Microwave and Millimeter-Wave Chips Based on Thin-Film Integrated Passive Device Technology

Yongle Wu · Weimin Wang

Microwave and Millimeter-Wave Chips Based on Thin-Film Integrated Passive Device Technology

Design and Simulation

電子工業出版社
PUBLISHING HOUSE OF ELECTRONICS INDUSTRY
http://www.phei.com.cn

Springer

Yongle Wu
School of Integrated Circuits
Beijing University of Posts
and Telecommunications
Beijing, China

Weimin Wang
School of Electronic Engineering
Beijing University of Posts
and Telecommunications
Beijing, China

ISBN 978-981-99-1454-8 ISBN 978-981-99-1455-5 (eBook)
https://doi.org/10.1007/978-981-99-1455-5

Jointly published with Publishing House of Electronics Industry
The print edition is not for sale in China (Mainland). Customers from China (Mainland) please order the print book from: Publishing House of Electronics Industry.

This Springer imprint is published by the registered company Springer Nature Singapore Pte Ltd.
The registered company address is: 152 Beach Road, #21-01/04 Gateway East, Singapore 189721, Singapore

Preface

With the arrival of the communication era of multi-frequency multi-mode microwave and millimeter-wave, both base station and mobile terminal sides bring new challenges to microwave and millimeter-wave devices in aspects of the low insertion loss, high suppression, miniaturized integration, and the low cost in batch. Therefore, the theoretical basis and application research of microwave and millimeter-wave devices and chips become one of the hot and difficult spots in the field of electronic science and technology.

Nowadays, there are many relevant courses and textbooks for students and engineers to learn about the design and simulation of microwave and millimeter-wave circuits and components based on the printed circuit board (PCB). By referring to them, beginners can master and get familiar with the basic principles and related theories of the core microwave and millimeter-wave circuits and components as well as the basic operations of relevant simulation software. During many years of teaching and research working, we found that many readers in electronic engineering major are confined to the design and implementation of the large-size PCB in radio frequency (RF) and microwave devices, while having little contact with the practical design and theoretical research of RF chips. At the same time, we also noticed that some readers have always thought that the procedure of RF chip design and implementation is very difficult and complicated, resulting in a fear of difficulties, which makes them flinch from the field of RF chips. Based on the years of studies and work experiences, we consider that providing beginners with a professional book, which is easy to learn and understand, convenient to get started, and close to the practical application, is a feasible method to solve the problems above. In addition, it is necessary for beginners to repeatedly imitate and ponder the design and simulation procedure of the representative passive chips, which is an essential process of the scientific research. However, some novices always expect their mentors to provide detailed guidance on this similar and repetitive procedure with specific time and efforts instead of doing it themselves, which is very laborious and inefficient. In view of the above realities, the authors have the initial motivation to write this book, hoping that beginners can eliminate their fear in the research field of RF devices and chips by learning this book.

This book adopts the latest academic achievements of microwave and millimeter-wave chips based on the thin-film integrated passive device technology in our research group as specific cases, shows the coherent and cohesive procedure of its basic theory and design implementation, provides comprehensive and systematic steps of its design and simulation, and forms a complete system from design theories, circuit simulations, full-wave electromagnetic simulations, and manufacturing implementations to chip measurements. This book is highly representative in subject matter, full and accurate in content, and has high theoretical value and practical guiding significance.

This book is divided into five chapters. Professor Wu Yongle's research team provided innovative and practical chip design cases and was responsible for the planning and adjustment of the book's structure and content. Weimin Wang was responsible for the unified draft of the whole book. Furthermore, Huiting Yu, Yuhao Yang, and Mengdan Kong also participated in the compilation of this book. I would like to appreciate for the hard work of all authors.

This book was supported in part by the Innovative Research Group Project in the National Natural Science Foundation of China (NSFC) and the Beijing Outstanding Youth Science Fund Project. The final completion of this book is also thanks to the good working environment and conditions provided by the School of Integrated Circuits, Beijing University of Posts and Telecommunications (BUPT) for our team. Moreover, I would like to express my special thanks to the academic colleagues who have given our guidance and help in the process of scientific research for many years. It is impossible to have the opportunity to produce innovative design ideas and specific cases related to this book without their strong support.

In the process of writing this book, the authors refer and cite to the original technical data of the Advanced Design System (ADS) software and Origin software, so hereby we would like to express our sincere thanks to the original authors of the technical data and related software companies.

Due to our limited abilities of the compilation, omissions in this book may be inevitable, so we gratefully welcome the suggestions and corrections from readers. In addition, if readers have any doubts or questions during the reading of this book, please contact the author (E-mail: wuyongle138@gmail.com) for discussion.

Beijing, China
September 2021

Yongle Wu

About This Book

This book mainly aims at the design and simulation of microwave and millimeter-wave chips based on the thin-film integrated passive device technology. On the basis of specific cases, the complete procedure including theories, designs, simulations, optimizations, fabrications, and measurements of representative microwave and millimeter-wave passive chips, such as the balanced bandpass filter chip, the millimeter-wave microstrip bandpass filter chip, the input-absorptive bandstop filter chip, the impedance-transforming power divider chip, and the bandpass filtering Marchand balun chip, is introduced in detail.

This book is suitable for professional technicians who are engaged in the design and engineering application of microwave and millimeter-wave device chips. It can also be used in colleges and universities as a textbook for electronic science and technology, electromagnetic field and microwave technology, electronic engineering, radar engineering, integrated circuit, and other related majors.

Introduction

1. Overview of Integration Technique

With the updating of intelligent mobile terminals, their internal RF millimeter-wave modules become increasingly complex. In order to ensure the portability and endurance of mobile terminals at the same time, the design space left for RF devices and chips is more and more limited on the premise of ensuring diversified analog functions. Therefore, the researches on miniaturized RF devices and chips with high performances are of great significance for the future development of wireless communications.

At present, many integrated technologies, including surface mounted devices (SMD), low-temperature co-fired ceramics (LTCC), complementary metal oxide semiconductors (CMOS), and thin-film integrated passive devices (TFIPD), have been widely used in the design and manufacturing of RF chips. Among them, the SMD technology has low production cost and little manufacturing difficulty; however, its process accuracy is limited, and the circuit size is relatively large. The LTCC technology belongs to the thick-film processing technology, which brings the risk of warping and cracking to the substrate because of the usage of the multilayer structure. The CMOS technology is mainly used for the realization of active chips. Moreover, the TFIPD technology is a kind of specialized semiconductor processing technology to construct passive microwave and millimeter-wave chips. Metal layers are deposited on thin-film substrate materials such as the gallium arsenide (GaAs) and high resistivity silicon, which can achieve micron-level process accuracy and provide fine spacing characteristics and good tolerance control for the layout design of the internal chip. Compared with other frequently used technologies, the TFIPD technology also has great advantages in improvements of the performance, design flexibility, integration, and compatibility. Therefore, this book mainly focuses on the design of microwave and millimeter-wave chips using the TFIPD technology.

2. Overview of Microwave and Millimeter-Wave Chip

Microwave and millimeter-wave devices and chips are the core components of RF circuits and modules, which play an indispensable role. Among them, microwave and millimeter-wave active chips have been widely concerned. In multi-frequency and multi-mode intelligent mobile terminals, microwave and millimeter-wave passive chips occupy a large proportion of space, so its basic research of ultra-miniaturization and high-performance is particularly important. The commonly used microwave and millimeter-wave passive chips mainly include filters, couplers, phase shifters, duplexers (multiplexers), power dividers, and balun chips, etc., which complete specific functions from one- or multi-port input RF signals to one- or multi-port output RF signals. For example, filters can achieve the function of selecting the specific spectrum by attenuating the specific frequency signal that is not needed in the input RF signal and completing the low loss transmission of the required frequency signal; phase shifters can complete the phase control function of the specific spectrum by adjusting the phase of the input RF signal and outputting the specified phase difference of the RF signal; duplexers (multiplexers) are used to distribute different frequencies of RF signals to different channels, which can complete the distribution or combination and the duplexer (multiplexer) function of different frequencies signals; power dividers can divide the one-way input RF signal into two or multiple outputs in an equal or unequal way, which can also be used as a combiner to synthesize two or multiple RF signals into one channel and complete the adjustment of signal amplitude; Balund is used to complete the conversion of signal type between the balanced circuit and the unbalanced circuit [1–3].

This book takes five kinds of representative microwave and millimeter-wave passive chips based on the TFIPD technology as examples to illustrate the complete process of the theory, design, simulation, manufacturing, and the measurement, which includes the balanced filter chip, microstrip bandpass filter chip, absorptive filter chip, power divider chip, and balun chip. The simulating software used in this book is the Advanced Design System (ADS). This book shows the methods of using the ADS to build, simulate, and optimize the ideal simulated models and the corresponding full-wave electromagnetic simulated models of the microwave and millimeter-wave passive chips in detail, so as to obtain the expected design performance. Finally, the manufactured chips are measured, and the measured results are given to verify the design theory and simulation results. In addition, this book also demonstrates the method of using the Origin software to plot data figures, and the process is simple and easy to learn. The figures are succinct and beautiful, which can greatly improve the ability of scientific researchers or engineers to write excellent academic papers and professional project summaries.

3. Introduction of Software

[Introduction of ADS] ADS is the electronic design automation (EDA) software launched by Agilent in the early stage of the USA, which can complete time domain and frequency domain simulations, the digital/analog hybrid simulation, and linear and nonlinear simulations at device level, circuit level, and even system level. The software is a 2.5D electromagnetic field simulator momentum based on the Method of Moment (MoM) to simplify the third dimension, which is very suitable for the simulation of the PCB multilayer board, passive device, and millimeter-wave integrated circuit with uniform changes in the third dimension. It shows a fast-speed and high-accuracy simulation. In addition, the 3D electromagnetic field simulator based on the finite element method (FEM) algorithm in the software can also simulate structures with non-uniform extension in the third dimension, such as antennas. At present, ADS software has become a very popular and widely used RF EDA software in academia and industry because of its powerful simulating functions.

This book introduces the usage of the ADS software to simulate and optimize the ideal simulated models and the corresponding full-wave electromagnetic simulated models of five representative microwave and millimeter-wave chips (the balanced filter chip, microstrip bandpass filter chip, absorptive filter chip, power divider chip, and the balun chip) and demonstrates how to view the S-parameter, amplitude, and phase information of these microwave millimeter-wave passive chips. Moreover, the method and process of calculating the actual physical sizes of the circuit distributed parameters using the LineCalc tool in ADS are also shown. The version of the ADS software used in this book is ADS 2020. For more detailed information about the ADS software, please refer to the references [4–7].

[Introduction of Origin] Origin is a scientific function drawing software developed by OriginLab Company in the USA. Its simple operation interface enables scientific researchers and engineers to visually draw figures in 2D, 3D, and multi-dimension. The software supports the data import in various formats, including ASCII, Excel, NI TDM, etc. At the same time, the graphics output formats are also diverse, such as JPEG, GIF, EPS, and TIFF. In Chap. 5, this book takes the simulated and measured results of balun chips as examples to introduce the detailed drawing steps from the data import, data processing, drafting beautification to graphics output based on the Origin software. The version of the Origin software used in this book is Origin 2019b. For more detailed information about the Origin software, please refer to the references [8] and [9].

4. Arrangement of Book

This book is characterized by the realization of the whole processes of the microwave and millimeter-wave chips based on the TFIPD technology from the design theory analysis to the use of the ADS software for the simulation, optimization, and measurement. Balanced bandpass filter chips are introduced in Chap. 1, millimeter-wave microstrip bandpass filter chips are introduced in Chap. 2, input-absorptive bandstop filter chips are introduced in Chap. 3, impedance-transforming power divider chips are introduced in Chap. 4, and bandpass filtering Marchand balun chips are introduced in Chap. 5. The complete design and implementation processes of the five representative passive chips are shown in detail. The cases introduced in this book are all from the academic papers published by the author's research group and the invention patents applied for. For other details of all the cases, please refer to the full text of the corresponding academic papers and invention patents.

As a reference book that combines theoretical basis with engineering practice closely, this book not only provides original innovative ideas, has a high level of academic value, but also focuses on engineering realization and practical application, which has important practical significance. Authors hope that by carefully learning and imitating this book, readers can understand and master the basic theories, performance indexes, and the theoretical design process of microwave and millimeter-wave filter chips, power dividers, balun chips, and other passive chips, and be familiar with the complete process of the design, simulation, optimization, and measurement of microwave and millimeter-wave chips, lay a solid foundation for further research, design innovation, and independent completion of chip research projects. Furthermore, authors also hope that the guidance of this book can make readers eliminate their fear, become interested, and have the ambition to make a new achievement in the field of microwave and millimeter-wave chips, so as to promote the RF chip industry to flourish forward.

References

1. X. Xie, Y. Xu, L. Xia. *Microwave Integrated Circuits* (Publishing House of Electronics Industry, Beijing, 2018)
2. Q. Gu, J. Xiang, X. Yuan. *Microwave Integrated Circuit Design* (People's Posts and Telecommunications Press, Beijing, 1978)
3. R. Li. *Key Issues in RF/RFIC Circuit Design* (Higher Education Press, Beijing, 2007)
4. X. Xu. *ADS2011 RF Circuit Design and Simulation Examples* (Publishing House of Electronics Industry, Beijing, 2014)
5. Y. Huang. *Design Basis and Typical Application of ADS RF Circuit,* 2nd edn. (Posts and Telecommunications Press, Beijing, 2015)
6. C. Chen. *Design and Simulation of ADS RF Circuit from Entry-Level to Proficient* (Publishing House of Electronics Industry, Beijing, 2013)
7. X. Feng, X. Kou. *Introduction and Application Examples of ADS RF Circuit Design and Simulation* (Publishing House of Electronics Industry, Beijing, 2014)
8. A. Fang, W. Ye. *Origin 8.0 Practical Guide* (China Machine Press, Beijing, 2009)
9. J. Zhang. *Origin 9.0 Technology Mapping and Data Analysis Super Learning Manual* (People's Posts and Telecommunications Press, Beijing, 2014)

Contents

Chapter 1
Design and Simulation of Balanced Bandpass Filter

As a kind of key frequency selection device, filters are widely used in circuit systems. Since communication systems are often disturbed by environmental noises, it is necessary to apply balance filters with excellent common-mode (CM) suppression to improve the anti-interference ability of the system and the quality of the communication. In addition, the application of the thin-film integrated passive device (TFIPD) technology in microwave passive devices can reduce the circuit size and realize the miniaturization of devices. In this chapter, we are going to introduce the basic principles of lumped-element balanced bandpass filters [1] based on the TFIPD technology, and the methods of using the ADS software to establish, simulate, and optimize the ideal parameter simulated model of the balanced bandpass filter and the corresponding full-wave electromagnetic simulated model.

1.1 Overview of Balanced Bandpass Filter

1.1.1 Theoretical Basis

1. Introduction

Filters are two-port devices, which are used to realize the low loss transmission of signals in the passband, achieve the substantial attenuation of signals in the stopband, and complete the selection of the specific spectrum. In essence, it is a frequency selection device.

In communication systems, radar systems, and measurement systems, filters are indispensable. According to the transmission characteristics, filters can be divided into lowpass filters, highpass filters, bandpass filters, and bandstop filters; According to the design methods, filters can be divided into Butterworth filters, Chebyshev

Y. Wu and W. Wang, *Microwave and Millimeter-Wave Chips Based on Thin-Film Integrated Passive Device Technology*,
https://doi.org/10.1007/978-981-99-1455-5_1

filters, and Elliptic filters, etc.; According to the structural elements, filters can be divided into lumped-element filters and distributed-element filters, etc.

2. Performance Index

The performance indexes of filters include the insertion loss, return loss, and group delay, etc. Assume port 1 as the input port of the filter and port 2 as the output port of the filter, its performance index parameters can be expressed as follow.

a. Insertion Loss (IL): IL (dB) $= -20\log_{10}|S_{21}|$. The insertion loss is the ratio of the output power to the input power when the signal is input from port 1 and output from port 2. It represents the power loss caused by the insertion of the filter. Theoretically, in the case of $S_{21} = 0$, the frequencies f_z of transmission zeros (TZs) can be obtained.
b. Return Loss (RL): RL (dB) $= -20\log_{10}|S_{11}|$. The return loss is the ratio of the incident power to the reflected power of the filter. Theoretically, when $S_{11} = 0$, the frequencies f_p of transmission poles (TPs) can be obtained.

1.1.2 Propaedeutics

Compared with the traditional single-ended filter, the balanced filter can reduce the influence of noises significantly on the communication system because of its high ability in the CM suppression. Figure 1.1 shows the lumped-element circuit model of the proposed balanced bandpass filter [1]. Considering the influence of 200-μm and 400-μm wire bonds on the performance of balanced bandpass filters in the wire bonding technology used in the chip packaging, the equivalent inductances $L_2 = 0.3$ nH and $L_3 = 0.6$ nH of wire bonds are introduced in the circuit design process.

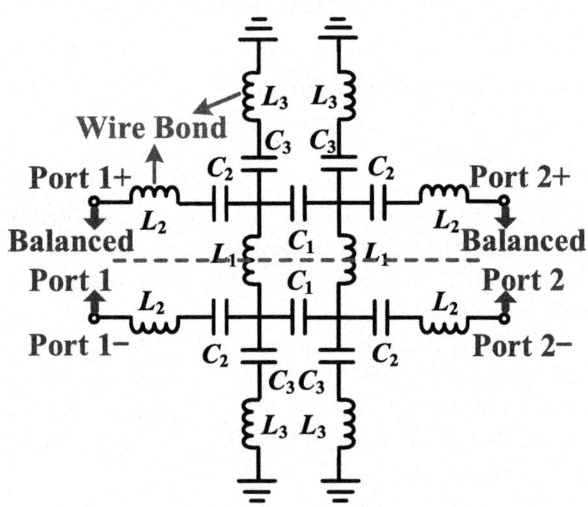

Fig. 1.1 Lumped-element circuit model of the proposed balanced bandpass filter

- Explanation

  The following methods can be used to create a new workspace:

  - Execute menu commands [File] → [New] → [Workspace…];
  - Click the [Create A New Workspace] icon in the toolbar.

  If the following content relates to the creating of a new workspace, use one of the methods mentioned above.

(3) The [Folder View] in the ADS main interface can display the name and working path of the created workspace, as shown in Fig. 1.5. The workspace name is "Balanced_Bandpass_Filter_wrk", the corresponding working path is "D:\ADS\Balanced_Bandpass_Filter_wrk", and a subfolder named "Balanced_Bandpass_Filter_wrk" in the ADS folder of disk D can be found in your computer.

2. New Circuit Model

(1) Create a new schematic. Execute menu commands [File] → [New] → [Schematic…] to open the [New Schematic] dialog box shown in Fig. 1.6. Modify the name of "Cell" to "circuit structure" and click the [Create Schematic] button to complete the creation of a new schematic, as shown in Fig. 1.7.

- Explanation

  The following methods can be used to create a new schematic:

- Execute menu commands [File] → [New] → [Schematic…];

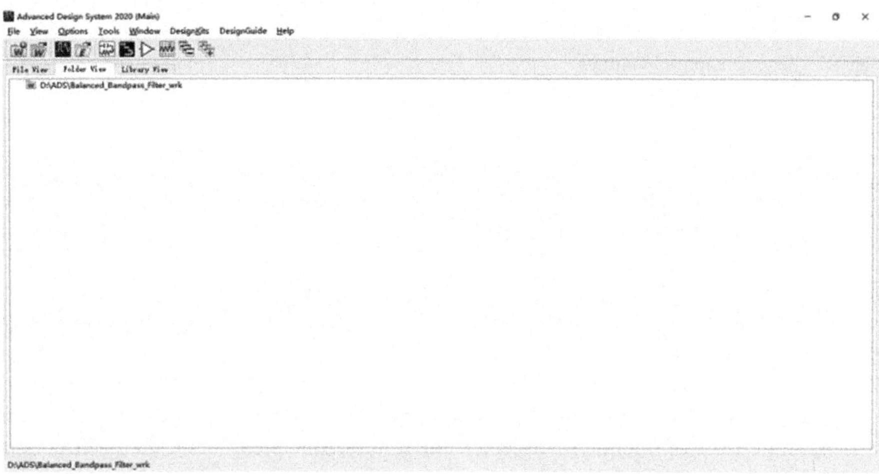

Fig. 1.5 Folder view of the new workspace

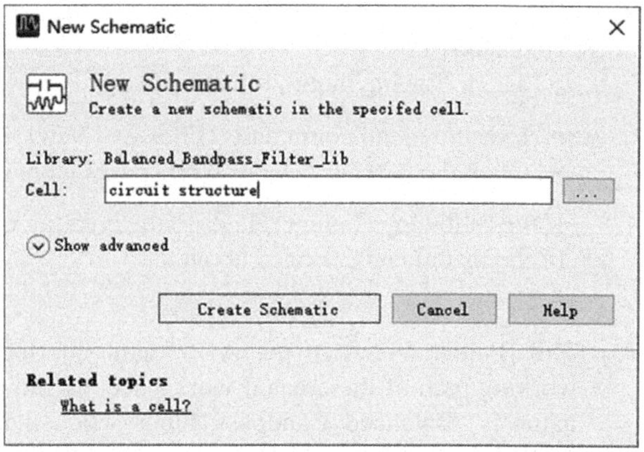

Fig. 1.6 [New schematic] dialog box

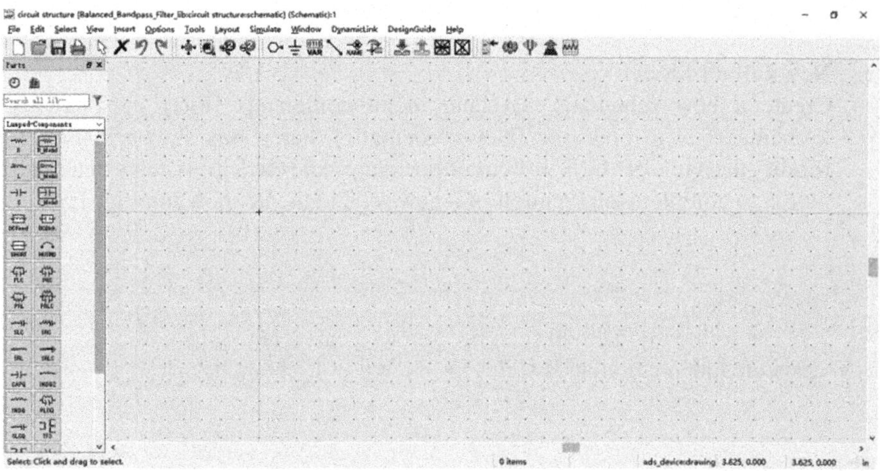

Fig. 1.7 New schematic window

- Click the [New Schematic Window] icon in the toolbar.

  If the following content relates to the creating of a new schematic, use one of the methods mentioned above.

(2) Insert and rotate capacitors. Select the [Lumped-Components] from the drop-down menu in the left component palette list, as shown in Fig. 1.8, which includes some commonly used ideal lumped-element models, such as the capacitor, inductor, and the resistor, etc. Click the [Capacitor] icon under the [Lumped-Components] to insert ten capacitors, then press "Esc" on the keyboard to exit, as shown in Fig. 1.9.

Fig. 1.8
[Lumped-Components] list

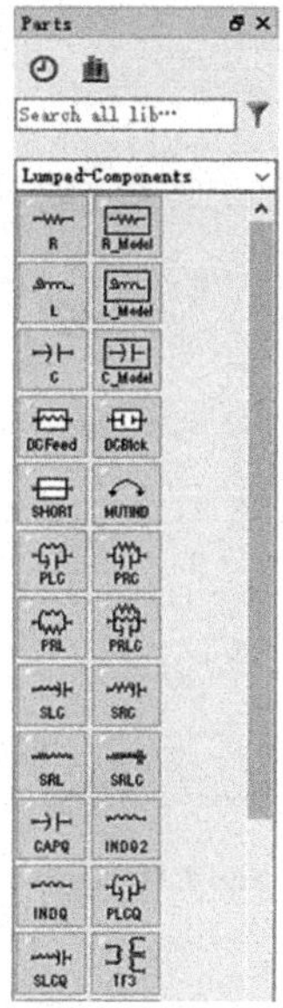

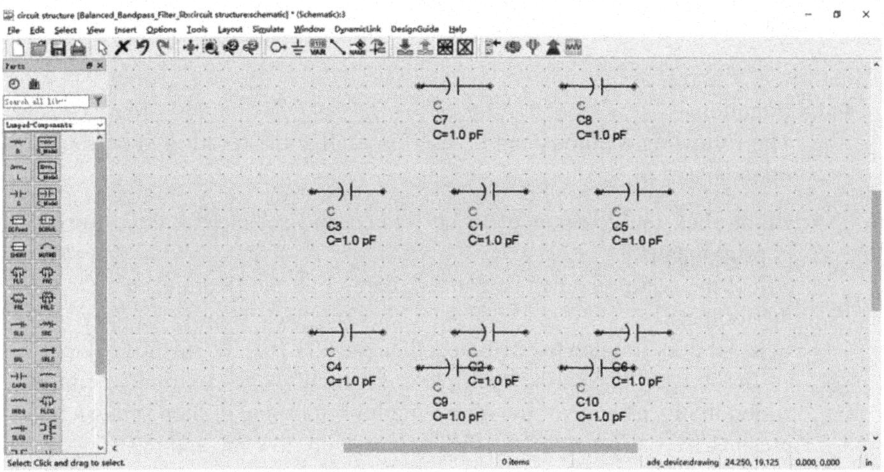

Fig. 1.9 Insertion of capacitors

- Explanation

  If the component palette list is not displayed in the left of the schematic, execute menu commands [View] → [Docking Windows] → [Component Palette] to make it displayed. If similar situation occurs in the following content, this method can be used to solve it.

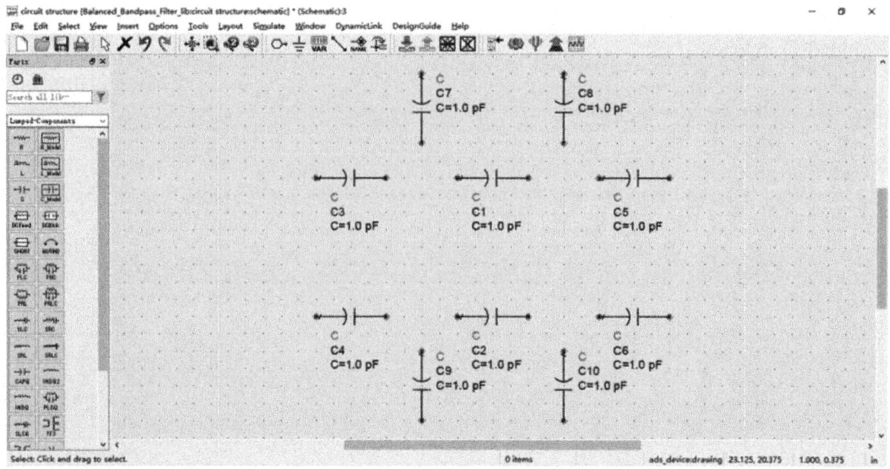

Fig. 1.10 Rotation of capacitors

Right-click the capacitors to be rotated, select [Rotate] from the pop-up menu to rotate them in 90° clockwise. Capacitors have been inserted and rotated, as shown in Fig. 1.10.

- Explanation

  The following methods can be used to realize the rotating operation of components:

  - Right-click the component to be rotated and select [Rotate] from the pop-up menu;
  - Select the component to be rotated, then click the [Rotate] icon in the toolbar;
  - Select the component to be rotated, then press "Ctrl + R" on the keyboard.

    In the case of rotating multiple components, one component can be added firstly, use one of the above methods to rotate it, then select it and press "Ctrl + C" and "Ctrl + V" successively to copy and paste.

    If the following content relates to the rotating operation of components, use one of the methods mentioned above.

(3) Insert and rotate inductors. Click the [Inductor] icon under the [Lumped-Components] to insert ten inductors, then press "Esc" on the keyboard to exit, as shown in Fig. 1.11. Right-click the inductors to be rotated, select [Rotate] from the pop-up menu to rotate them in 90° clockwise. Inductors have been inserted and rotated, as shown in Fig. 1.12.

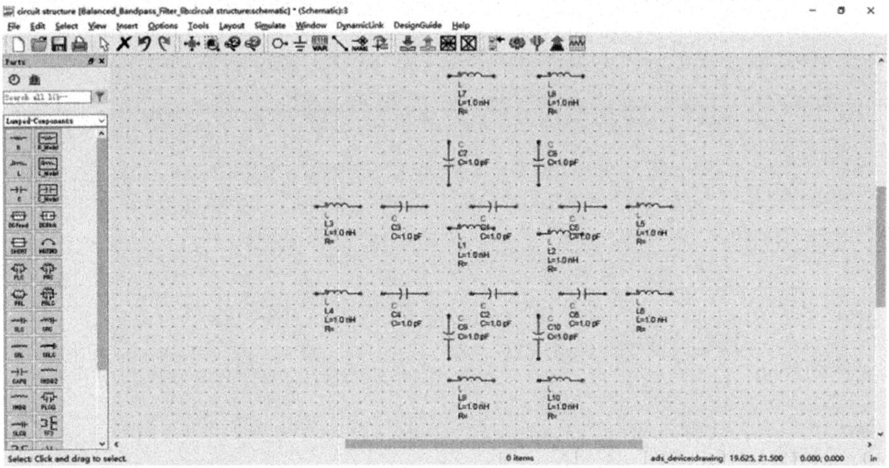

Fig. 1.11 Insertion of inductors

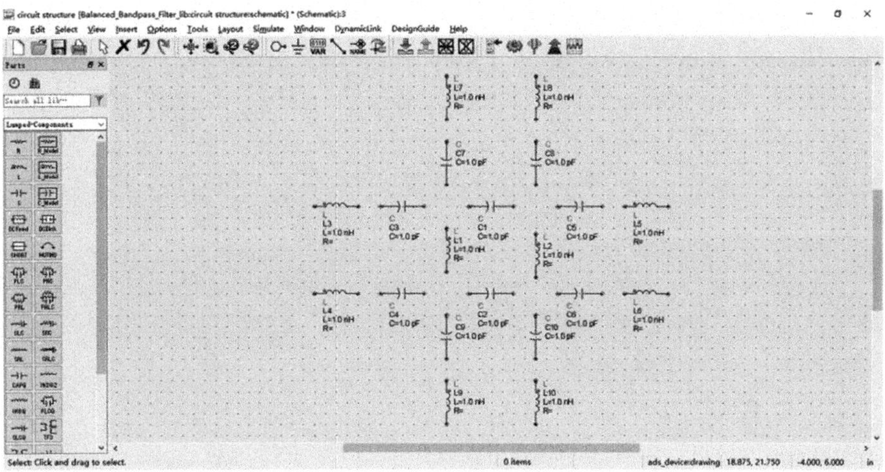

Fig. 1.12 Rotation of inductors

(4) Place and rotate grounds. Execute menu commands [Insert] → [GROUND] to place four grounds, then press "Esc" on the keyboard to exit, as shown in Fig. 1.13. Right-click the ground symbols to be rotated, select [Rotate] from the pop-up menu and operate it twice to rotate them in 180° clockwise. Ground symbols have been inserted and rotated, as shown in Fig. 1.14.

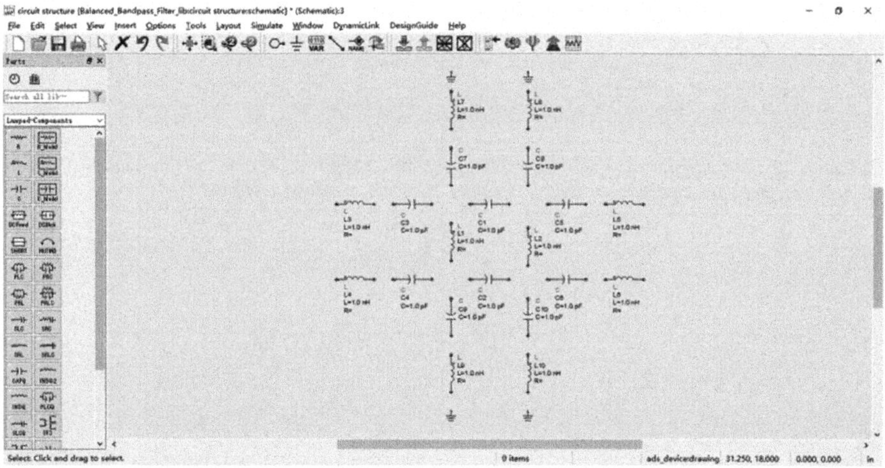

Fig. 1.13 Placement of grounds

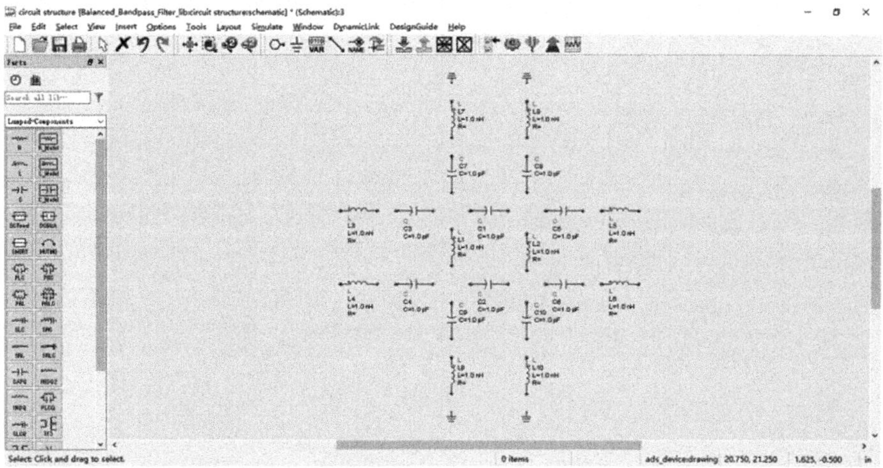

Fig. 1.14 Rotation of grounds

- Explanation

  The following methods can be used to place grounds:

- Execute menu commands [Insert] → [GROUND];
- Click the [Insert GROUND] icon in the toolbar.

  If the following content relates to the placement of grounds, use one of the methods mentioned above.

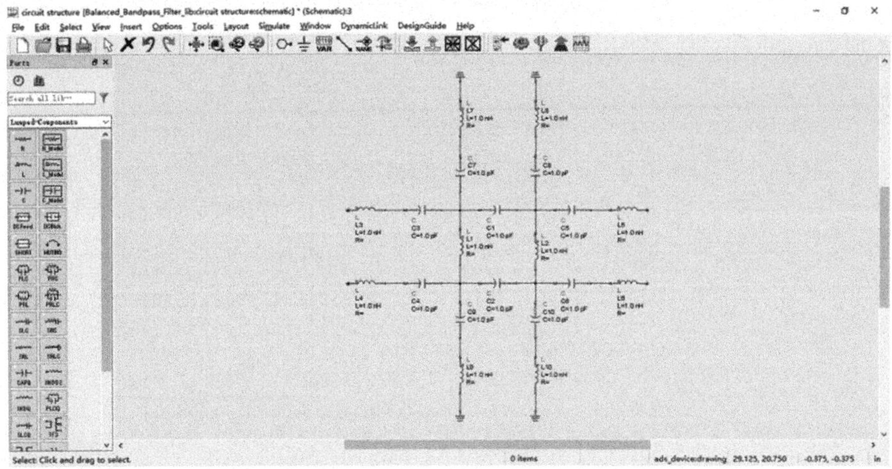

Fig. 1.15 Connection of components

(5) Connect components. Execute menu commands [Insert] → [Wire] to connect each component according to the circuit model shown in Fig. 1.1. Figure 1.15 shows the circuit schematic after the components are connected.

- Explanation

  The following methods can be used to realize the connecting operation of components:

  - Execute menu commands [Insert] → [Wire];
  - Click the [Insert Wire] icon  in the toolbar;
  - Use the shortcut "Ctrl + W".

  If the following content relates to the connecting operation of components, use one of the methods mentioned above.

(6) Modify circuit model parameters. Double-click the capacitor C1 and modify its parameter value to C1 (pF) in the pop-up [Edit Instance parameters] dialog box, as shown in Fig. 1.16. Click the [OK] button to save the parameter and close the dialog box. Similarly, modify the parameter value of C2 to C1 (pF), the parameter values of C3, C4, C5, and C6 to C2 (pF) (taking C3 as an example, as shown in Fig. 1.17), the parameter value of C7, C8, C9, and C10 to C3 (pF) (taking C7 as an example, as shown in Fig. 1.18), the parameter values of L1 and L2 to L1 (nH) (taking L1 as an example, as shown in Fig. 1.19), the parameter values of L3, L4, L5, and L6 to L2 (nH) (taking L3 as an example, as shown in Fig. 1.20), the parameter values of L7, L8, L9, and L10 to L3 (nH) (taking L7 as an example, as shown in Fig. 1.21).

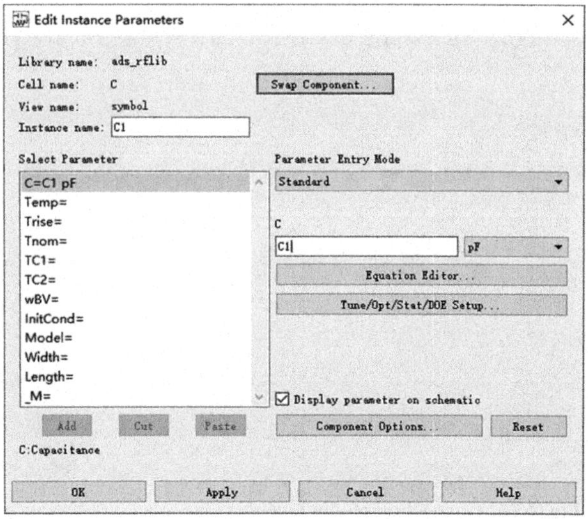

Fig. 1.16 Parameter setting of the capacitor C1

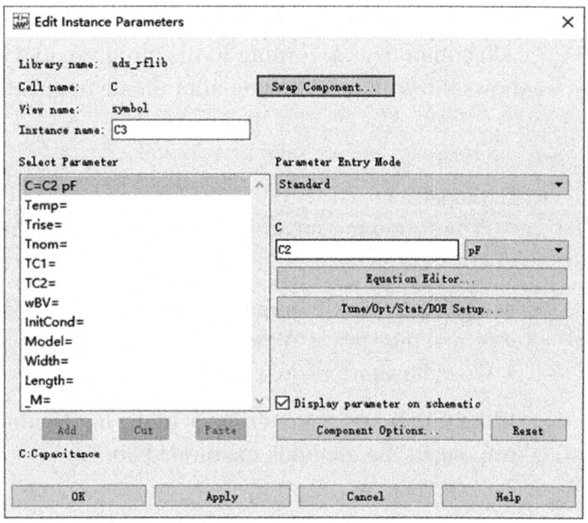

Fig. 1.17 Parameter setting of the capacitor C3

- Note

  The parameter unit is selected from the drop-down menu behind the input column. Ensure that the parameter units are set consistently, otherwise errors will occur during the simulation.

(7) Define parameter values of variables. Click the [Insert VAR: Variable Equations] icon 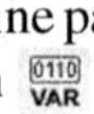 in the toolbar to insert a VAR item in the schematic. Double-click the

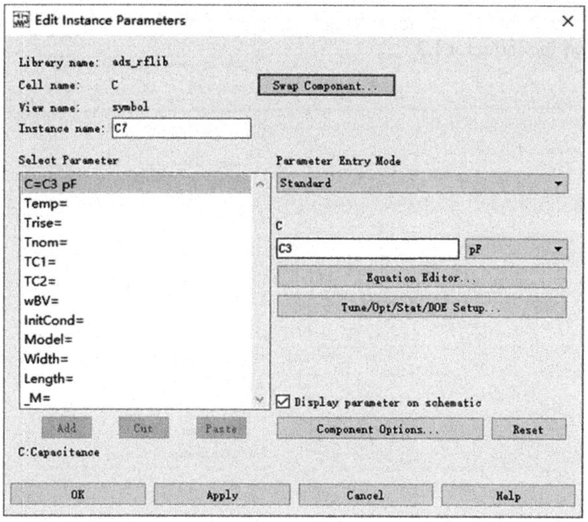

Fig. 1.18 Parameter setting of the capacitor C7

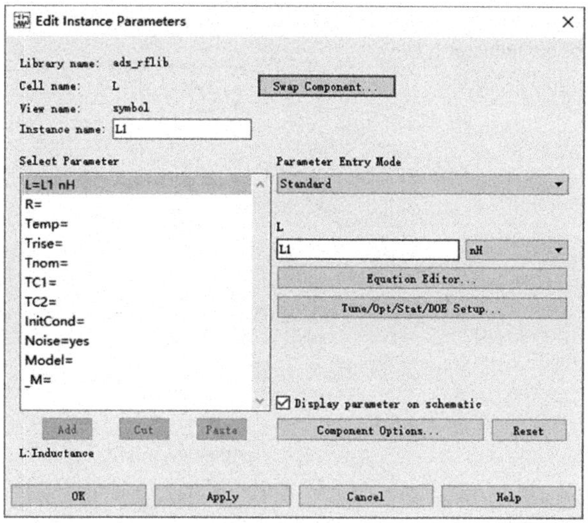

Fig. 1.19 Parameter setting of the inductor L1

variable component 

to open the [Edit Instance Parameter] dialog box, the "Variable or Equation Entry Mode" column is "Standard" by default. Enter the variable name "C1" in the "Name" column and the variable value "0.56" in the "Variable Value" column. Click the [Apply] button, and the dialog box after the setting is shown in Fig. 1.22. If you click [OK], the dialog box can be closed directly. Then define other variables in turn, including C2 = 0.89, C3 = 1.00, L1 = 0.68, L2 = 0.3, and L3 = 0.6, as shown in Fig. 1.23. While defining other variables, click the [Add] button to add them. If you click the [Apply] button, the selected variable in the parameter list on the left can be replaced directly.

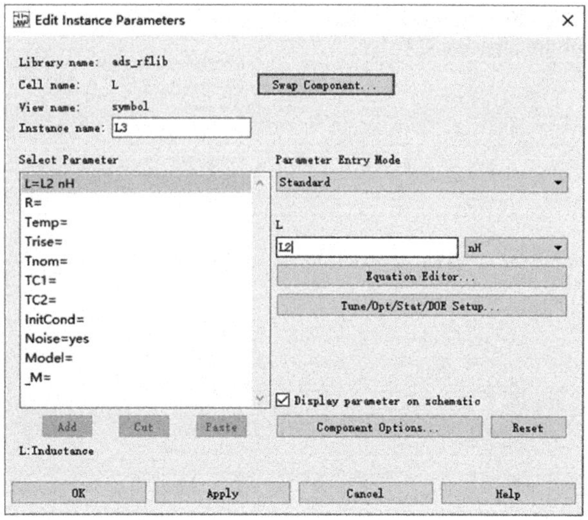

Fig. 1.20 Parameter setting of the inductor L3

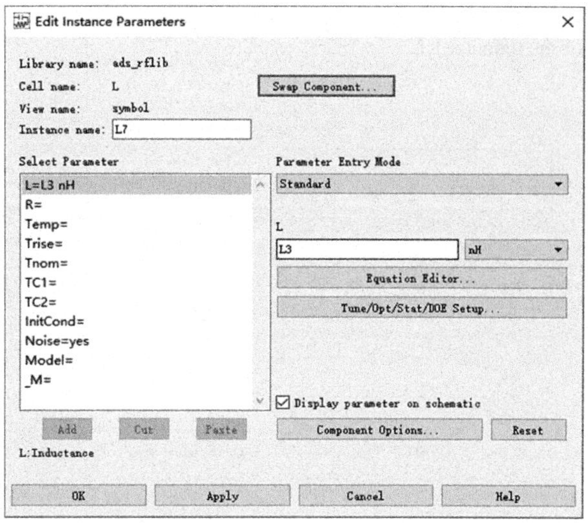

Fig. 1.21 Parameter setting of the inductor L7

In addition, there is no need to set units because the unit of each variable has already been defined in the component model.

(8) The VAR item provides three methods to add variables, including "Standard", "Name = Value", and "File Based". You can select the most suitable method in the "Variable or Equation Entry Mode" column. The second method is introduced below. Select the "Name = Value" mode, as shown in Fig. 1.24. Input "C1 = 0.56" under the "Variable Value" column, then click the [Apply] button. Similarly, when defining other variables, click the [Add] button to add them.

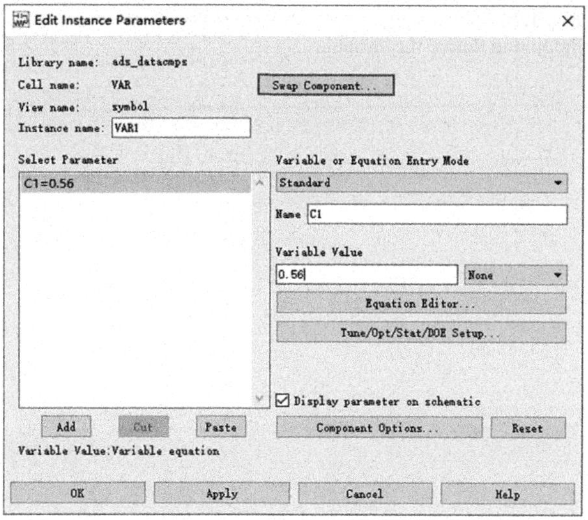

Fig. 1.22 Definition of the variable C1

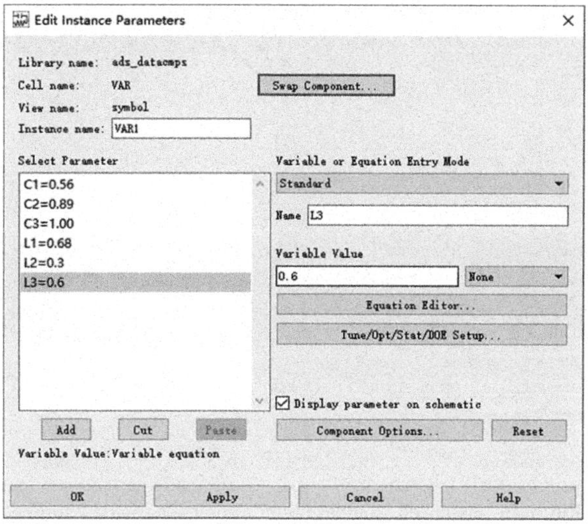

Fig. 1.23 Definition of all variables in the circuit

(9) Complete parameter definitions. After the definition of all variables, click the [OK] button, the final circuit schematic of the balanced bandpass filter with the defined parameters is shown in Fig. 1.25. Readers can compare the parameters in the lumped-element circuit schematic to check whether all the parameters are set correctly in detail.

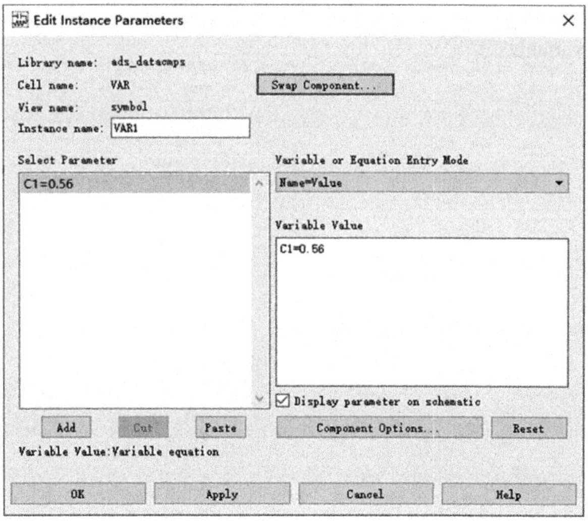

Fig. 1.24 The second method to define the variable C1

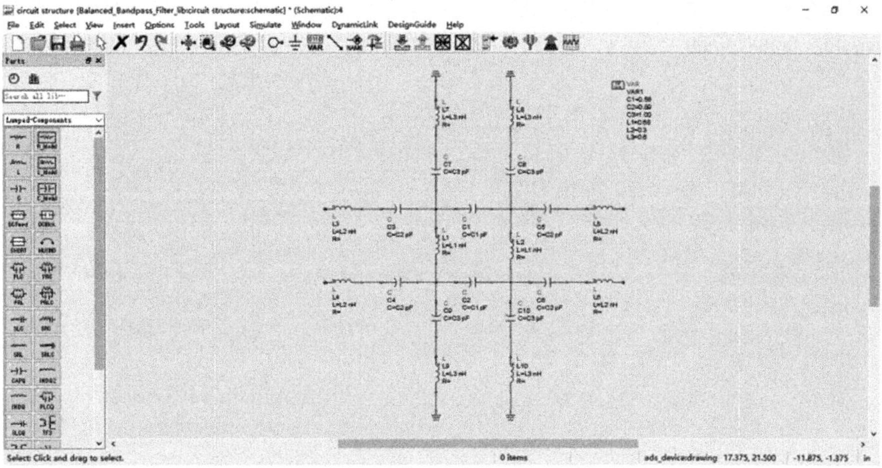

Fig. 1.25 Balanced bandpass filter circuit schematic with the defined parameters

1.2.2 Schematic Simulation

1. Set Simulation Parameters

(1) Insert the S-parameter simulator, term ports, and grounds. As shown in Fig. 1.26, select the [Simulation-S_Param] from the drop-down menu in the left component palette list and click the [S-parameter Simulator] icon to insert an S-parameter simulator. Click the [Port Impedance Termination for S-Parameters] icon to insert four term ports, then press "Esc" on the keyboard to exit.

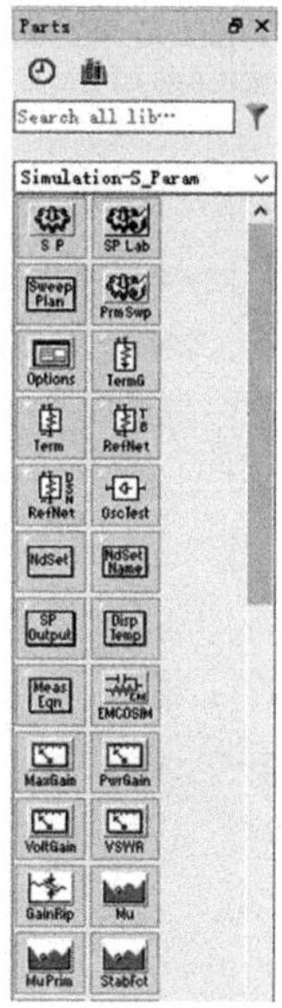

Fig. 1.26
[Simulation-S_Param] list

After that, execute menu commands [Insert] → [GROUND] to place four grounds (Or click the [TermG Port Impedance Termination for S-Parameters with Grounded Reference] icon to insert four term ports with the grounded reference). Execute menu commands [Insert] → [Wire] to connect all the inserted components. After finishing all the steps, press "Esc" to exit.

(2) Set simulation frequency. Double-click the S-parameter simulator S-PARAMETERS. Complete the settings shown in Fig. 1.27, in which the start frequency is 0 GHz, the stop frequency is 8 GHz, and the step-size is 0.01 GHz, then click the [OK] button. The final ideal parameter simulated circuit model of the balanced bandpass filter is obtained, as shown in Fig. 1.28.

2. View Simulated Results

(1) Execute menu commands [Simulate] → [Simulate]. After the simulation completes, the data display window will be opened automatically, as shown in Fig. 1.29.

- Explanation

  The following methods can be used to realize the operation of simulation:

  - Execute menu commands [Simulate] → [Simulate];
  - Click the [Simulate] icon in the toolbar;
  - Use the shortcut "F7".

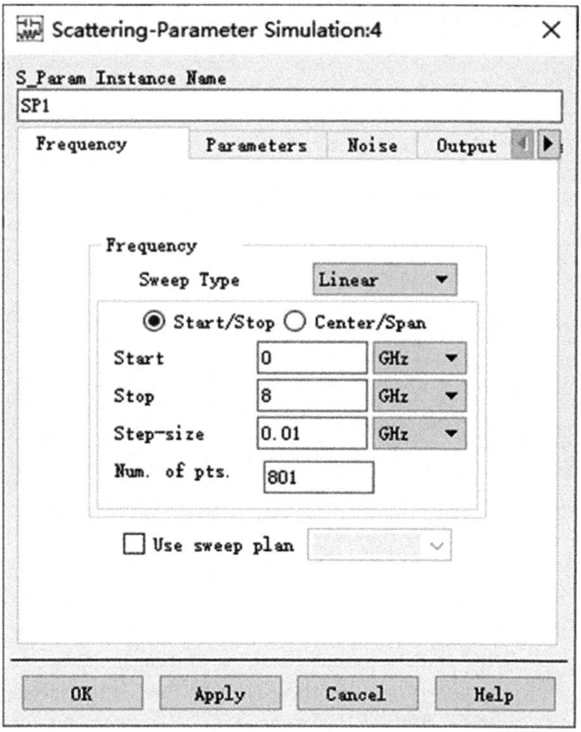

Fig. 1.27 Frequency settings of the S-parameter simulation

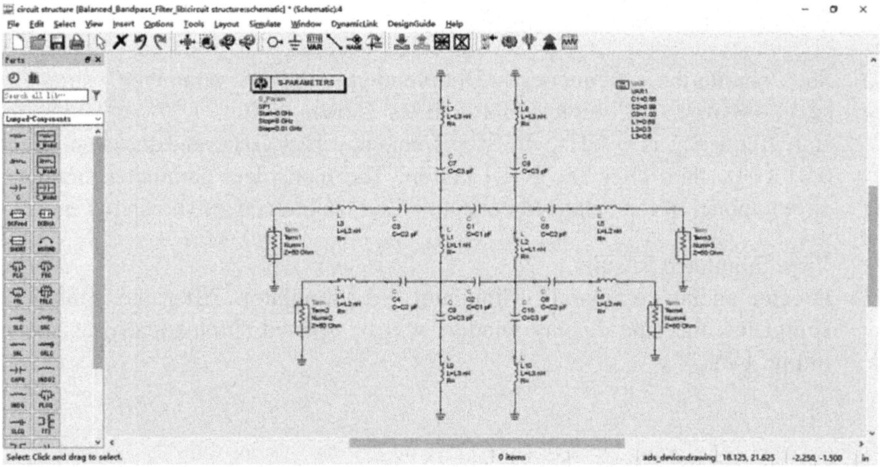

Fig. 1.28 Final ideal circuit simulated model of the balanced bandpass filter

Fig. 1.29 Data display window

> If the following content relates to the operation of simulation, use one of the methods mentioned above.

(2) Click the [Equation] icon Eqn in the left [Palette]. Click the left mouse button in the data display area to open the [Eenter Equation] dialog box. Write the equation in Fig. 1.30 under the [Enter equation here] to calculate the DM return loss, then click the [OK] button to close the dialog box. Similarly, the equations in Figs. 1.31, 1.32, and 1.33 can be entered to calculate the DM insertion loss, CM return loss, and CM insertion loss, respectively.

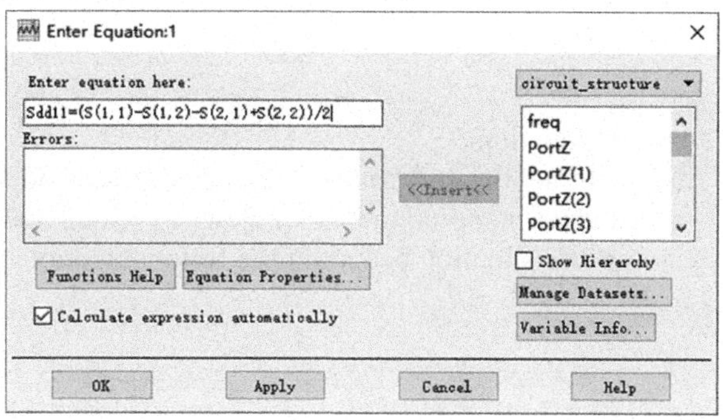

Fig. 1.30 Operation of writing the DM return loss calculating equation

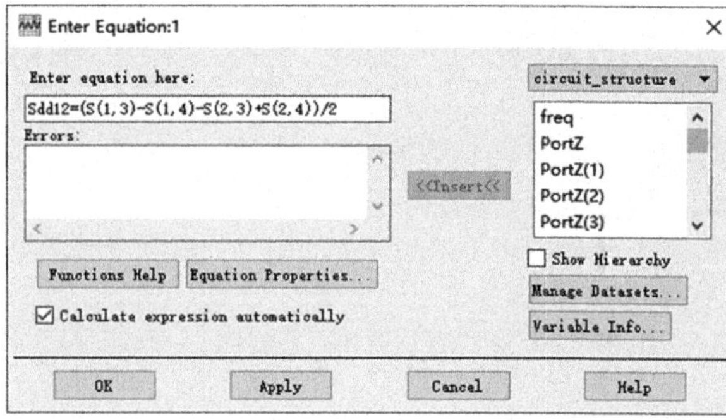

Fig. 1.31 Operation of writing the DM insertion loss calculating equation

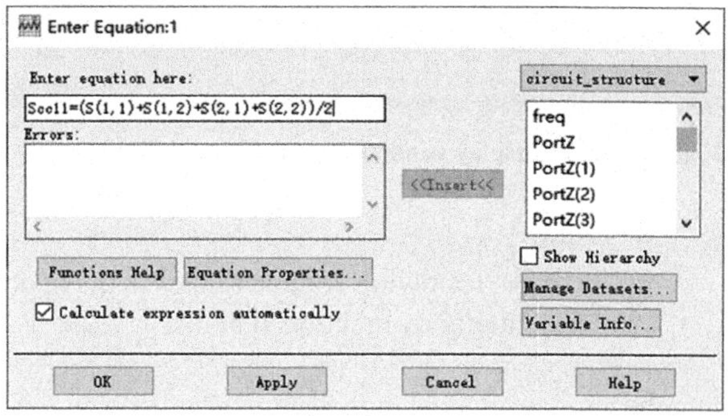

Fig. 1.32 Operation of writing the CM return loss calculating equation

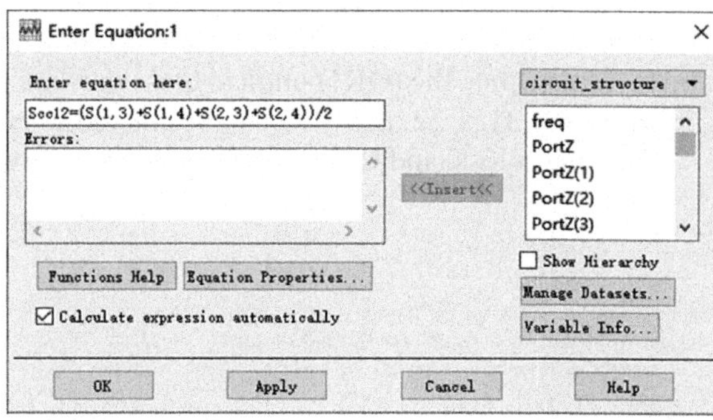

Fig. 1.33 Operation of writing the CM insertion loss calculating equation

- Explanation

  If the [Palette] is not displayed on the left side of the data display window, execute menu commands [View] → [Item Palette] to make it display. If similar situation occurs in the following content, use this method to solve it.

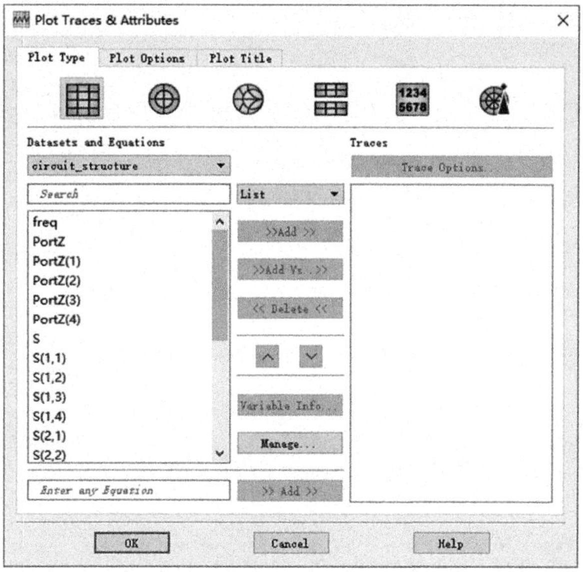

Fig. 1.34 [Plot Traces and Attributes] dialog box

(3) Click the [Rectangular Plot] icon  in the [Palette], then click the left mouse button at the blank data display area to open the [Plot Traces & Attributes] dialog box shown in Fig. 1.34, where the parameter traces can be set and plotted.

(4) Select "Equations" in the "Datasets and Equations" column, long press the "Ctrl" key to select the defined Sdd11 and Sdd12 successively, and click the [>> Add >>] button. Choose [dB] in the pop-up dialog box shown in Fig. 1.35, then click the [OK] button. It can be seen that dB(Sdd11) and dB(Sdd12) have been added to the [Traces] list box, as shown in Fig. 1.36.

(5) Click the [OK] button in Fig. 1.36, the traces of dB(Sdd11) and dB(Sdd12) can be displayed in the data display area, the ordinate is the dB value, as shown in Fig. 1.37.

(6) Similarly, the traces of dB(Scc11) and dB(Scc12) can also be displayed, as shown in Fig. 1.38.

3. Handle Traces

Next, traces dB(Sdd11) and dB(Sdd12) are taken as examples to introduce the curves handling. Similar steps can also be applied in other traces.

(1) By changing the frequency position of the marker, the value of any point on the curve can be read. Execute menu commands [Marker] → [New…] to open the dialog box shown in Fig. 1.39. Select a trace and click the left mouse button to insert a marker on it, as shown in Fig. 1.40. Similarly, insert a marker for another trace dB(Sdd12). In addition, you can long press the left mouse button to move the position of the data frame.

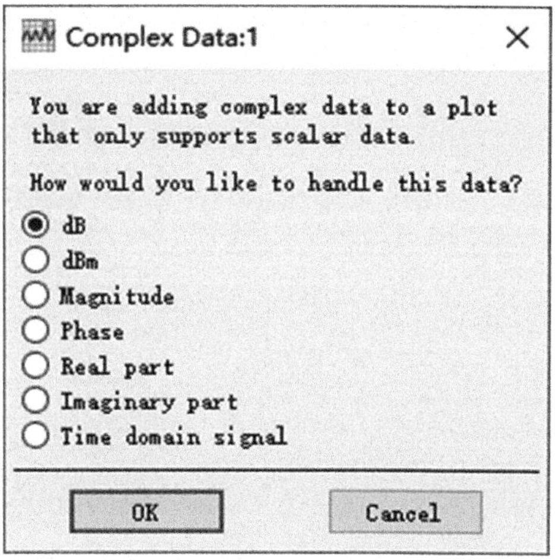

Fig. 1.35 Handling of the data

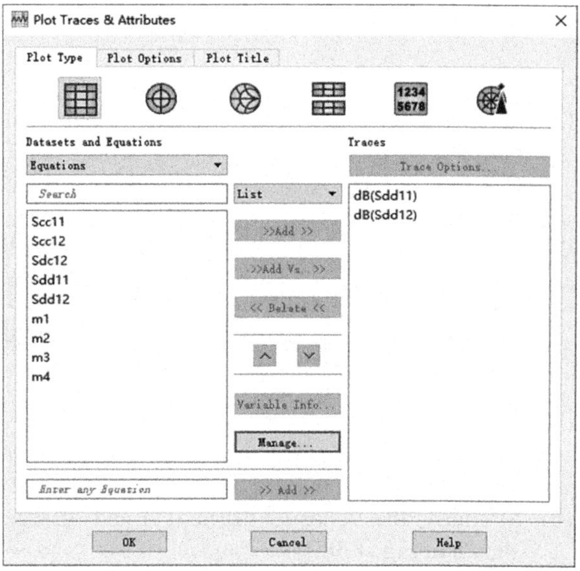

Fig. 1.36 Add of dB(Sdd11) and dB(Sdd12)

- Explanation

  The following methods can be used to insert markers:

  - Execute menu commands [Marker] → [New…];
  - Click the [Insert A New Marker] icon in the toolbar;
  - Use the shortcut "Ctrl + M".

  If the following content relates to the insertion of markers, use one of the methods mentioned above.

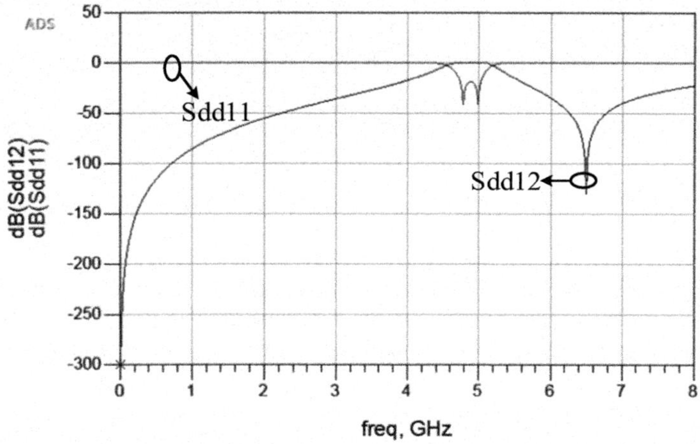

Fig. 1.37 Traces of dB(Sdd11) and dB(Sdd12)

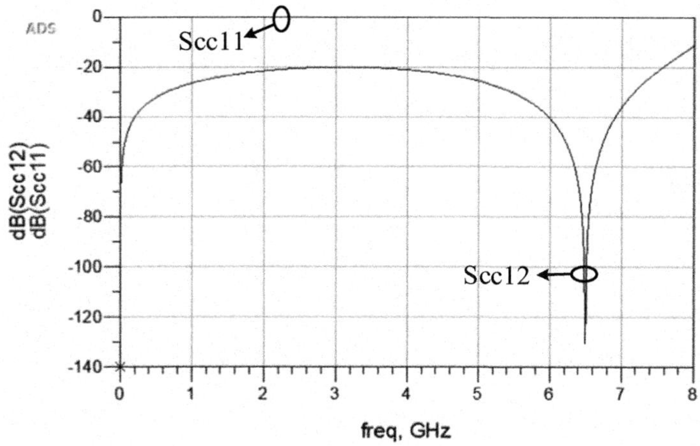

Fig. 1.38 Traces of dB(Scc11) and dB(Scc12)

Fig. 1.39 [Insert Marker] dialog box

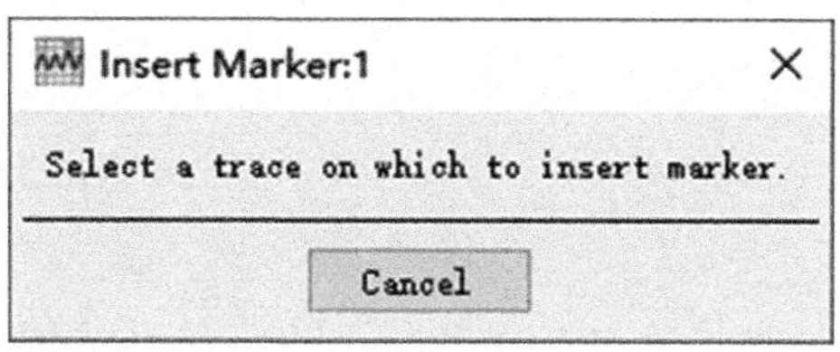

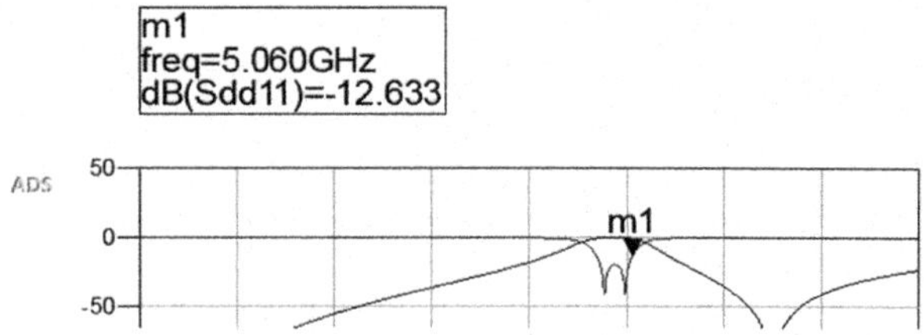

Fig. 1.40 Insertion of the marker on the trace

(2) After selecting the inserted marker, the left and right arrow keys on the keyboard can be used to adjust the position of the abscissa (freq). You can also click the position shown in Fig. 1.41 with the left mouse button to modify the specific frequency you want to view directly. Here, the values of dB(Sdd11) and dB(Sdd12) at the center frequency of 4.9 GHz are displayed, as shown in Fig. 1.42.

- Explanation

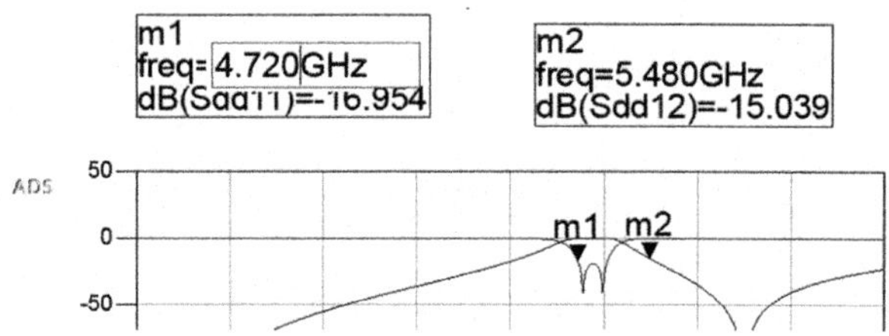

Fig. 1.41 Modification of the abscissa (freq) value

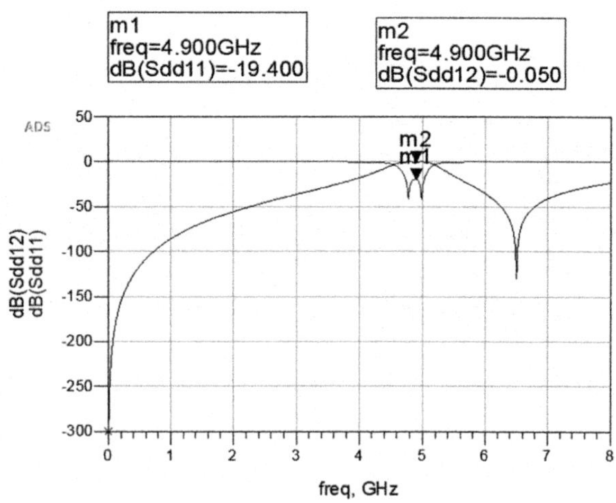

Fig. 1.42 Values of dB(Sdd11) and dB(Sdd12) at the center frequency

There are other types of markers: Execute menu commands [Marker] → [New Peak…] to view the peak value of a trace; Execute menu commands [Marker] → [New Valley…] to view the valley value of a trace; Execute menu commands [Marker] → [New Max..] to view the maximum value of a trace; Execute menu commands [Marker] → [New Min..] to view the minimum value of a trace; Execute menu commands [Marker] → [New Line..] to insert a vertical line which intersects with multiple traces, so multiple values in one frequency can be displayed at the same time.

(3) The editing function of the data display is introduced below. Take the modification and beautification of the Y axis as examples. Double-click the simulated S-parameters figure, then the [Plot Traces and Attributes] dialog box pops up. Click the [Plot Options] tab to deselect the "Auto Scale" (The automatic adjustment scale of the ADS software is not used). Adjust the Y axis scale shown in Fig. 1.43, then the adjusted S-parameter results in Fig. 1.44 can be obtained.

(4) In addition, the type, color, thickness, and the symbol of the trace can also be modified. Double-click the dB(Sdd11) trace to open the [Trace Options] dialog box and modify the trace, as shown in Fig. 1.45 (Keep the trace color in default). Similarly, modify the dB(Sdd12) trace shown in Fig. 1.46. The final dB(Sdd11) and dB(Sdd12) traces are shown in Fig. 1.47.

(5) Similarly, modify dB(Scc11) and dB(Scc12) traces. The final dB(Scc11) and dB(Scc12) traces can be obtained, as shown in Fig. 1.48.

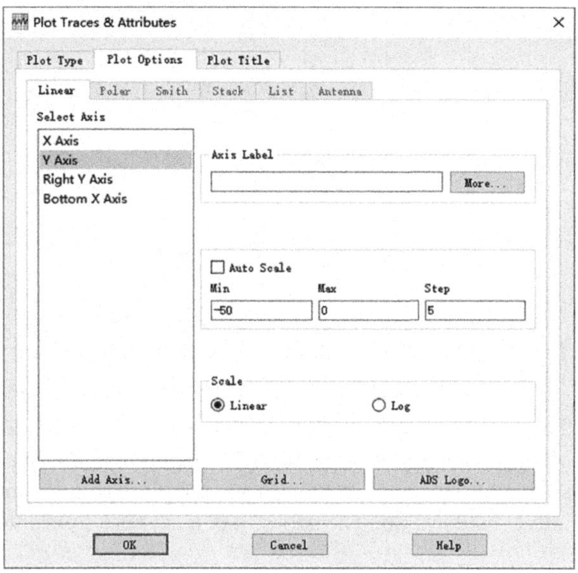

Fig. 1.43 Adjustment of the Y axis scale

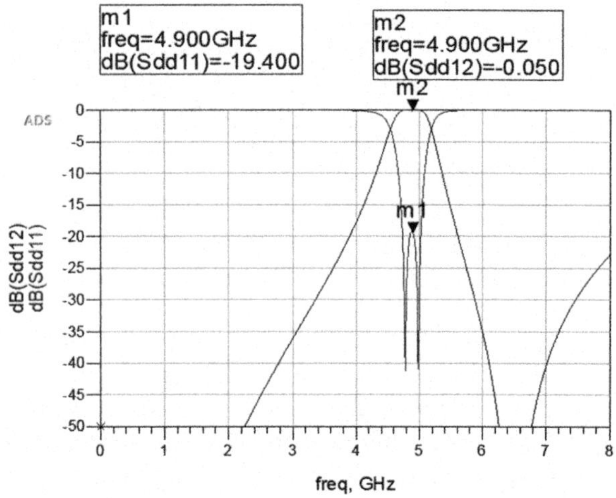

Fig. 1.44 The adjusted S-parameter results

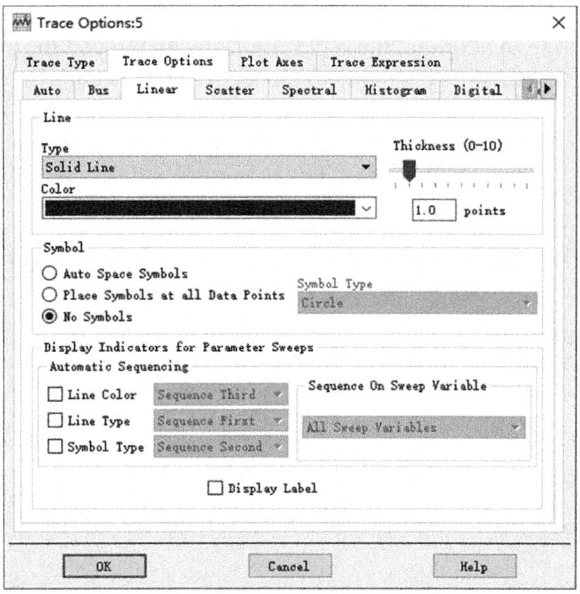

Fig. 1.45 Modification of the dB(Sdd11) trace

1.3 Layout of Balanced Bandpass Filter

The simulation of the schematic is carried out in an ideal state, while the performance of the actual circuit often has some differences comparing with the theoretical results, so the influences of interferences, couplings, and some other factors should be taken into consideration. Therefore, it is necessary to use the ADS software for the layout

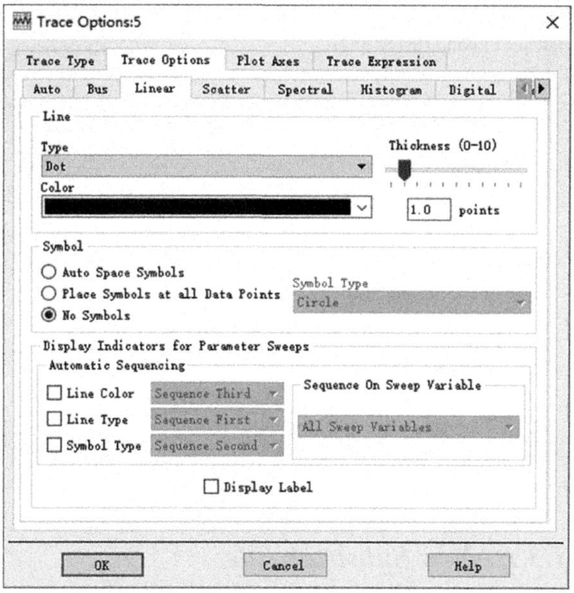

Fig. 1.46 Modification of the dB(Sdd12) trace

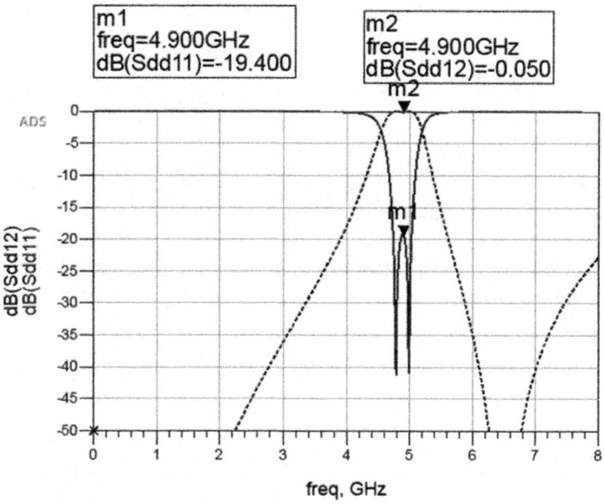

Fig. 1.47 The final dB(Sdd11) and dB(Sdd12) traces

simulation. ADS can use the Method of Moment (MOM) to analyze the electromagnetic simulation of the circuit, making the simulated results more accurate than those in the schematic. Besides, co-simulation of the schematic and layout can also be carried out. After completing the layout simulation and the co-simulation of the schematic and layout in ADS, the performance of the final manufactured chips can be more consistent with the expectation.

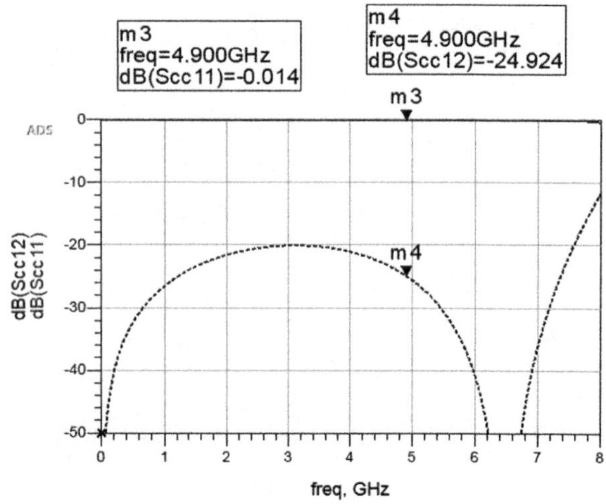

Fig. 1.48 The final dB(Scc11) and dB(Scc12) traces

1.3.1 New Substrate File

All circuit elements are constructed on the gallium arsenide (GaAs) substrate with the thickness of 200 μm, relative dielectric constant of 12.85, and the loss tangent of 0.006. The top and bottom 5-μm-thick metals and the 0.2-μm-thick intermediate silicon nitride (Si_3N_4) material construct the metal-insulate-metal (MIM) capacitor.

Before drawing the layout, the substrate file needs to be set according to the used TFIPD technology.

1. New Layout

Return to the "Balanced_Bandpass_Filter_wrk" main interface of the workspace. Execute menu commands [File] → [New] → [Layout...] to open the [New Layout] dialog box in Fig. 1.49. Modify the "Cell" name to "MIM capacitor", then click the [Create Layout] button. After that, the [Choose Layout Technology] dialog box shown in Fig. 1.50 will pop up. Here, select "Standard ADS Layers, 0.001 micron layout resolution", which is 0.001 μm (note that the unit is μm in this chapter), click [Finish], and the new layout window will pop up, as shown in Fig. 1.51.

- Explanation

  The following methods can be used to create a new layout:

  - Execute menu commands [File] → [New] → [Layout...];
  - Click the [New Layout Window] icon in the toolbar.

  If the following content relates to the creating of a new layout, use one of the methods mentioned above.

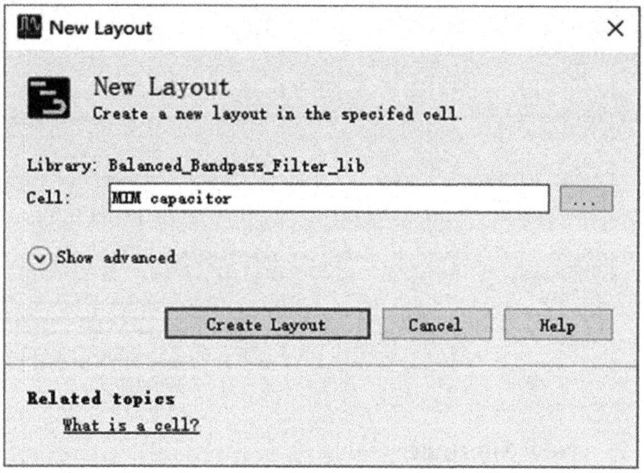

Fig. 1.49 [New Layout] dialog box

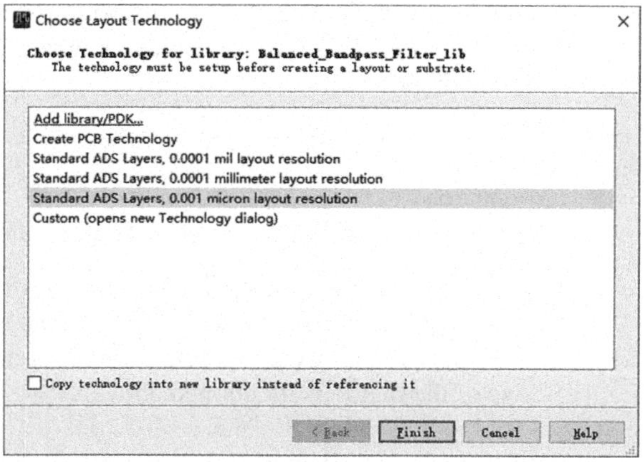

Fig. 1.50 [Choose Layout Technology] dialog box

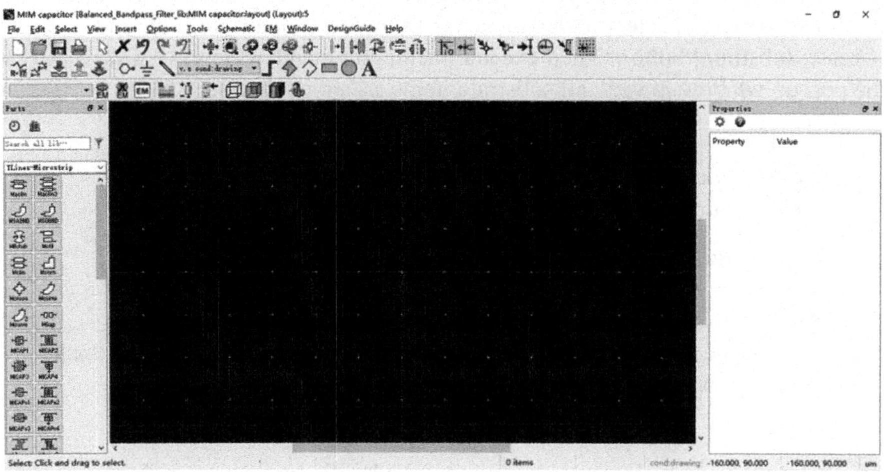

Fig. 1.51 New layout window

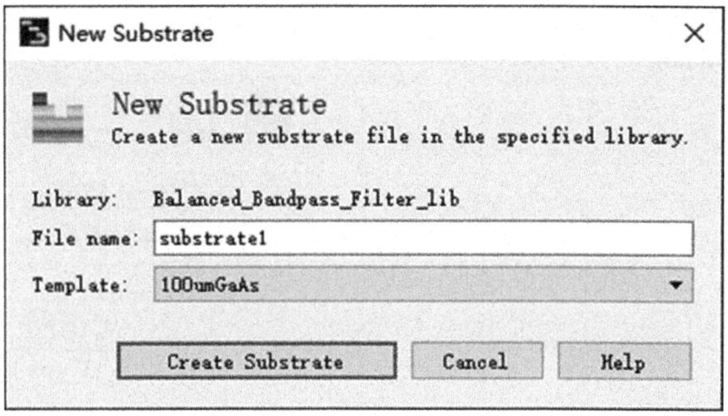

Fig. 1.52 [New Substrate] dialog box

2. New Substrate

(1) Create a new substrate. Execute menu commands [EM] → [Substrate…] in the layout window, then click [OK] in the pop-up dialog box. The [New Substrate] dialog box in Fig. 1.52 will be displayed, in which "File name" and "Template" can be modified.

- Explanation

  The following methods can be used to create a new substrate:

  - Execute menu commands [EM] → [Substrate…];
  - Click the [Substrate Editor] icon in the toolbar.

  If the following content relates to the creating of a new substrate, use one of the methods mentioned above.

Keep the default file name here. Because the used TFIPD technology is based on the GaAs substrate, select "100μmGaAs" in the "Template" column and click the [Create Substrate] button. Then execute menu commands [View] → [View All] in the pop-up window to view all substrate items, as shown in Fig. 1.53.

- Explanation

  The following methods can be used to view all substrate items:

  - Execute menu commands [View] → [View All];
  - Click the [View All] icon in the toolbar;
  - Use the shortcut "F".

  If the following content relates to the view all substrate items, use one of the methods mentioned above.

(2) Add conductors. Execute menu commands [Technology] → [Material Definitions…] to open the [Material Definitions] dialog box shown in Fig. 1.54.

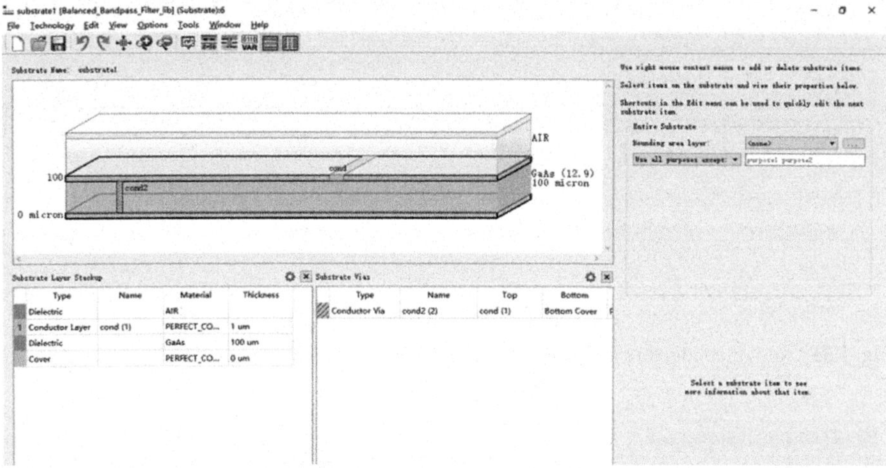

Fig. 1.53 View of all substrate items

Select the [Conductors] tab and define the relevant conductors in Fig. 1.55. The specific steps are listed as follows: Click the [Add From Database…] button in the lower right corner of Fig. 1.55. If the conductor to be added exists in the pop-up [Add Materials From Database] dialog box shown in Fig. 1.56, select the conductor and click [OK] to add it. If the conductor to be added does not exist in the [Add Materials From Database] dialog box, return to the [Material Definitions] dialog box and click the [Add Conductor] button to add a conductor and modify its properties. In addition, you can remove unwanted conductors by clicking the [Remove Conductor] button in the [Material Definitions] dialog box. After finishing all the steps, click the [Apply] button.

(3) Add dielectrics. Select the [Dielectrics] tab and define the relevant dielectrics in Fig. 1.57. The specific steps are listed as follows: Click the [Add From Database…] button in the lower right corner of Fig. 1.57. If the dielectric to be

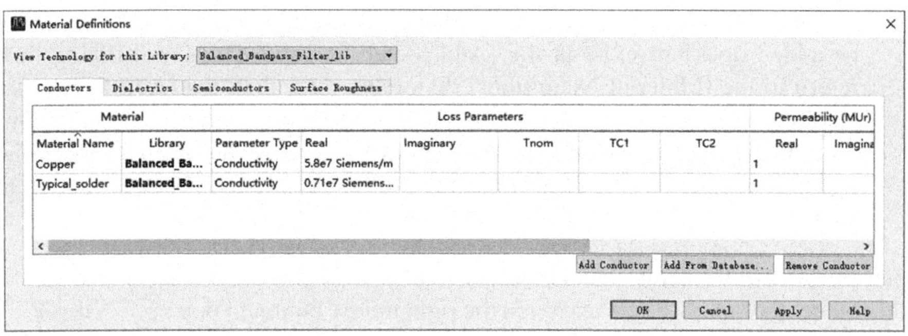

Fig. 1.54 [Material Definitions] dialog box

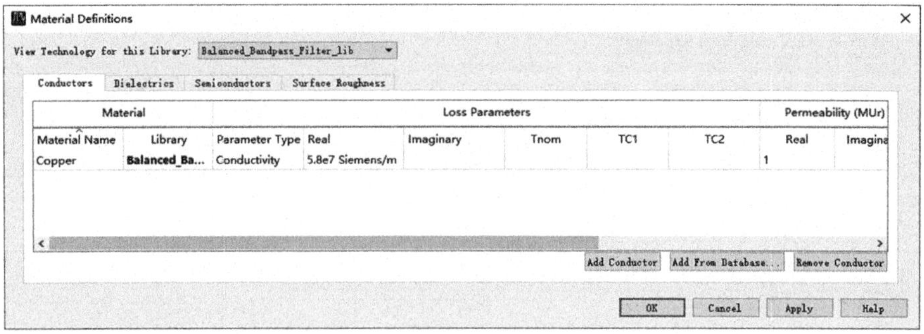

Fig. 1.55 Add of conductors and modification of their properties

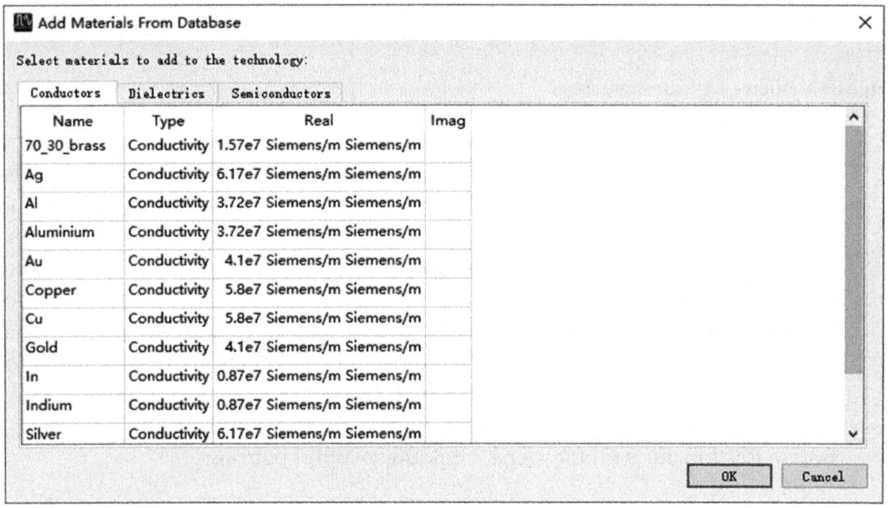

Fig. 1.56 [Conductors] tab of the [Add Materials From Database] dialog box

added exists in the pop-up [Add Materials From Database] dialog box shown in Fig. 1.58, select the dielectric and click [OK] to add it. If the dielectric to be added does not exist in the [Add Materials From Database] dialog box, return to the [Material Definitions] dialog box and click the [Add Dielectric] button to add a dielectric and modify its properties. In addition, you can remove unwanted dielectrics by clicking the [Remove Dielectric] button in the [Material Definitions] dialog box. After finishing all the steps, click the [OK] button to close the [Material Definitions] dialog box.

3. Edit Substrate Items

(1) Select the "cond" layer and click the right mouse button to pop up Unmap . Click the "Unmap" to delete this layer (Or select the "cond" layer, then press

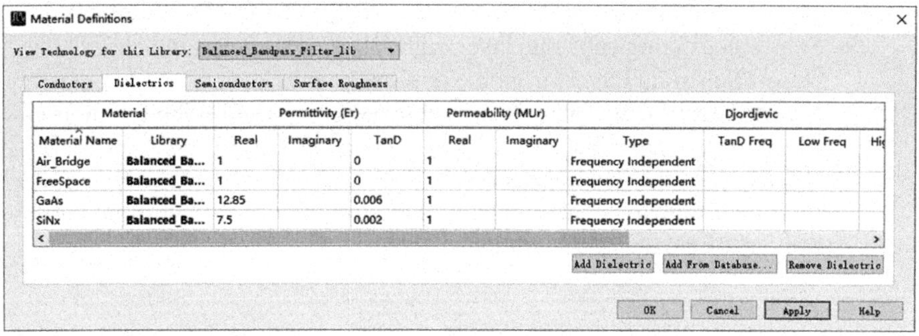

Material		Permittivity (Er)			Permeability (MUr)		Djordjevic			
Material Name	Library	Real	Imaginary	TanD	Real	Imaginary	Type	TanD Freq	Low Freq	Hi
Air_Bridge	Balanced_Ba...	1		0	1		Frequency Independent			
FreeSpace	Balanced_Ba...	1		0	1		Frequency Independent			
GaAs	Balanced_Ba...	12.85		0.006	1		Frequency Independent			
SiNx	Balanced_Ba...	7.5		0.002	1		Frequency Independent			

Fig. 1.57 Add of dielectrics and modification of their properties

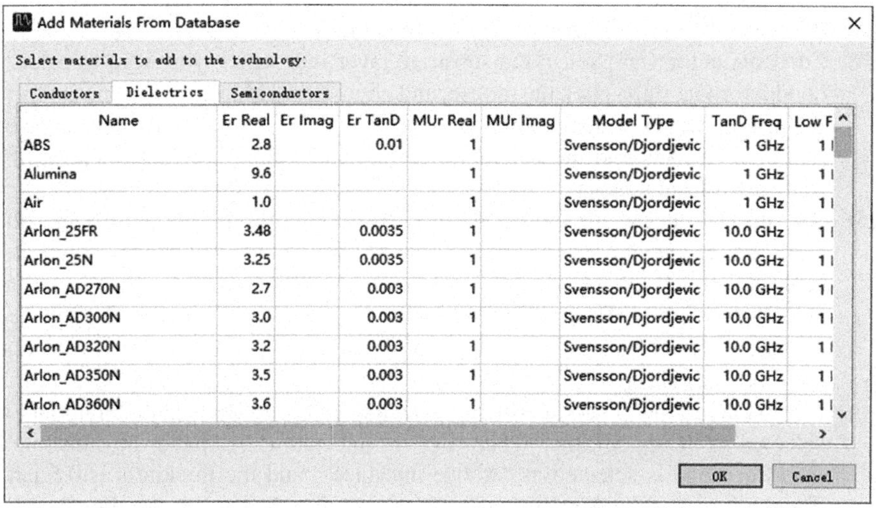

Name	Er Real	Er Imag	Er TanD	MUr Real	MUr Imag	Model Type	TanD Freq	Low F
ABS	2.8		0.01	1		Svensson/Djordjevic	1 GHz	1
Alumina	9.6			1		Svensson/Djordjevic	1 GHz	1
Air	1.0			1		Svensson/Djordjevic	1 GHz	1
Arlon_25FR	3.48		0.0035	1		Svensson/Djordjevic	10.0 GHz	1
Arlon_25N	3.25		0.0035	1		Svensson/Djordjevic	10.0 GHz	1
Arlon_AD270N	2.7		0.003	1		Svensson/Djordjevic	10.0 GHz	1
Arlon_AD300N	3.0		0.003	1		Svensson/Djordjevic	10.0 GHz	1
Arlon_AD320N	3.2		0.003	1		Svensson/Djordjevic	10.0 GHz	1
Arlon_AD350N	3.5		0.003	1		Svensson/Djordjevic	10.0 GHz	1
Arlon_AD360N	3.6		0.003	1		Svensson/Djordjevic	10.0 GHz	1

Fig. 1.58 [Dielectrics] tab of the [Add Materials From Database] dialog box

"Delete" on the keyboard to delete it). Delete the "cond2" layer in the same way.

(2) Edit substrate layers. Select an existing substrate layer, right-click the mouse, and choose the [Insert Substrate Layer] from the pop-up menu to insert a new substrate layer. Select the substrate layer to be modified to view and modify its properties under the [Substrate Layer] at the right side of the window.

- Explanation

  The following methods can be to insert a substrate layer:

- Select an existing substrate layer, right-click the mouse, and choose the [Insert Substrate Layer] from the pop-up menu;
- Select the surface of an existing substrate layer, right-click the mouse, and choose [Insert Substrate Layer Above] or [Insert Substrate Layer Below] from the pop-up menu.

If the following content relates to the insertion of the substrate layer, use one of the methods mentioned above.

(3) Edit conductor layers. Select the surface of the substrate layer on which you want to insert a conductor layer, right-click the mouse, and choose the [Map Conductor Layer] from the pop-up menu to insert a new conductor layer. Select the conductor layer to be modified to view and modify its properties under the [Conductor Layer] at the right side of the window.

(4) Edit conductor vias. Select the substrate layer into which you want to insert a conductor via, right-click the mouse, and choose the [Map Conductor Via] from the pop-up menu to insert a new conductor via. Select the conductor via to be modified to view and modify its properties under the [Conductor Via] at the right side of the window.

(5) All substrate items are shown in Fig. 1.59. The bottom layer is "Cover", the thickness of the GaAs substrate layer is 200 μm, and the thickness of the first SiNx layer is 0.1 μm. For the bond layer, the "Process Role" is selected as "Conductor", the "Material" is selected as "Copper", the "Operation" is selected as "Expand the substrate", the "Position" is selected as "Above interface", and the thickness is 5 μm. The thickness of the second SiNx layer is 0.2 μm. For the text layer, the "Process Role" is selected as "Conductor", the "Material" is selected as "Copper", the "Operation" is selected as "Expand the substrate", the "Position" is selected as "Above interface", and the thickness is 0.5 μm. The thickness of the Air_Bridge layer is 3 μm. For the leads layer, the "Process Role" is selected as "Conductor", the "Material" is selected as "Copper", the "Operation" is selected as "Intrude the substrate", the "Position" is selected as "Above interface", and the thickness is 5 μm. The top layer is the "FreeSpace" layer. For the symbol layer, the "Process Role" is selected as "Conductor Via" and the "Material" is selected as "Copper". For the packages layer, the "Process Role" is selected as "Conductor Via" and the "Material" is selected as "Copper".

(6) Moreover, execute menu commands [Technology] → [Layer Definitions…] to open the [Layer Definitions] dialog box shown in Fig. 1.60, where you can modify the display properties of each layer such as color and pattern, etc. Here, keep all the settings in default.

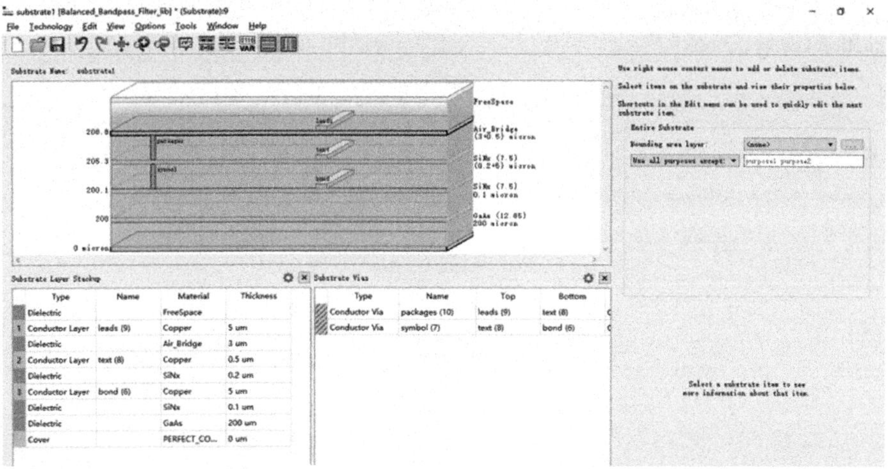

Fig. 1.59 Substrate layer stackup and substrate vias

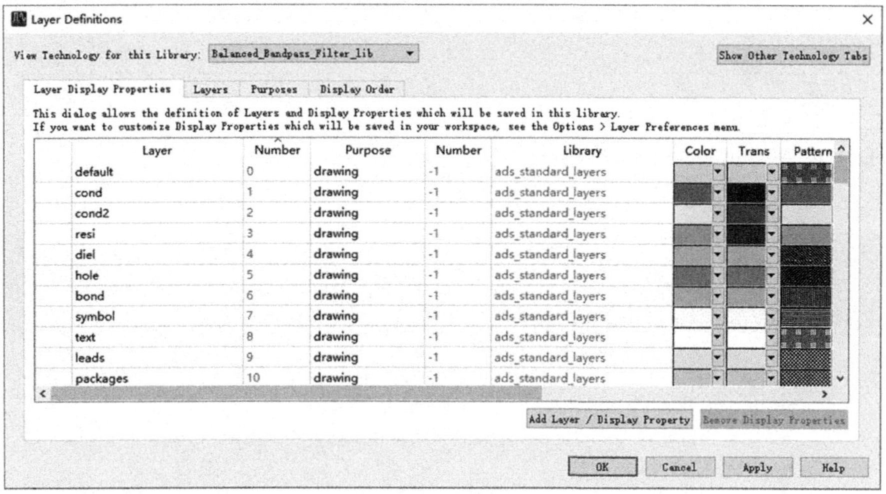

Fig. 1.60 [Layer Display Properties] tab of the [Layer Definitions] dialog box

1.3.2 Layout of MIM Capacitor

To facilitate the layout drawing, select the function keys in the box of Fig. 1.61 in the "MIM capacitor" cell. Execute menu commands [Options] → [Preferences…], select the [Grid/Snap] tab in the pop-up [Preferences for Layout] dialog box to modify the display grid in the layout. Set the "Snap Grid Distance (in layout units)", "Snap Grid Per Minor Display Grid", and the "Minor Grid Per Major Display Grid" to appropriate values, as shown in Fig. 1.62.

Fig. 1.61 Selection of the function key in the box

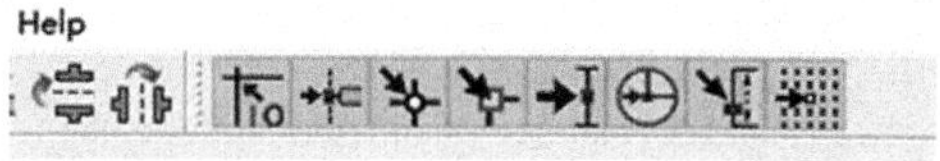

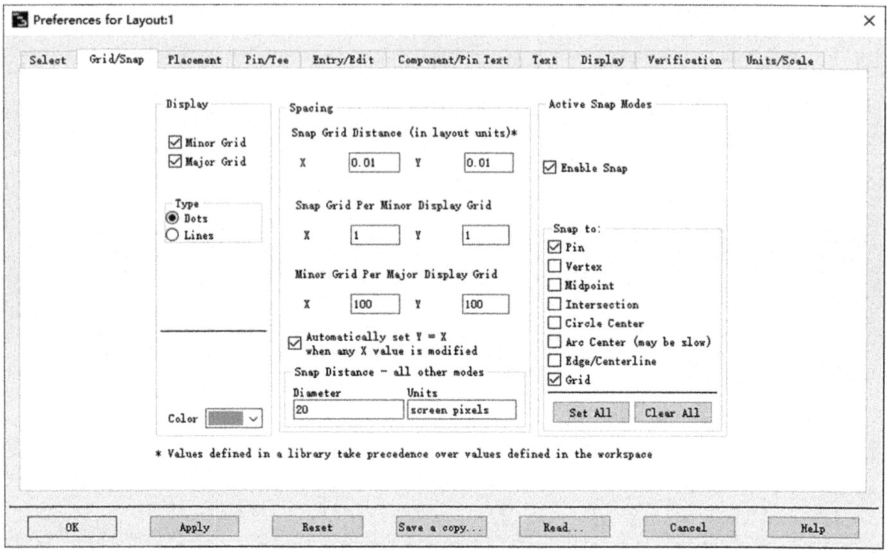

Fig. 1.62 Modification of the display grid in the layout

- Explanation

  The following methods can be used to realize the modification of the displayed grid in the layout:

  - Select the [Grid/Snap] tab in the [Preferences for Layout] dialog box, then set the "Snap Grid Distance (in layout units)", "Snap Grid Per Minor Display Grid", and the "Minor Grid Per Major Display Grid" to appropriate values;
  - Right-click in the drawing area of the layout, select " < 0.1–1-100 > " under the "Grid Spacing..." from the menu that pops up;
  - Use the shortcut "Ctrl + Shift + 8".

If the following content relates to the modification of the displayed grid in the layout, use one of the methods mentioned above.

1. Layout Drawing of MIM Capacitor

Two 5-μm-thick copper layers (bond and leads layers) and a 0.2-μm-thick intermediate silicon nitride (Si_3N_4) layer are used to construct the MIM capacitors. The area

Fig. 1.63 [Copy to Layer] dialog box

and thickness of the intermediate dielectric layer determine the value of the MIM capacitor. The MIM capacitor with a capacitance of 0.89 pF is taken as an example to describe its drawing steps in detail below.

(1) MIM capacitor lamination. Execute menu commands [Insert] → [Rectangle] to insert a rectangle into the layout, then press "Esc" to exit. Select the inserted rectangle, select "bond: drawing" in the drop-down menu of the [All Shapes] → [Layer] under [Properties] at the right side of the window, and set the [Rectangles] → [Width] to 58 μm, [Height] to 56 μm. Execute menu commands [Edit] → [Copy/Paste] → [Copy To Layer…], select [text: drawing] in the pop-up [Copy to Layer] dialog box, as shown in Fig. 1.63. Click the [Apply] button to copy a text layer in the original position. Similarly, copy a leads layer and a packages layer in the original position, respectively. After copying all the layers, click the [Cancel] button to close this dialog box.

- Explanation

  The following methods can be used to insert a rectangle:

  - Execute menu commands [Insert] → [Rectangle];
  - Click the [Insert Rectangle] icon ▭ in the toolbar;
  - Use the shortcut "R".

If the following content relates to the insertion of a rectangle, use one of the methods mentioned above.

(2) The indentation of layers. There are different indentations between the layers of the MIM capacitor. Select the leads layer and execute menu commands [Edit] →

[Scale/Oversize] → [Oversize…]. Since the leads layer is indented by 1.5 μm compared with the bond layer, enter -1.5 in the "Oversize(+)/Undersize(-)" column of the pop-up [Oversize] dialog box and click the [Apply] button, as shown in Fig. 1.64. Similarly, indent the text and packages layers by 3 μm relative to the bond layer.

(3) Draw connection parts. Execute menu commands [Insert] → [Rectangle] to insert a rectangle in the layout, then press "Esc" to exit. Select the inserted rectangle, select "bond: drawing" in the drop-down menu of the [All Shapes] → [Layer] under [Properties] at the right side of the window, and set the [Rectangles] → [Width] to 20 μm, [Height] to 20 μm. Select and long press the modified rectangle by the left mouse button to move it to the middle position of the MIM capacitor and connect it with the original bond layer. Similarly, a leads layer rectangle is inserted in the opposite position and is connected to the original leads layer. The layout of a MIM capacitor has been drawn, as shown in Fig. 1.65.

2. Layout Simulation of MIM Capacitor
(1) Insert ports. Execute menu commands [Insert] → [Pin] and click the left mouse button to insert two pins into the I/O ports of the MIM capacitor, as shown in Fig. 1.66.

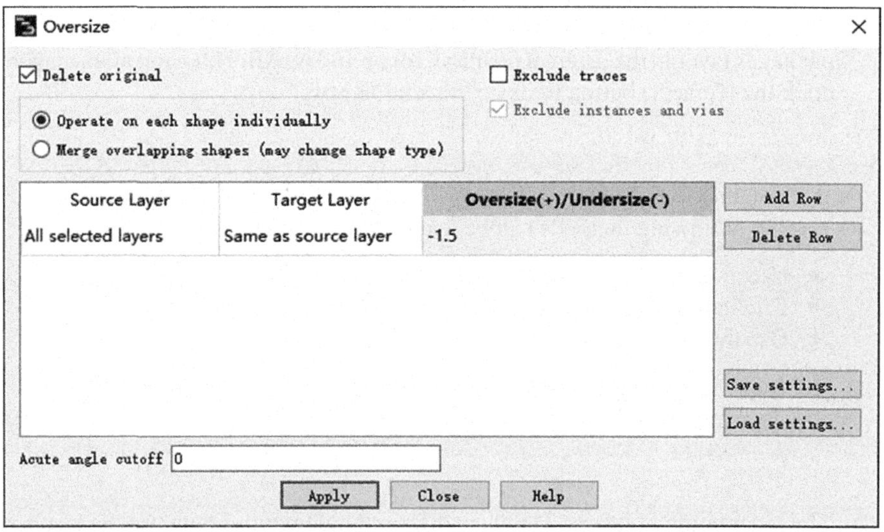

Fig. 1.64 [Oversize] dialog box

Fig. 1.65 Final layout of a MIM capacitor

Fig. 1.66 Insert pins on MIM capacitor

- Explanation

  The following methods can be used to insert a pin:

  - Execute menu commands [Insert] → [Pin];
  - Click the [Insert Pin] icon  in the toolbar.

  If the following content relates to the insertion of a pin, use one of the methods mentioned above.

(2) Modify simulation settings. Execute menu commands [EM] → [Simulation Settings…] and the [New EM Setup View] dialog box shown in Fig. 1.67 will pop up. Click the [Create EM Setup View] button and the [EM Setup for simulation] window will pop up, as shown in Fig. 1.68. Select the EM simulator first. The second method "Momentum Microwave" is used generally because this method runs fast and its high accuracy can also meet the application requirement (The first method "Momentum RF" is the fastest method with the least accuracy. The third method "FEM" has the highest accuracy, but it shows the slowest simulation speed, so it is mainly used for some complex three-dimensional structure simulations). Select the [Frequency Plan] tab to set the simulation frequency. Select "Adaptive" in the "Type" column, then set the "Fstart" to 0 GHz, the "Fstop" to 8 GHz, and the "Npts" to 50. Finally, select the [Options] tab. Click "Preprocessor", select the "User specified snap distance" option under

the "Heal the layout", and set the distance to 2.5 μm. Click "Mesh" and select the "Edge Mesh" option. Keep other settings in default. After finishing all the steps, close the window and click the [OK] button to save the settings.

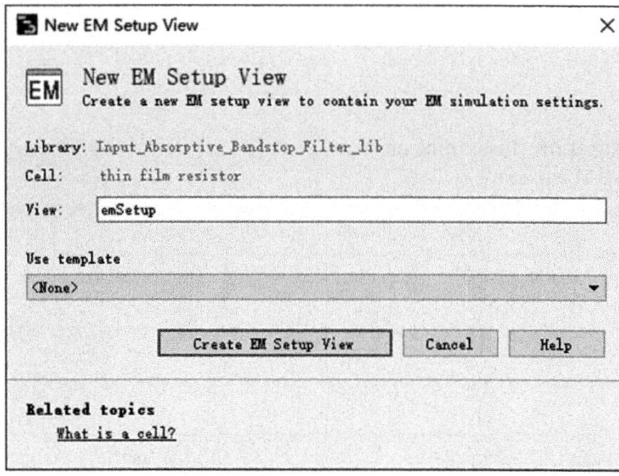

Fig. 1.67 [New EM Setup View] dialog box

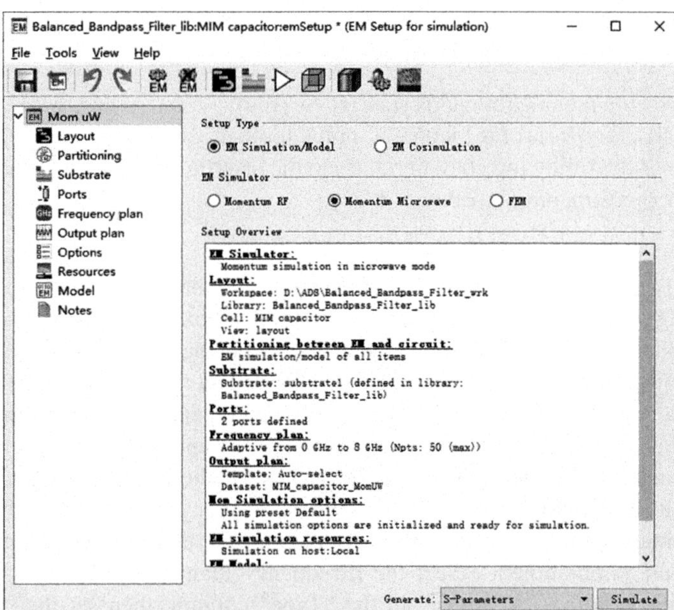

Fig. 1.68 [EM Setup for simulation] window

- Explanation

  The following methods can be used to achieve the operation of simulation settings:

  - Execute menu commands [EM] → [Simulation Settings…];
  - Click the [EM Simulation Setup] icon EM in the toolbar;
  - Use the shortcut "F6".

  If the following content relates to the operation of simulation settings, use one of the methods mentioned above.

(3) Layout simulation. Execute menu commands [EM] → [Simulate] to simulate the layout of the MIM capacitor. During the simulation, a status window will pop up to display the current item progress. After the finishing of the simulation, the data display window will be opened automatically. Refer to the steps in Sect. 1.2.2 to view and handle traces dB(S(1,1)) and dB(S(2,1)), the final results are shown in Fig. 1.69.

- Explanation

  The following methods can be used to realize the layout simulation:

  - Execute menu commands [EM] → [Simulate];
  - Click the [Start EM Simulation] icon EM EM in the toolbar;
  - Use the shortcut "F7".

  If the following content relates to the layout simulation, one of the methods mentioned above.

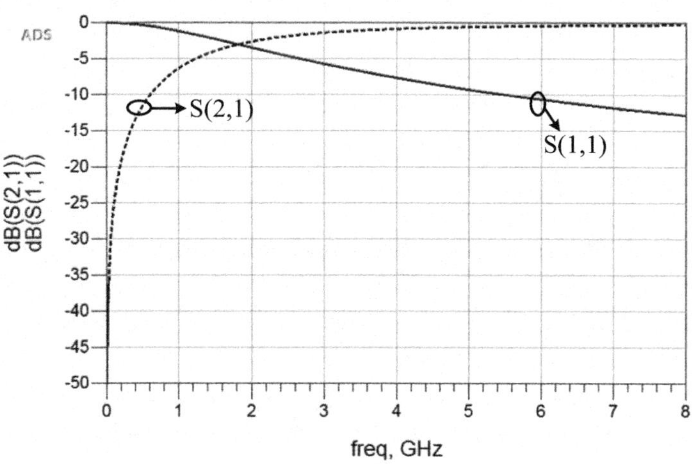

Fig. 1.69 Final simulated results of the drawn MIM capacitor

3. Co-Simulation of MIM Capacitor

To verify whether the capacitance of the drawn MIM capacitor is 0.89 pF, the co-simulation of the capacitor schematic and layout is required.

(1) Create EM model and symbol of the MIM capacitor. In the layout window, execute menu commands [EM] → [Component] → [Create EM Model And Symbol…], then click the [OK] button in the pop-up dialog boxes. Execute menu commands [Edit] → [Component] → [Update Component Definitions…], then click the [OK] button in the pop-up dialog box to complete the creation of the EM model and the symbol in the MIM capacitor.

(2) Create a new schematic and insert the MIM capacitor component. Return to the "Balanced_Bandpass_ Filter_wrk" main interface of the workspace. Execute menu commands [File] → [New] → [Schematic…], modify the "Cell" name to "MIM capacitor-cosimulation" in the opened [New Schematic] dialog box, and click the [Create Schematic] button to create a new schematic. Click the [Open the Library Browser] icon  at the left side of the schematic window, then select the "MIM capacitor" component under the [Workspace Libraries] in the displayed [Component Library] window, as shown in the Fig. 1.70. Click the right mouse button, then click the [Place Component] in the pop-up menu to insert a MIM capacitor component in the schematic. Press "Esc" to exit.

- Explanation
  The following methods can be used to insert a component:
  - Click the [Place Component] and click the left mouse button in the schematic window;
  - Drag and drop the component name into the schematic window.

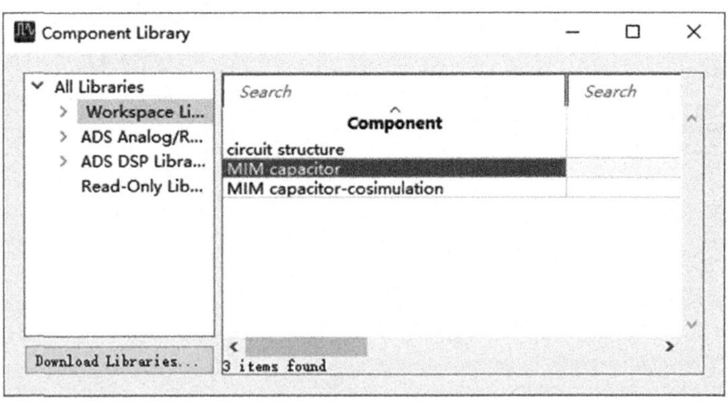

Fig. 1.70 MIM capacitor component in the [Component Library] window

If the following content relates to the insertion of a component, use one of the methods mentioned above.

(3) Insert the ideal capacitor. Select the [Lumped-Components] from the drop-down menu in the left component palette list and click the [Capacitor] icon 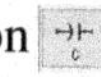 to insert a capacitor. Press "Esc" on the keyboard to exit. Double-click the capacitor and modify C = 0.89 pF in the [Edit Instance Parameter] pop-up dialog box (check whether the unit setting is consistent), then click the [OK] button to save the parameter.

(4) Insert the S-parameter simulator, term ports, and grounds. Select the [Simulation-S_Param] from the drop-down menu in the left component palette list and click the [S-parameter Simulator] icon  to insert an S-parameter simulator. Click the [Port Impedance Termination for S-Parameters] icon  to insert four term ports, then press "Esc" to exit. Execute menu commands [Insert] → [GROUND] to place four grounds (Or click the [TermG Port Impedance Termination for S-Parameters with Grounded Reference] icon  to insert four term ports with the grounded reference). Execute menu commands [Insert] → [Wire] to connect inserted components. After finishing all the steps, press "Esc" to exit.

(5) Set simulation frequency. Double-click the S-parameter simulator . Set the "Start" to 0 GHz, the "Stop" to 8 GHz, and the "Step-size" to 0.01 GHz, then click the [OK] button. The final capacitor co-simulation circuit is obtained, as shown in Fig. 1.71.

(6) Co-simulation. Execute menu commands [Simulate] → [Simulate]. After the simulation completes, the data display window will be opened automatically. Refer to the steps in Sect. 1.2.2 to view and handle traces of dB(S(1,1)), dB(S(2,1)), dB(S(3,3)), and dB(S(4,3)), the final results are shown in Fig. 1.72. It can be seen that traces dB(S(1,1)) and dB(S(3,3)), traces dB(S(2,1)) and dB(S(4,3)) are almost coincident, indicating that the capacitance of the drawn MIM capacitor is approximately equal to 0.89 pF. If the differences between these traces are large, go back to modify the MIM capacitor layout and repeat the steps above until the error of two traces is in the range of acceptance (within an acceptable range).

1.3.3 Layout of Balanced Bandpass Filter

(1) Draw MIM capacitors. Refer to the steps in Sect. 1.3.2 of this chapter to draw MIM capacitors with capacitances of 0.56 and 1.00 pF, respectively.

(2) Create a new layout and copy the corresponding component layouts. In the "Balanced_Bandpass_Filter_wrk" main interface of the workspace, execute menu

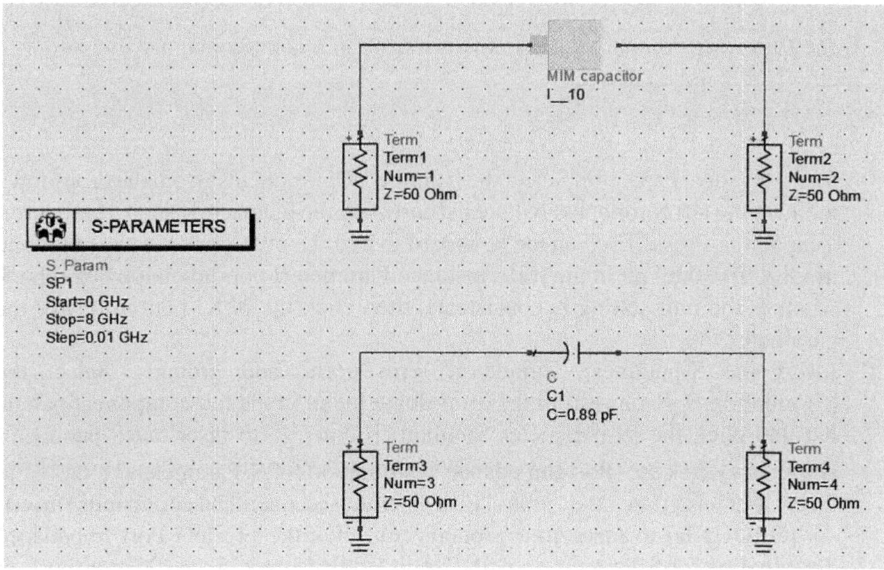

Fig. 1.71 Capacitor co-simulation circuit

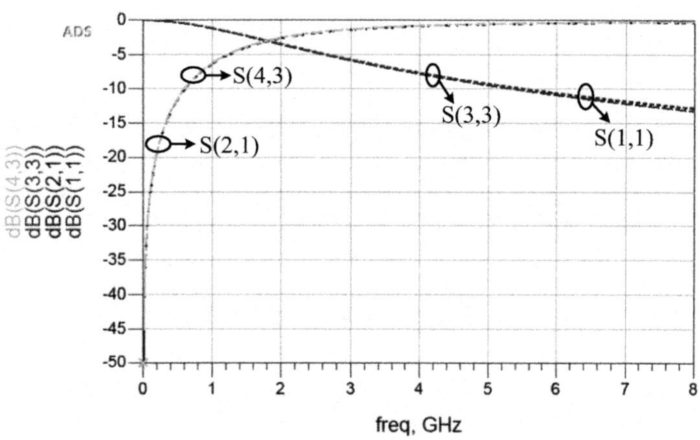

Fig. 1.72 S-parameter results of the capacitor co-simulation

commands [File] → [New] → [Layout…] and modify the "Cell" name to "balanced bandpass filter" in the opened [New Layout] dialog box. Click the [Create Layout] button, then a new layout window will pop up. Press "Ctrl + C" and "Ctrl + V" successively to copy the drawn MIM capacitor layout to this layout window. It can be observed from the circuit model of the balanced bandpass filter shown in Fig. 1.28 that there are two capacitors with a capacitance of 0.56 pF, four capacitors with the capacitance of 0.89 pF, and four capacitors with a capacitance of 1.00 pF. So the 0.56 pF MIM capacitor should be copied twice, the 0.89 pF MIM capacitor and 1.00 pF MIM capacitor should be copied for four times, respectively.

(3) Draw inductors. Execute menu commands [Insert] → [Path…], the [Path] dialog box shown in Fig. 1.73 will pop up. Select "bond: drawing" in the "Layer" column and set the "Width (μm)" column to 15, then the bottom metal of the inductor can be drawn. As shown in Fig. 1.74, the total length is 607 μm, the total width is 258 μm, and the length of two sections on the right side are both 70 μm. Execute menu commands [Edit] → [Copy/Paste] → [Copy To Layer…] and select [text: drawing] in the pop-up [Copy to Layer] dialog box. Click the [Apply] button to copy a text layer in the original position. Similarly, copy a leads layer, a symbol layer, and a packages layer in the original position, respectively. After copying all the needed layers, click the [Cancel] button to close this dialog box. Then indent each layer. Select the text layer and execute menu commands [Edit] → [Scale/Oversize] → [Oversize…]. Since the text layer is indented by 2 μm compared with the bond layer, enter -2 in the "Oversize(+)/Undersize(-) " column of the pop-up [Oversize] dialog box and click the [Apply] button. Similarly, indent the symbol and packages layers by 2 μm relative to the bond layer. Finally, the inductor layout drawing is completed. It is noteworthy that the chip package can bring in the equivalent inductors of wire bonds ($L_2 = 0.3$ nH and $L_3 = 0.6$ nH), so there is no need to draw the inductors L_2 and L_3 in the layout of the balanced bandpass filter.

(4) Draw pads. To facilitate the packaging and testing, pads should be added to the I/O and grounding ports. The size of the pad is fixed to 100 μm × 100 μm here. The drawing of pads is similar to that of inductors, except that the shape of pads in each metal layer is rectangular, which should execute menu commands

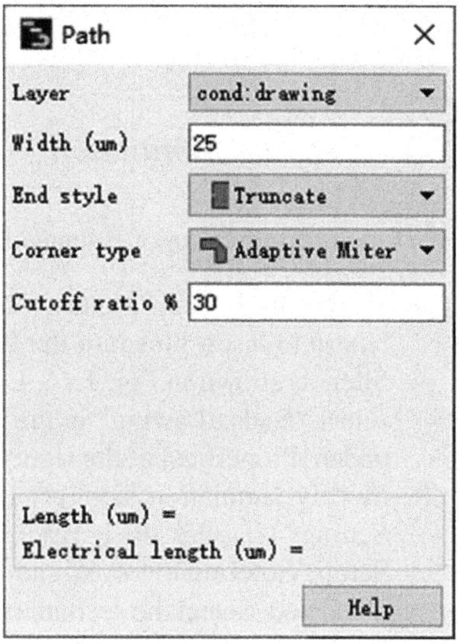

Fig. 1.73 [Path] dialog box

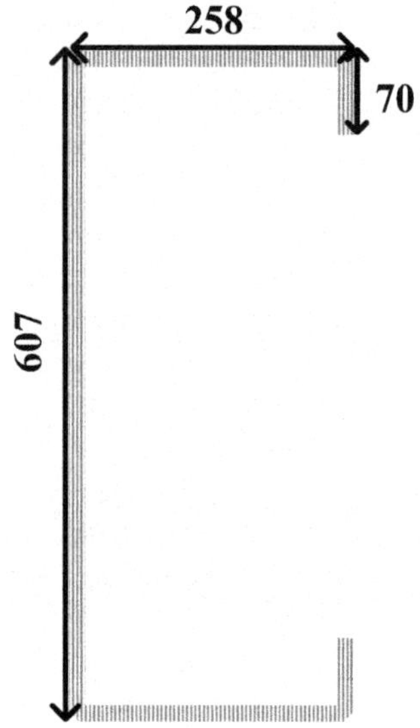

Fig. 1.74 Bottom metal of the inductor

[Insert] → [Rectangle] to insert the rectangle. For more details, please refer to step (3), which are not repeated here.

(5) Layout adjustment and component connection. The layout is adjusted in the consideration of the circuit size, layout beautification, and other factors. According to the simulated circuit model in Fig. 1.28, all components are connected by microstrip lines. The drawing of the microstrip line is similar to that of pads. For more details, please refer to step (3) and (4), which are not repeated here. Considering the influence of wire bonds in chip package on the performance of the balanced bandpass filter, the final layout is obtained after the optimizations in several times, as shown in Fig. 1.75 (unit: μm).

1.3.4 Layout Simulation

1. Layout Simulation of Balanced Bandpass Filter

(1) Insert pins. Execute menu commands [Insert] → [Pin], then click the left mouse button to insert pins into the I/O and grounded ports of the balanced bandpass filter, as shown in Fig. 1.76. Long press the "Ctrl" key and select all pins, then select "leads: drawing" in the drop-down menu of the [All Shapes] → [Layer] under [Properties] at the right side of the window.

(2) Modify simulation settings. Execute menu commands [EM] → [Simulation Settings…], click the [Create EM Setup View] button in the pop-up [New EM Setup View] dialog box, and the [EM Setup for simulation] window will be displayed. Select the second method "Momentum Microwave" of the EM simulator at first. Then select the [Frequency Plan] tab to set the simulation frequency.

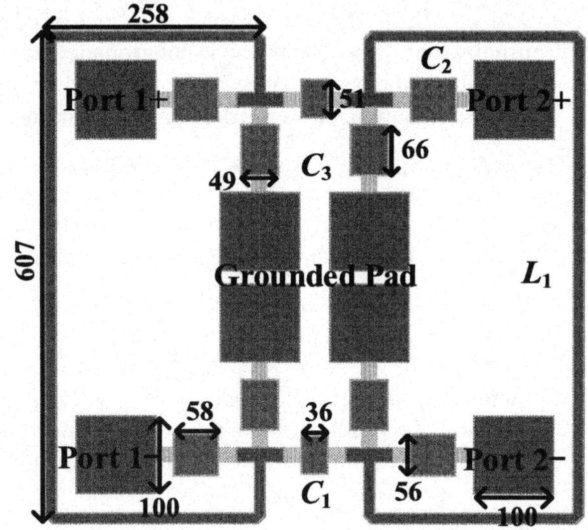

Fig. 1.75 Final layout of the balanced bandpass filter

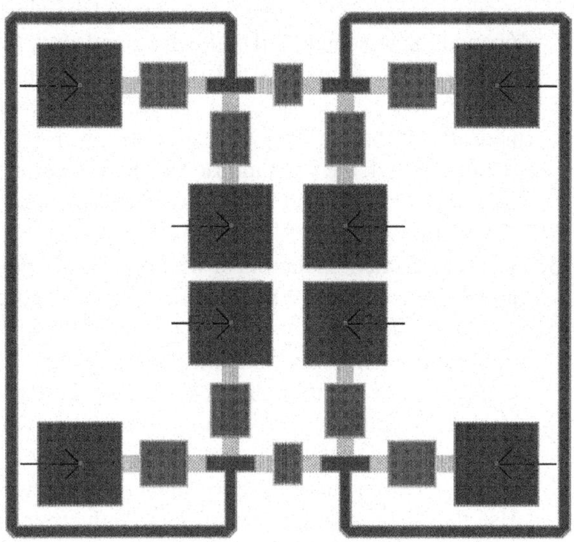

Fig. 1.76 Insert pins on the final layout

Select "Adaptive" in the "Type" column, then set the "Fstart" to 0 GHz, the "Fstop" to 8 GHz, and the "Npts" to 50. Finally, select the [Options] tab. Click "Preprocessor", select the "User specified snap distance" option under the "Heal the layout", and set the default distance to 2.5 μm. Click [Options] → [Mesh] and select the "Edge Mesh" option. Keep other settings in default. After finishing all the steps, close the window and click the [OK] button to save the settings.

(3) Layout simulation. Execute menu commands [EM] → [Simulate] to simulate the layout of the balanced bandpass filter. During the simulation, a status window

will pop up to display the current item's progress. The whole simulation is usually long. Since the layout is not grounded, the simulated results are not used to evaluate the performance of the balanced bandpass filter. After the finishing of the simulation, close the data display window.

2. Co-simulation of Balanced Bandpass Filter

In order to evaluate the performance of the balanced bandpass filter, the co-simulation of the balanced bandpass filter is carried out.

(1) Create EM model and symbol of the balanced bandpass filter. In the layout window, execute menu commands [EM] → [Component] → [Create EM Model And Symbol…], then click the [OK] button in the pop-up dialog boxes. Execute menu commands [Edit] → [Component] → [Update Component Definitions…] and click the [OK] button in the pop-up dialog box to complete the creation of the EM model and symbol of the balanced bandpass filter.
(2) Create a new schematic and insert the balanced bandpass filter component. Return to the "Balanced_Bandpass_ Filter_wrk" main interface of the workspace. Execute menu commands [File] → [New] → [Schematic…], modify the "Cell" name to "balanced bandpass filter-cosimulation" in the opened [New Schematic] dialog box, and click the [Create Schematic] button to create a new schematic. Click the [Open the Library Browser] icon  at the left side of the schematic window and select the "balanced bandpass filter" component under the [Workspace Libraries] in the displayed [Component Library] window. Click the right mouse button, then click the [Place Component] in the pop-up menu to insert a balanced bandpass filter component in the schematic. Press "Esc" to exit.
(3) Insert equivalent inductors of wire bonds. The equivalent inductors $L_2 = 0.3$ nH and $L_3 = 0.6$ nH of 200 um and 400 um wire bonds used in the chip package are inserted in the co-simulation circuit to make the co-simulation results be more consistent to the final measured results of the chip. Select the [Lumped-Components] from the drop-down menu in the left component palette list and click the [Inductor] icon 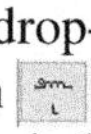 to insert an inductor. Press "Esc" on the keyboard to exit. Double-click the inductor and modify L = 0.3 nH in the pop-up [Edit Instance Parameter] dialog box, then click the [OK] button to save the parameter. Use the same method to insert three inductors of 0.3 nH and four inductors of 0.6 nH (Or insert two inductors and modify their inductances to 0.3 nH and 0.6 nH, respectively, then select the two inductors and press "Ctrl + C" and "Ctrl + V" to copy and paste).
(4) Insert the S-parameter simulator, term ports, and grounds. Select the [Simulation-S_Param] from the drop-down menu in the left component palette list and click the [S-parameter Simulator] icon  to insert an S-parameter simulator. Click the [Port Impedance Termination for S-Parameters] icon  to insert four term ports, then press "Esc" to exit. Execute menu commands [Insert] → [GROUND] to place eight grounds (Or click the [TermG Port Impedance Termination for S-Parameters with Grounded Reference] icon  to insert four term

ports with the grounded reference, then execute menu commands [Insert] → [GROUND] to place four grounds). Execute menu commands [Insert] → [Wire] to connect all components according to the simulated circuit model shown in Fig. 1.28. After finishing all the steps, press "Esc" to exit.

(5) Set simulation frequency. Double-click the S-parameter simulator 

. Set the "Start" to 0 GHz, the "Stop" to 8 GHz, and the "Step-size" to 0.01 GHz, then click the [OK] button. The final balanced bandpass filter co-simulation circuit is obtained, as shown in Fig. 1.77.

(6) Co-simulation. Execute menu commands [Simulate] → [Simulate] to complete the simulation. After the simulation completes, the data display window will be opened automatically. Refer to the steps in Sect. 1.2.2 to view and handle traces of dB(Sdd11), dB(Sdd12), dB(Scc11), and dB(Scc12), the final results are shown in Figs. 1.78 and 1.79.

(7) Analysis of simulated results. By comparing Figs. 1.47, 1.48, 1.78 and 1.79, it can be seen that the theoretical simulated results are in good agreement with the EM simulated results of this balanced bandpass filter. The DM return loss $|S_{dd11}|$ is better than 18 dB from 4.8 to 5.0 GHz in the EM simulation, while the insertion loss $|S_{dd12}|$ is below than 2.3 dB. Furthermore, the CM $|S_{cc12}|$ is larger than 20 dB from 0 to 7.5 GHz in the EM simulation.

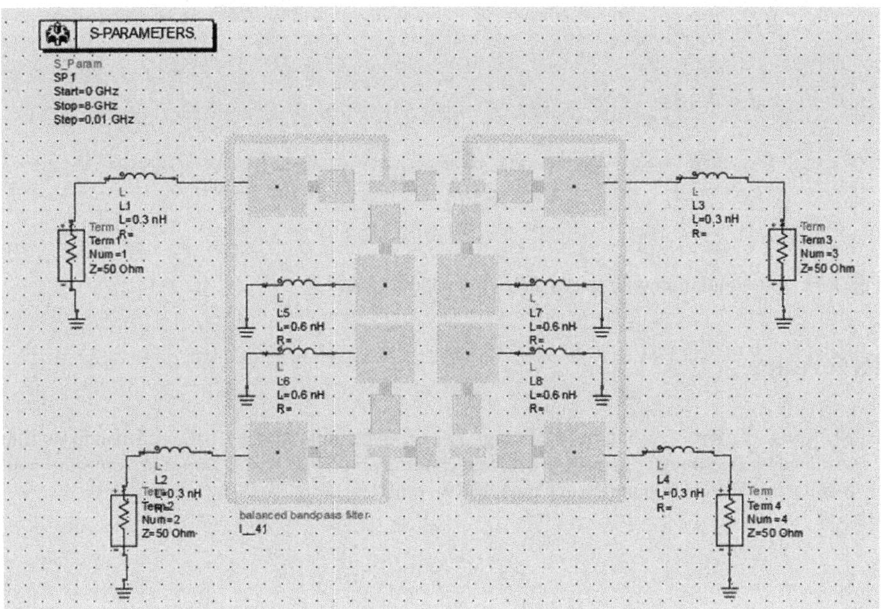

Fig. 1.77 Co-simulation circuit of the balanced bandpass filter

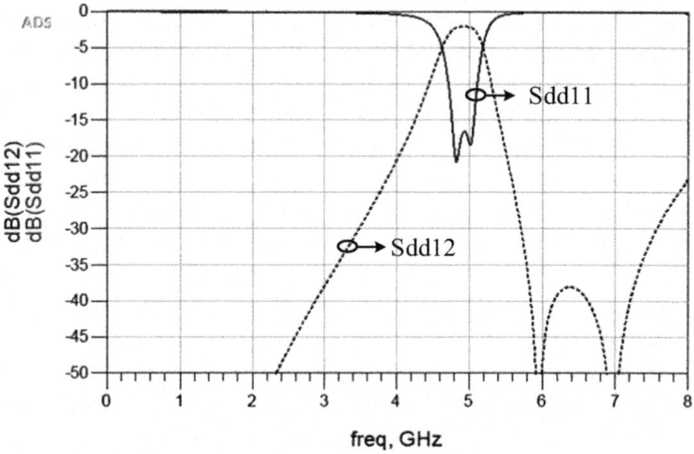

Fig. 1.78 Co-simulation with traces dB(Sdd11) and dB(Sdd12)

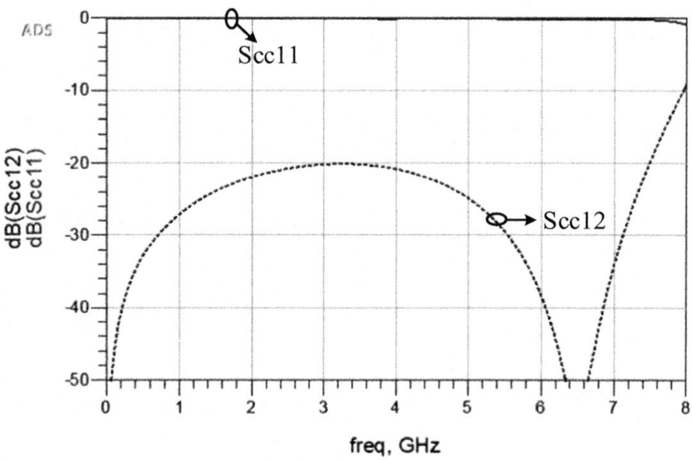

Fig. 1.79 Co-simulation with traces dB(Scc11) and dB(Scc12)

Reference

1. M. Kong, Y. Wu, Z. Zhuang, W. Wang, Y. Liu, Ultra-miniaturized balanced bandpass filter using gas-based integrated passive device technology, in *IEEE MTT-S International Microwave Symposium Digest* (Guangzhou, China, 2019), pp. 1–3

Chapter 2
Design and Simulation of Millimeter-Wave Microstrip Bandpass Filter

As an important component of microwave millimeter wave circuit systems, filters are widely used. In recent years, with the rapid development of the 5G mobile communication system, the millimeter-wave spectrum has attracted more and more attention. Researchers are committed to designing filters operating at the millimeter-wave band, so as to promote the development and construction of communication networks [1–3]. Inspired by References [4–8], a millimeter-wave microstrip bandpass filter based on the TFIPD technology is introduced in this chapter. The corresponding Chinese patent (No. 202010228778.3) has been pending. This filter with a simple structure only consists of two coupled lines and a stepped-impedance open-circuited stub. This chapter is going to introduce the basic theories of the millimeter-wave microstrip bandpass filter and how to use the ADS software to build, simulate, and optimize its ideal simulated model and full-wave electromagnetic simulated model.

2.1 Overview of Microstrip Filter

2.1.1 Theoretical Basis

Filters can be divided into lumped-element filters and distributed-element filters according to the used components. The millimeter-wave microstrip bandpass filter introduced in this chapter belongs to the distributed-element filter.

Y. Wu and W. Wang, *Microwave and Millimeter-Wave Chips Based on Thin-Film Integrated Passive Device Technology*,
https://doi.org/10.1007/978-981-99-1455-5_2

2.1.2 Propaedeutics

The circuit structure of the millimeter-wave microstrip bandpass filter introduced in this chapter is shown in Fig. 2.1 [4]. The filter is composed of two coupled lines and a stepped-impedance open-circuited stub. The even- and odd-mode analysis method can be applied to analyze the symmetric circuit model of the proposed bandpass filter. The even- and odd-mode equivalent circuits are displayed in Figs. 2.2 and 2.3, respectively.

To construct an ideal microstrip bandpass filter with the center frequency f_0, the S-parameters of the filter can be expressed by the even- and odd-mode input admittance Y_{ino}, Y_{ine}, and the standard characteristic admittance $Y_0 = 1/Z_0$:

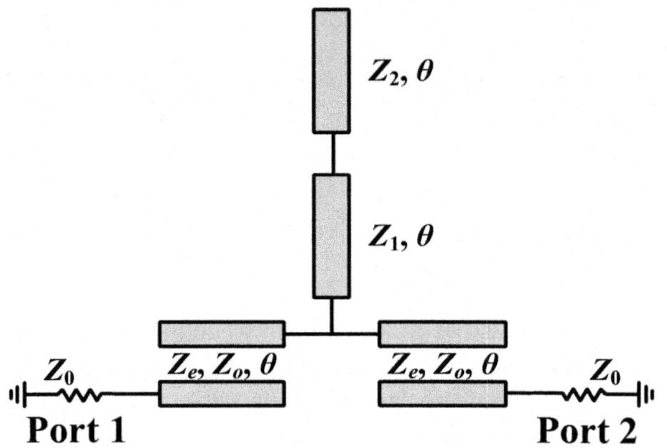

Fig. 2.1 Circuit structure of the millimeter-wave microstrip bandpass filter

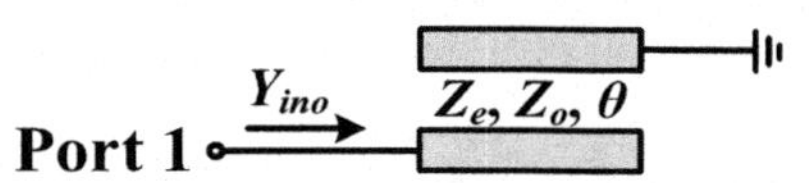

Fig. 2.2 Odd-mode equivalent circuit of the proposed filter

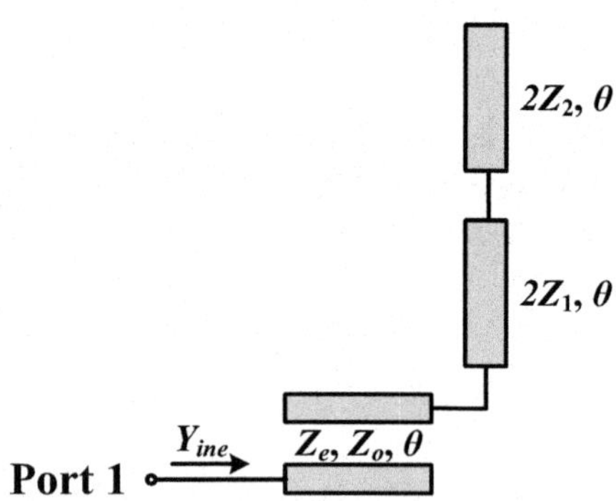

Fig. 2.3 Even-mode equivalent circuit of the proposed filter

$$S_{11} = \frac{Y_0^2 - Y_{ine}Y_{ino}}{(Y_0 + Y_{ine})(Y_0 + Y_{ino})} \tag{2.1}$$

$$S_{21} = \frac{Y_0(Y_{ine} - Y_{ino})}{(Y_0 + Y_{ine})(Y_0 + Y_{ino})} \tag{2.2}$$

Set $S_{11} = 0$ to obtain the frequencies of transmission poles f_{p} and set $S_{21} = 0$ to obtain the frequencies of transmission zeros f_{z}. According to the desired performance of the millimeter-wave microstrip bandpass filter, the appropriate values of Z_{e}, Z_{o}, and Z_i ($i = 1$ and 2) are chosen and the corresponding f_{p} and f_{z} can be determined.

2.2 Schematic of Millimeter-Wave Microstrip Bandpass Filter

ADS can realize the parameterized model simulation. Take a millimeter-wave microstrip bandpass filter with the center frequency $f_0 = 28$ GHz as an example to introduce how to build and simulate its ideal and microstrip circuit models below. The even- and odd-mode impedances of the coupled lines are Z_{e} and Z_{o}, respectively, and the characteristic impedances of the stub Z_i ($i = 1$ and 2) can be parameterized in the circuit model.

2.2.1 *New Workspace and Circuit Model*

1. New Workspace

(1) Double-click the ADS shortcut icon, and click the [OK] in the displayed dialog box to start the ADS software. After ADS runs, the [Get Started] dialog box will pop up automatically. Click the [Close] button in the lower right corner to enter the main interface [Advanced Design System 2020 (Main)], as shown in Fig. 2.4.

(2) Create a new workspace to store all files of the design simulation. Execute menu commands [File] → [New] → [Workspace…] to open the new workspace dialog box shown in Fig. 2.5, where you can set the workspace name [Name] and the working path [Create in]. Here, change the workspace name to "Millimeter_Wave_Microstrip_Bandpass_Filter_wrk" and keep the working path in the default setting. Click the [Create Workspace] button to complete the creation of the new workspace.

(3) The [Folder View] in the ADS main interface can display the name and working path of the created workspace, as shown in Fig. 2.6. The workspace name is "Millimeter_Wave_Microstrip_Bandpass_Filter_wrk", the corresponding path is "D:\ADS\Millimeter_Wave_Microstrip_Bandpass_Filter_wrk", and a

Fig. 2.4 Main interface of ADS

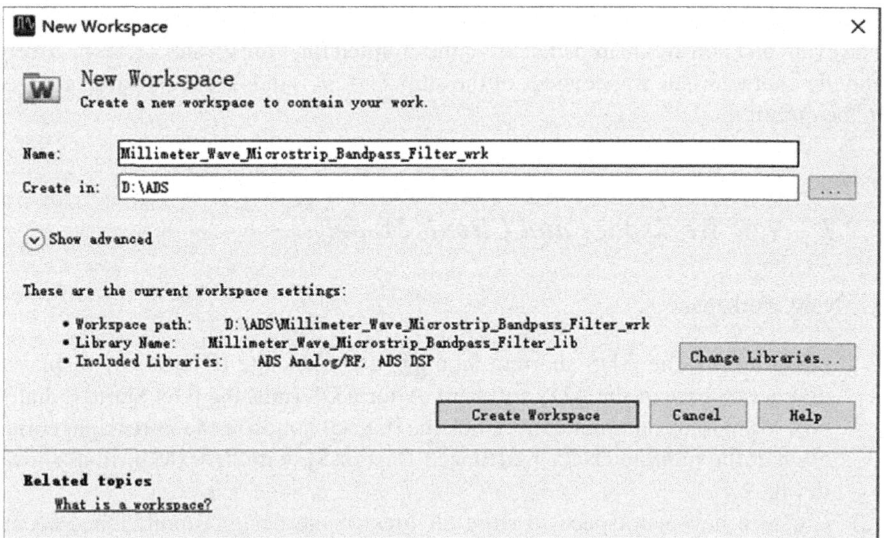

Fig. 2.5 [New workspace] dialog box

subfolder named "Millimeter_Wave_Microstrip_Bandpass_Filter_wrk" in the ADS folder of disk D can be found in your computer.

2. New Circuit Model

(1) Create a new schematic. Execute menu commands [File] → [New] → [Schematic…] to open the [New Schematic] dialog box shown in Fig. 2.7.

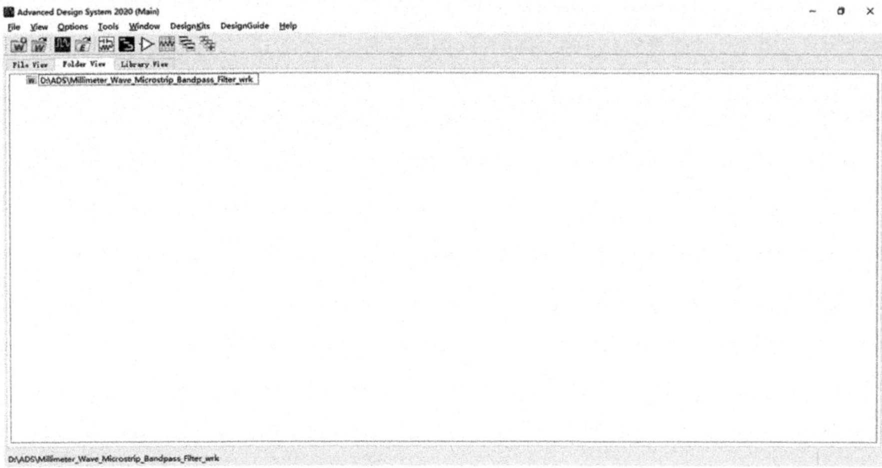

Fig. 2.6 Folder view of the new workspace

Modify the name of "Cell" to "circuit structure" and click the [Create Schematic] button to complete the creation of a new schematic, as shown in Fig. 2.8.

(2) Insert coupled transmission lines. Select the [TLines-Ideal] from the drop-down menu in the left component palette list, as shown in Fig. 2.9, which includes some

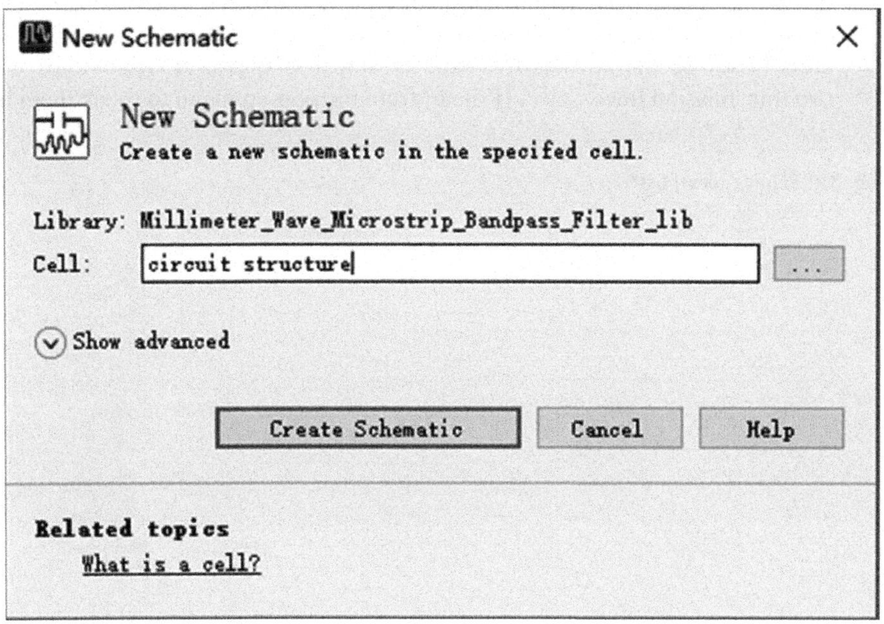

Fig. 2.7 [New schematic] dialog box

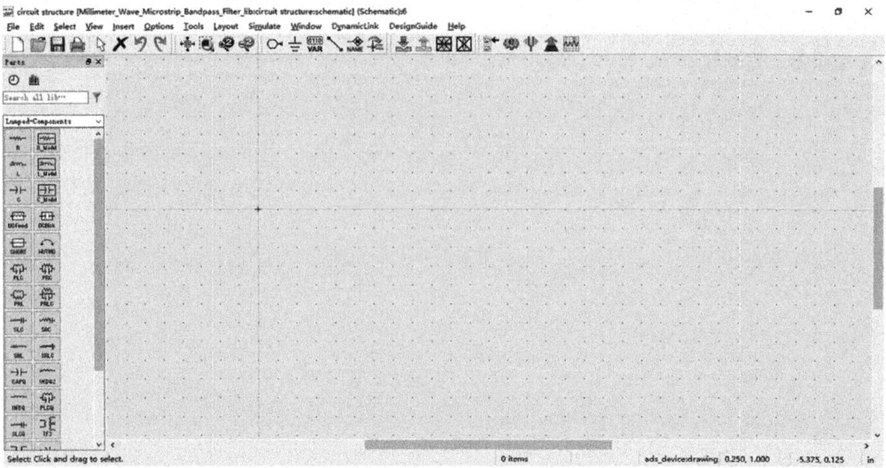

Fig. 2.8 New schematic window

commonly used ideal distributed-element models, such as the transmission line and the coupled transmission line, etc. Click the [Ideal Coupled Transmission Lines] icon  under the [TLines-Ideal] to insert two ideal coupled transmission lines, then press "Esc" on the keyboard to exit. Coupled transmission lines have been inserted, as shown in Fig. 2.10.

(3) Insert and rotate transmission lines. Click the [Ideal Terminal Transmission Line] icon  under the [TLines-Ideal] to insert two transmission lines, then press "Esc" on the keyboard to exit, as shown in Fig. 2.11. Right-click the two transmission lines, select [Rotate] from the pop-up menu to rotate them in

Fig. 2.9 [Tlines-Ideal] list

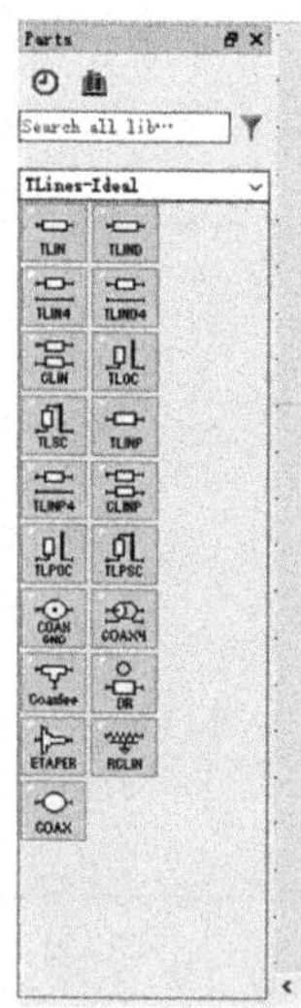

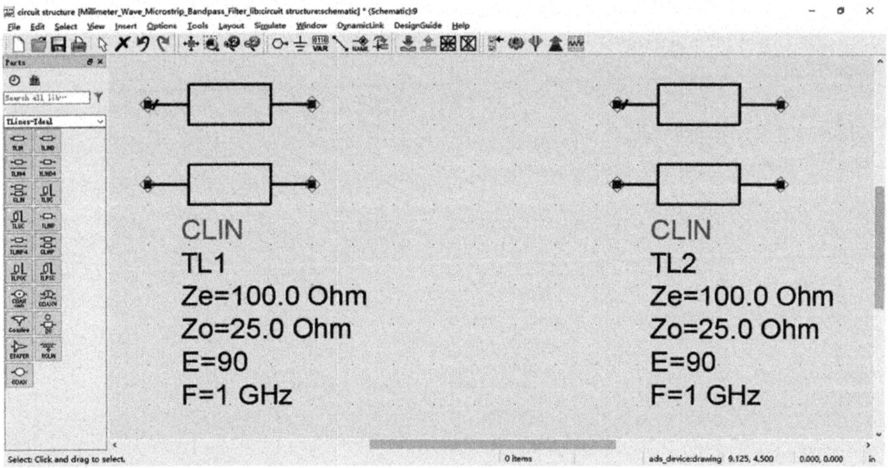

Fig. 2.10 Insertion of coupled transmission lines

90° clockwise. Transmission lines have been inserted and rotated, as shown in Fig. 2.12.

(4) Connect components. Execute menu commands [Insert] → [Wire] to connect each component according to the circuit structure shown in Fig. 2.1. Figure 2.13 shows the circuit structure after the components are connected.

(5) Modify circuit model parameters. Double-click the coupled transmission line TL1, then modify the parameter value to Ze (Ohm), Zo (Ohm), SitaT (deg), and f0 (GHz) in the pop-up [Edit Instance parameters] dialog box, as shown in Fig. 2.14 (Check whether the unit settings are consistent). Click the [OK] button

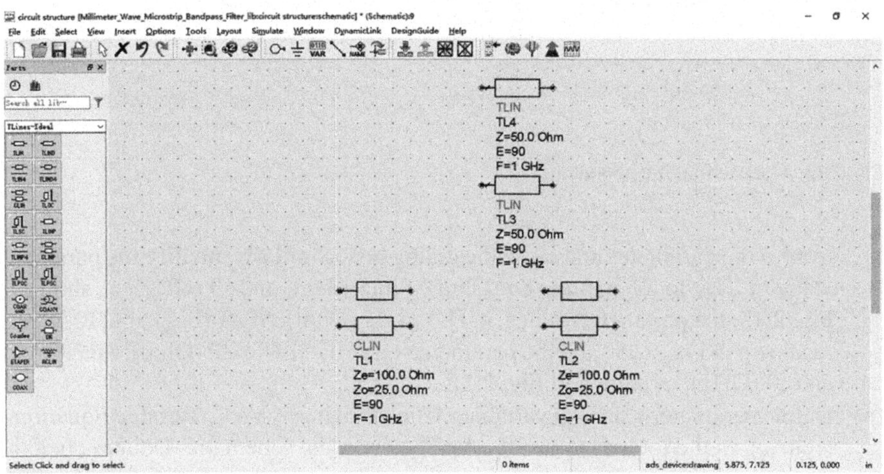

Fig. 2.11 Insertion of transmission lines

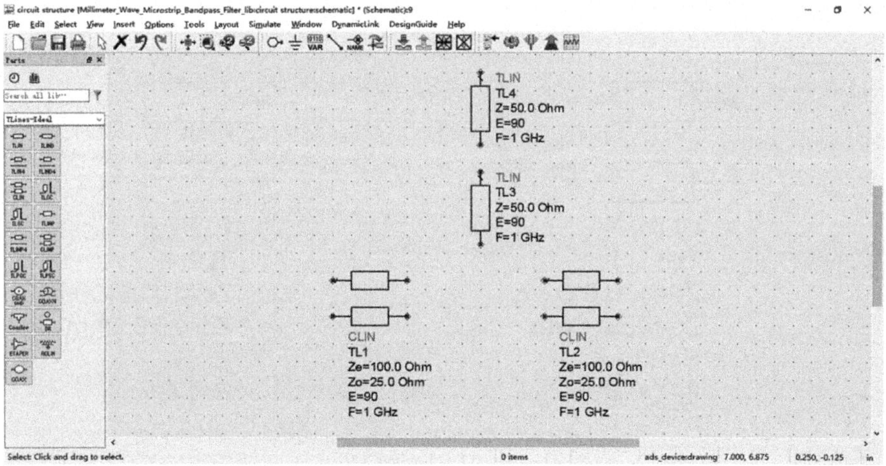

Fig. 2.12 Rotation of transmission lines

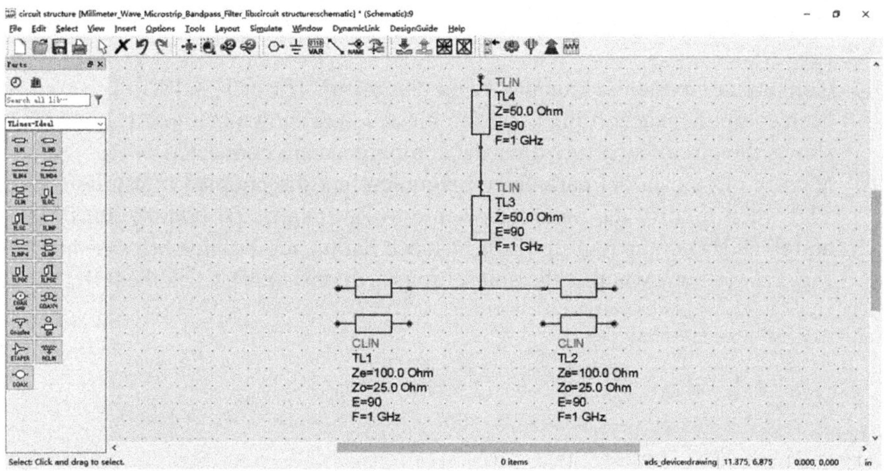

Fig. 2.13 Connection of components

to save the parameter and close the dialog box. Similarly, modify the parameter value of TL2 to Ze (Ohm), Zo (Ohm), SitaT (deg), and f0 (GHz), as shown in Fig. 2.15, the parameter values of TL3 to Z1 (Ohm), SitaT (deg), and f0 (GHz), as shown in Fig. 2.16, and the parameter value of TL4 to Z2 (Ohm), SitaT (deg), and f0 (GHz), as shown in Fig. 2.17.

(6) Define parameter values of variables. Click the [Insert VAR: Variable Equations] icon in the toolbar to insert a VAR item in the schematic. Double-click the variable component to open the [Edit Instance Parameter] dialog box, the "Variable or Equation Entry Mode" is "Standard" by default. Input the variable

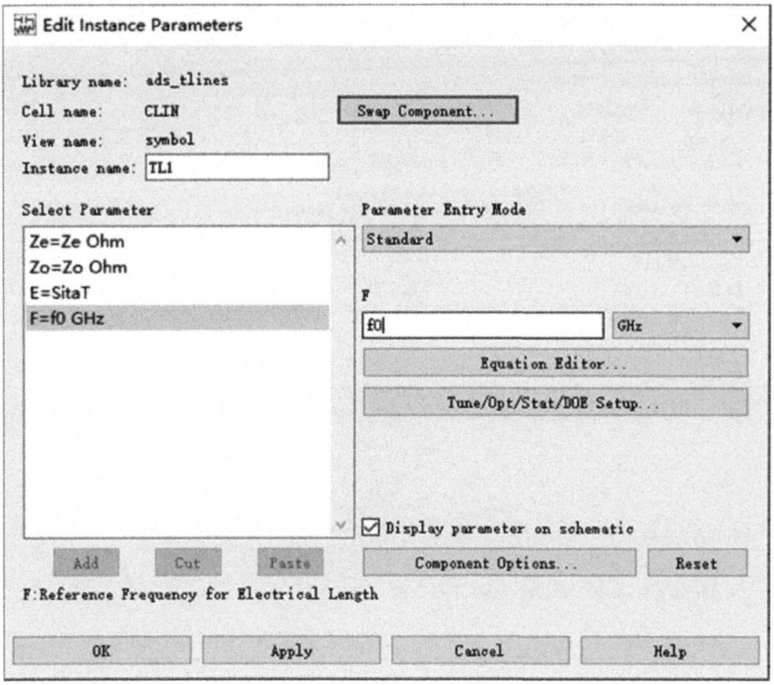

Fig. 2.14 Ideal parameter settings of TL1

name “Ze” in the “Name” column and the variable value “120” in the “Variable Value” column. Click the [Apply] button, and the dialog box after the setting is shown in Fig. 2.18. If you click [OK], the dialog box can be closed directly. Then define other variables in turn, including Zo = 56, Z1 = 30, Z2 = 120, SitaT = 90, and f0 = 28, as shown in Fig. 2.19. While defining other variables, click the [Add] button to add them. If you click the [Apply] button, the selected variable in the parameter list on the left can be replaced directly. In addition, there is no need to set units because the unit of each variable has already been defined in the component model.

(7) The VAR item provides three methods to add variables, including “Standard”, “Name = Value”, and “File Based”. You can select the most suitable method in the “Variable or Equation Entry Mode” column. The second method is introduced below. Select the “Name = Value” mode, as shown in Fig. 2.20. Input “Ze = 120” under the “Variable Value” column, then click the [Apply] button. Similarly, when defining other variables, click the [Add] button to add them.

(8) Complete parameter definitions. After the definition of all variables, click the [OK] button, the final circuit model of the millimeter-wave microstrip bandpass filter with the defined ideal parameters is shown in Fig. 2.21. Readers can compare the parameters in the circuit structure to check whether all parameters are set correctly in detail.

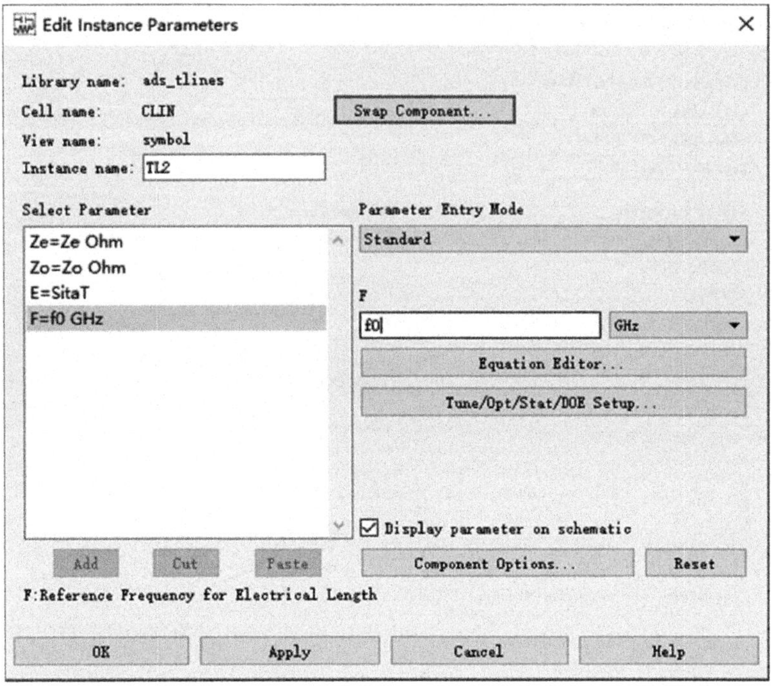

Fig. 2.15 Ideal parameter settings of TL2

2.2.2 Schematic Simulation

1. Set Simulation Parameters

(1) Insert the S-parameter simulator, term ports, and grounds. As shown in Fig. 2.22, select the [Simulation-S_Param] from the drop-down menu in the left component palette list and click the [S-parameter Simulator] icon to insert an S-parameter simulator. Click the [Port Impedance Termination for S-Parameters] icon to insert two term ports, then press "Esc" on the keyboard to exit. After that, execute menu commands [Insert] → [GROUND] to place two grounds (Or click the [TermG Port Impedance Termination for S-Parameters with Grounded Reference] icon to insert two term ports with the grounded reference). Execute menu commands [Insert] → [Wire] to connect all the inserted components. After finishing all the steps, press "Esc" to exit.

(2) Set simulation frequency. Double-click the S-parameter simulator S-PARAMETERS. Complete the settings shown in Fig. 2.23, in which the start frequency is 0 GHz, the stop frequency is 50 GHz, and the step-size is 0.05 GHz, then click the [OK] button. The final ideal parameter simulated circuit model of the millimeter-wave microstrip bandpass filter is obtained, as shown in Fig. 2.24.

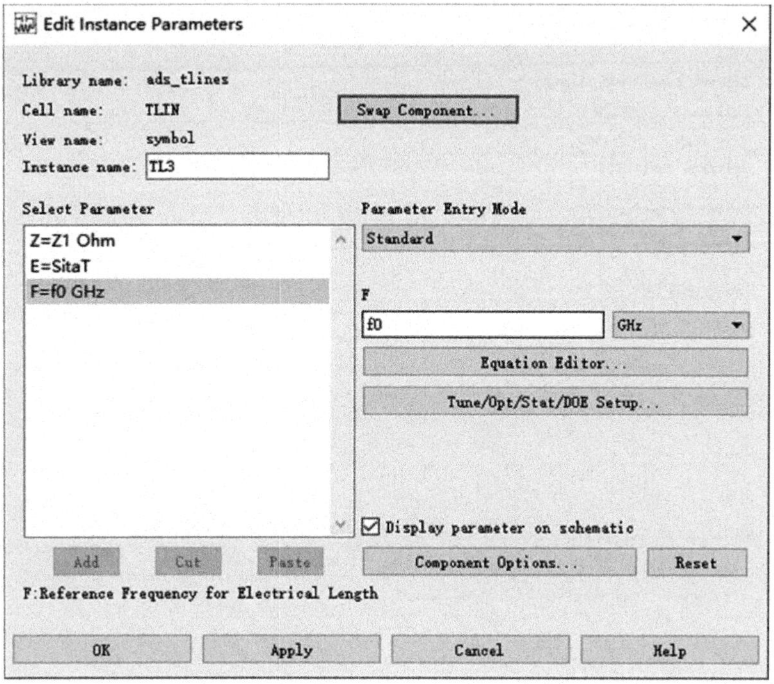

Fig. 2.16 Ideal parameter settings of TL3

2. View simulated results

(1) Execute menu commands [Simulate] → [Simulate]. After the simulation completes, the data display window will be opened, as shown in Fig. 2.25.
(2) Click the [Rectangular Plot] icon  in the [Palette], then click the left mouse button at the blank data display area to open the [Plot Traces & Attributes] dialog box shown in Fig. 2.26, where the parameter traces can be set and plotted.
(3) Long press the "Ctrl" key to select S(1,1) and S(2,1) successively, then click the [>>Add> >] button. Choose [dB] in the pop-up dialog box shown in Fig. 2.27, then click the [OK] button. It can be seen that dB(S(1,1)) and dB(S(2,1)) have been added to the [Traces] list box, as shown in Fig. 2.28.
(4) Click the [OK] button in Fig. 2.28, the traces of dB(S(1,1)) and dB(S(2,1)) can be displayed in the data display area, the ordinate is the dB value, as shown in Fig. 2.29.

3. Handle Traces

Next, traces dB(S(1,1)) and dB(S(2,1)) are taken as examples to introduce the curves handling. Similar steps can also be applied in other traces.

(1) By changing the frequency position of the marker, the values of any point on the trace can be read. Execute menu commands [Marker] → [New…] to open

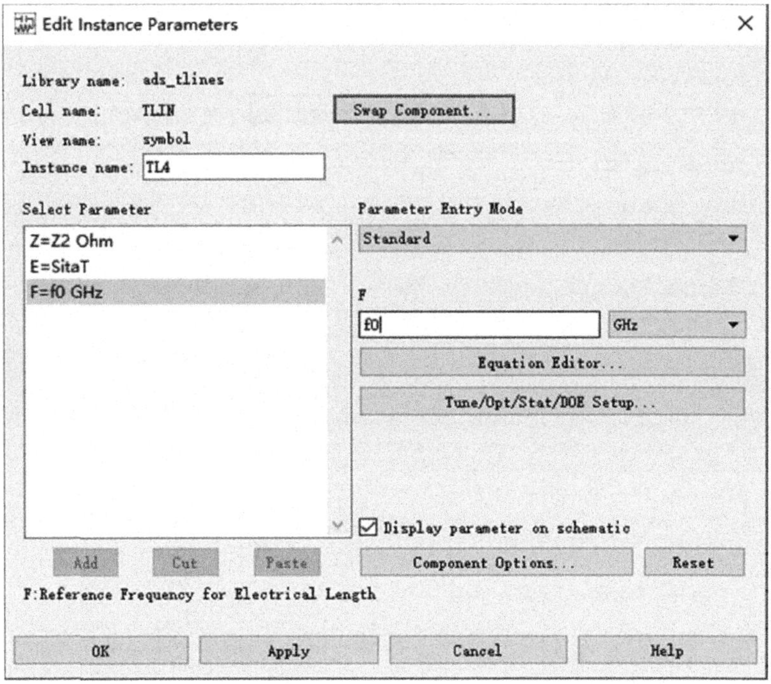

Fig. 2.17 Ideal parameter settings of TL4

the dialog box shown in Fig. 2.30. Select a trace and click the left mouse button to insert a marker on it, as shown in Fig. 2.31. Similarly, insert a marker for another trace dB(S(2,1)). In addition, you can long press the left mouse button to move the position of the data frame.

(2) After selecting the inserted marker, the left and right arrow keys on the keyboard can be used to adjust the position of the abscissa (freq). You can also click the position shown in Fig. 2.32 with the left mouse button to modify the specific frequency you want to view directly. Here, the values of dB(S(1,1)) and dB(S(2,1)) at the center frequency of 28 GHz are displayed, as shown in Fig. 2.33.
(3) The editing function of the data display is introduced below. Take the modification and beautification of the Y axis as examples. Double-click the simulated parameters figure, then the [Plot Traces & Attributes] dialog box pops up. Click the [Plot Options] tab to deselect the "Auto Scale" (The automatic adjustment scale of the ADS software is not used). Adjust the Y axis scale shown in Fig. 2.34, then the adjusted S-parameter results in Fig. 2.35 can be obtained.
(4) In addition, the type, color, thickness, and symbol of the trace can also be modified. Double-click the dB(S(1,1)) trace to open the [Trace Options] dialog box and modify the trace, as shown in Fig. 2.36 (Keep the trace color as default).

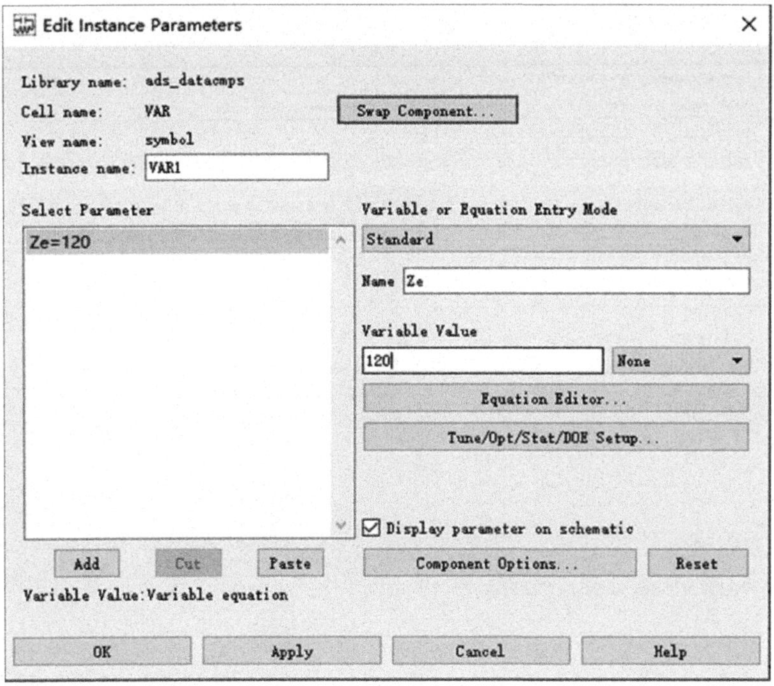

Fig. 2.18 Definition of the variable Ze

Similarly, modify the dB(S(2,1)) trace shown in Fig. 2.37. The final dB(S(1,1)) and dB(S(2,1)) traces are shown in Fig. 2.38.

2.2.3 *Microstrip Circuit Model*

The fixed parameters of the millimeter-wave microstrip bandpass filter are: $Z_e = 120$ Ω, $Z_o = 56$ Ω, $Z_1 = 30$ Ω, $Z_2 = 120$ Ω, SitaT = 90°, and $f_0 = 28$ GHz. The final chip is processed using the TFIPD technology, whose substrate is 100-μm-thick GaAs with the relative dielectric constant of 12.9 and the loss tangent of 0.001. Besides, the metal thickness of the microstrip line is 1.065 μm.

1. Calculate physical sizes

The "LineCalc" function in ADS can be used to solve the corresponding physical sizes of components at a certain frequency.

(1) Close the data display window, return to the schematic window. Execute menu commands [Tools] → [LineCalc] → [Start LineCalc], then the [LineCalc] window shown in Fig. 2.39 will pop up. Under the [Substrate Parameter], set the relative dielectric constant of substrate "Er" to 12.9, the thickness

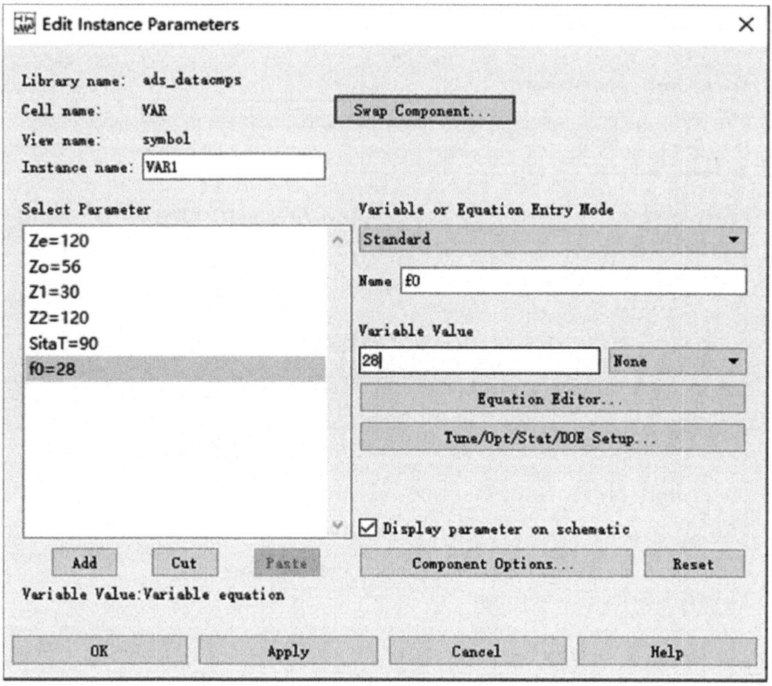

Fig. 2.19 Definitions of all variables in the ideal circuit

"H" to 100 μm, the loss tangent to 0.001, and the thickness of microstrip line to 1.065 μm (Check whether the unit settings are consistent). Keep other parameters in the default values, the modified results are shown in Fig. 2.40.

(2) Solve the physical sizes of the coupled line TL1 with the even-mode impedance $Z_e = 120\ \Omega$, the odd-mode impedance $Z_o = 56\ \Omega$, and the electrical length SitaT $= 90°$. Select the component type as " MCLIN " from the [Component] → [Type] drop-down menu. Under the [Component Parameters], set the operating frequency "Freq" to 28 GHz. Since the units in this chapter are μm, the units of physical lengths "W", "S", and "L" should be selected as μm from their drop-down list boxes under the [Physical]. Then set the even-mode impedance "ZE" to 120 Ω, the odd-mode impedance "ZO" to 56 Ω, and the electrical length "E_Eff" to 90 deg under the [Electrical], as shown in Fig. 2.41. Click the [Synthesize] button to solve the physical sizes of TL1, as shown in Fig. 2.42 (Figs. 2.41 and 2.42 are corresponding to the changes in size before and after clicking the [Synthesize] button, respectively). The circuit parameters of coupled lines TL2 and TL1 are the same, so their physical sizes are also the same.

(3) Solve the physical sizes of the microstrip line TL3 with the characteristic impedance $Z_1 = 30\ \Omega$ and the electrical length SitaT $= 90°$. Select the component type as " MLIN " from the [Component] → [Type] drop-down menu.

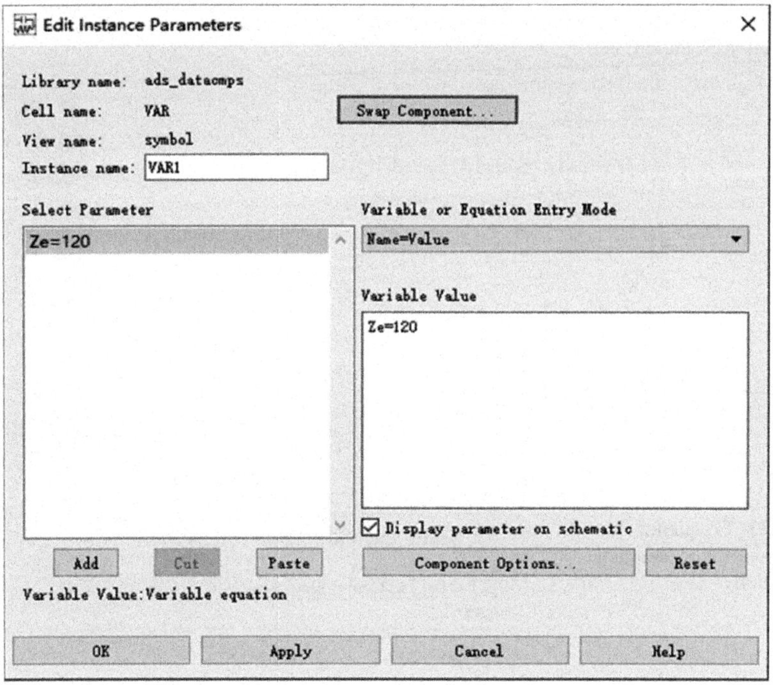

Fig. 2.20 The second method to define the variable Ze

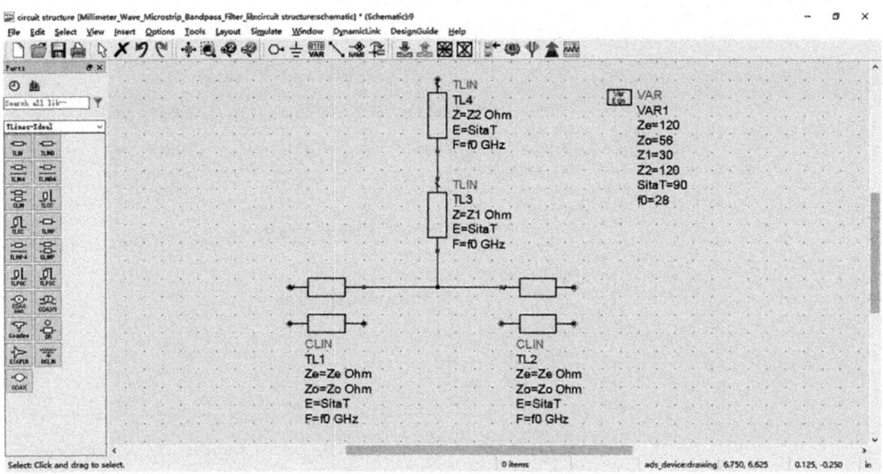

Fig. 2.21 Millimeter-wave microstrip bandpass filter circuit schematic with the defined ideal parameters

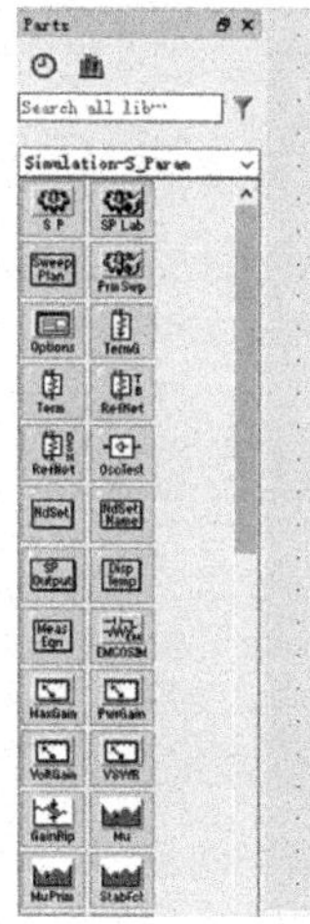

Fig. 2.22 [Simulation-S_Param] list

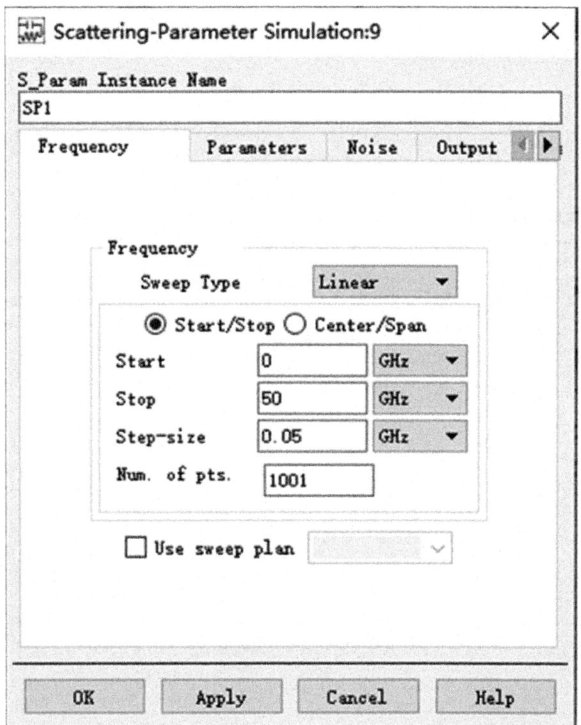

Fig. 2.23 Frequency settings of the S-parameter simulation

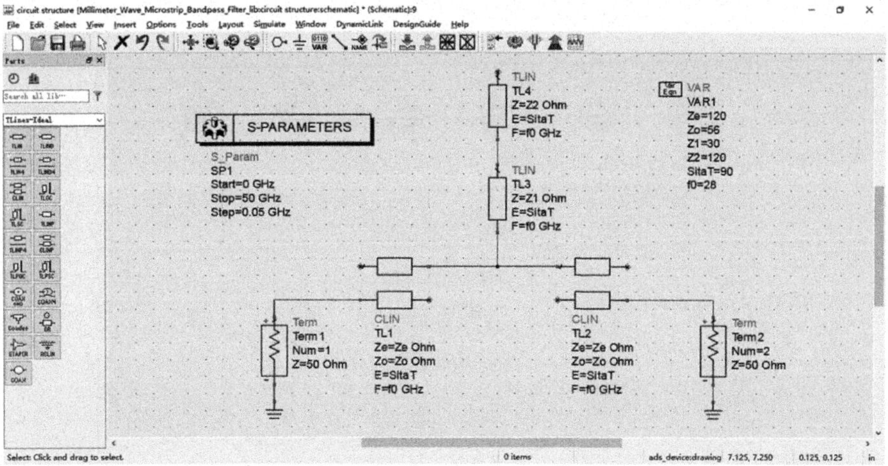

Fig. 2.24 Final ideal circuit simulated model of the millimeter-wave microstrip bandpass filter

Fig. 2.25 Data display window

Set the operating frequency "Freq" to 28 GHz under the [Component Parameters] and the units of "W", "S", and "L" to μm from their drop-down list boxes under the [Physical]. Then set the characteristic impedance "Z0" to 30 Ω and the electrical length "E_Eff" to 90 deg under the [Electrical]. Click the [Synthesize] button to solve the physical sizes of TL3, as shown in Fig. 2.43. Similarly, the corresponding physical sizes of microstrip line TL4 can be solved and the results are shown in Fig. 2.44.

2. Circuit model simulation

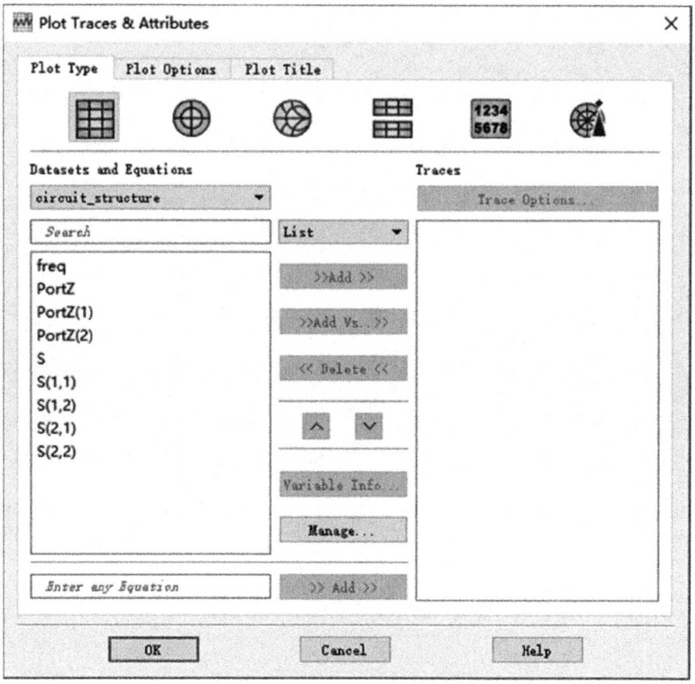

Fig. 2.26 [Plot Traces and Attributes] dialog box

Fig. 2.27 Handling of the data

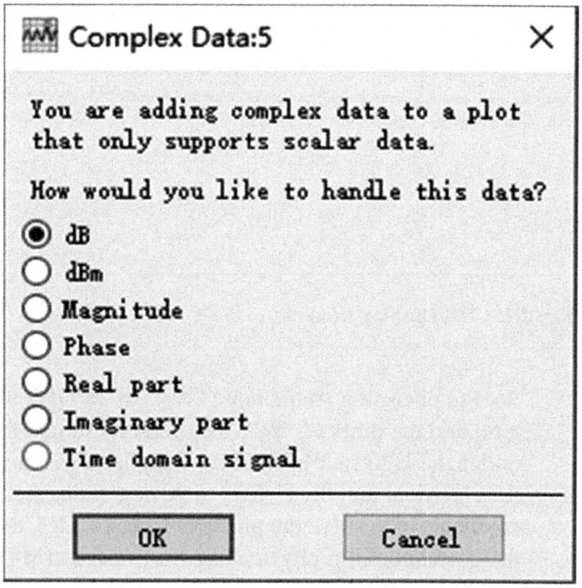

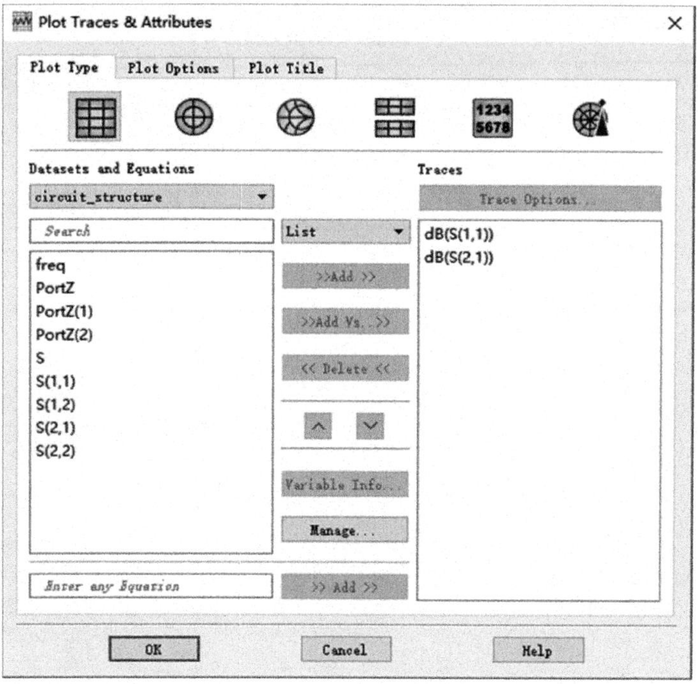

Fig. 2.28 Add of S(1,1) and S(2,1)

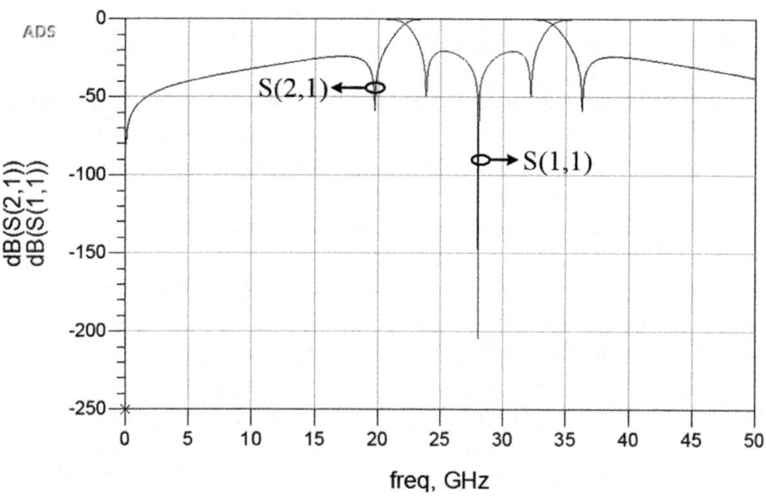

Fig. 2.29 Traces of dB(S(1,1)) and dB(S(2,1))

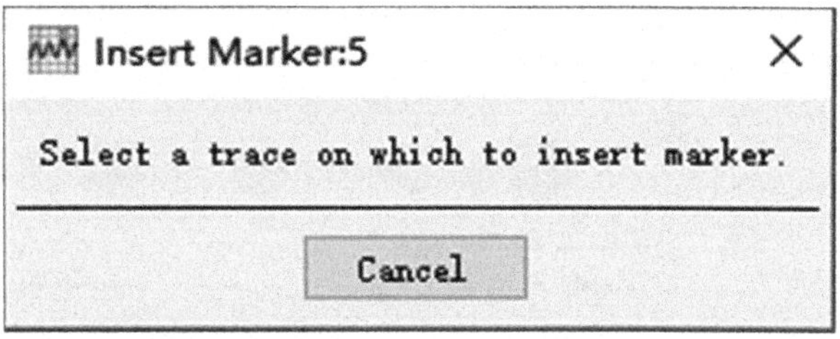

Fig. 2.30 [Insert Marker] dialog box

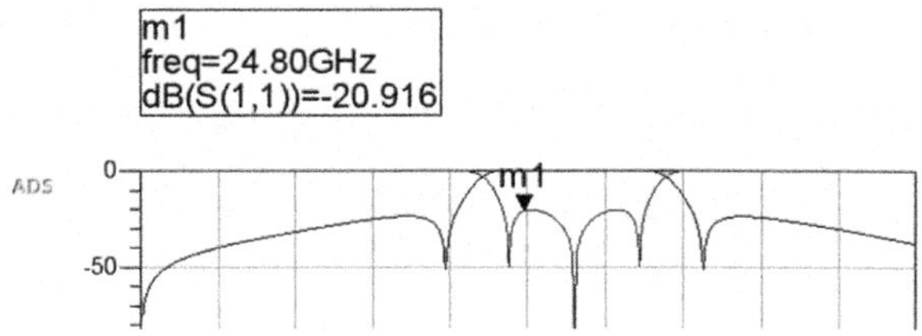

Fig. 2.31 Insertion of the marker on the trace

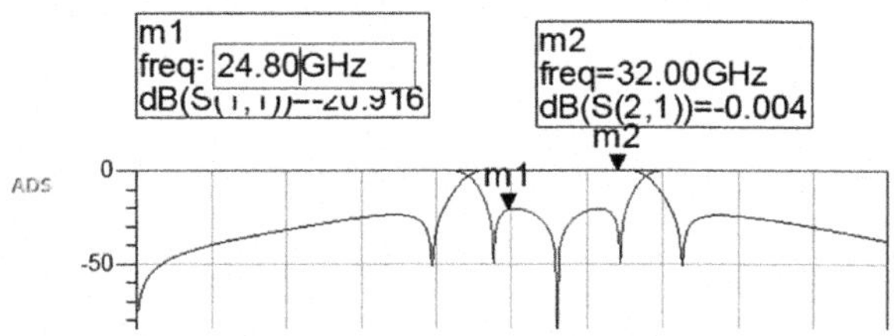

Fig. 2.32 Modification of the abscissa (freq) value

(1) Save the schematic. Return to the schematic window and save the schematic "circuit structure" as "microstrip circuit structure". Execute menu commands [File] → [Save As…], then modify the name of "Cell" to "microstrip circuit structure" in the pop-up [Save Design As] dialog box, as shown in Fig. 2.45. Click the [OK] button to save the circuit schematic.

(2) Delete ideal transmission lines and coupled transmission lines. The ideal transmission lines and coupled transmission lines are deleted from the schematic "microstrip circuit structure", as shown in Fig. 2.46.

(3) Insert microstrip coupled lines. Select the [TLines-Microstrip] from the drop-down menu in the left component palette list, as shown in Fig. 2.47, which includes some commonly used microstrip distributed-element models, such as microstrip line and microstrip coupled lines, etc. Click the [Microstrip Coupled Lines] (with physical size parameters) icon under the [TLines-Microstrip] to insert two microstrip coupled lines at the locations of the original ideal coupled transmission lines, then press "Esc" on the keyboard to exit, as shown in Fig. 2.48.

(4) Insert and rotate microstrip lines. Click the [Microstrip Line] (with physical size parameters) icon under the [TLines-Microstrip] to insert two microstrip

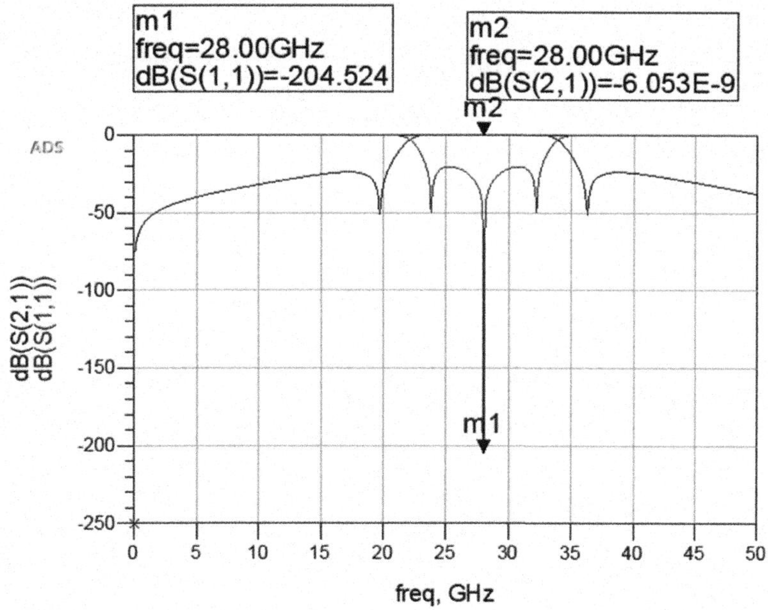

Fig. 2.33 Values of dB(S(1,1)) and dB(S(2,1)) at the center frequency

lines at the locations of the original ideal transmission lines, then press "Esc" on the keyboard to exit, as shown in Fig. 2.49. Right-click the two microstrip lines, select [Rotate] from the pop-up menu to rotate them in 90° clockwise. Microstrip lines have been inserted and rotated, as shown in Fig. 2.50.

(5) Set microstrip substrate. Click the [Microstrip Substrate] icon under the [TLines-Microstrip] and the [Choose Layout Technology] dialog box will pop up, as shown in Fig. 2.51. Choose "Standard ADS Layers, 0.001 micron layout resolution", which is 0.001 μm (note that the unit in this chapter is μm), then click the [Finish] button to insert a microstrip substrate model. Double-click the icon to edit substrate and conductor parameters: substrate thickness H = 100 μm, relative dielectric constant Er = 12.9, dielectric loss tangent TanD = 0.001, and conductor thickness T = 1.065 μm, as shown in Fig. 2.52. Click the [OK] to save and the schematic window is shown in Fig. 2.53.

(6) Modify circuit model parameters. Double-click the microstrip coupled line CLin1 and modify its parameter values to W (μm), S (μm), and L (μm) in the pop-up [Edit Instance parameters] dialog box, as shown in Fig. 2.54 (Check whether the unit settings are consistent). Click the [OK] button to save the parameter and close the dialog box. Similarly, modify the parameter values of CLin2 to W (μm), S (μm), and L (μm), as shown in Fig. 2.55, the parameter values of TL1 to Ws1 (μm), and Ls1 (μm), as shown in Fig. 2.56, and the parameter values of TL2 to Ws2 (μm), and Ls2 (μm), as shown in Fig. 2.57.

Fig. 2.34 Adjustment of the Y axis scale

(7) Define parameter values of variables. Select the original variable component in the schematic and press "Delete" to delete it. Click the [Insert VAR: Variable Equations] icon in the toolbar to insert a VAR item in the schematic (Or you can not delete the original variable component, insert a new variable component directly). Double-click the variable component to open the [Edit Instance Parameter] dialog box. Refer to the steps of the variables definitions in Sect. 2.2.1, and define all parameter values successively according to Figs. 2.42, 2.43, and 2.44, respectively, including W = 11.01, S = 33.67, L = 999.31, Ws1 = 192.92, Ls1 = 876.20, Ws2 = 2.43, and Ls2 = 1012.87 (all are accurate to 0.01 μm), as shown in Fig. 2.58. Click the [OK] button to close the dialog box, the final microstrip circuit model of the millimeter-wave microstrip bandpass filter is shown in Fig. 2.59.

(8) Microstrip circuit simulation. Execute menu commands [Simulate] → [Simulate]. After the simulation completes, the data display window can be opened automatically. Refer to the steps in Sect. 2.2.2 to view and handle traces of dB(S(1,1)) and dB(S(2,1)), the final results are shown in Fig. 2.60.

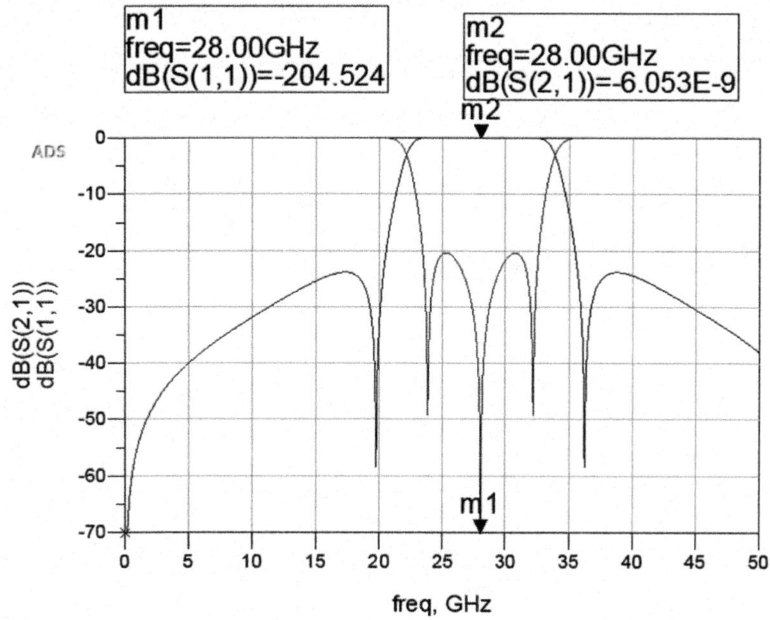

Fig. 2.35 The adjusted S-parameter results

(9) Analysis of simulated results. It can be found that there are differences between the microstrip circuit simulated results and ideal circuit simulated results of the millimeter-wave bandpass filter by comparing Figs. 2.38 and 2.60. The performance of the microstrip circuit deteriorates within a tolerable limit.

2.2.4 Parameter Tuning

In order to achieve better performances of the millimeter-wave bandpass filter, it is necessary to tune the parameter values of its microstrip circuit. Tuning can realize the continuous change of one or more design parameters for optimized values, and the corresponding effect of the output traces can be seen without resimulating the entire design.

(1) Set tunable parameter values. Double-click the variable component , and the first variable "W" has been selected. Click the [Tune/Opt/Stat/DOE Setup…] and the [Setup] dialog box shown in Fig. 2.61 will pop up. Select "Enabled" in the "Tuning Status" column, then set the "Minimum Value", "Maximum Value", and "Step Value" to proper values, as shown in Fig. 2.62. Click the [OK] button to close the dialog box, the variable "W" has been set to a tunable parameter value. Similarly, set the other variables to tunable parameter values. After completing all the settings, click the [OK] button in the [Edit Instance

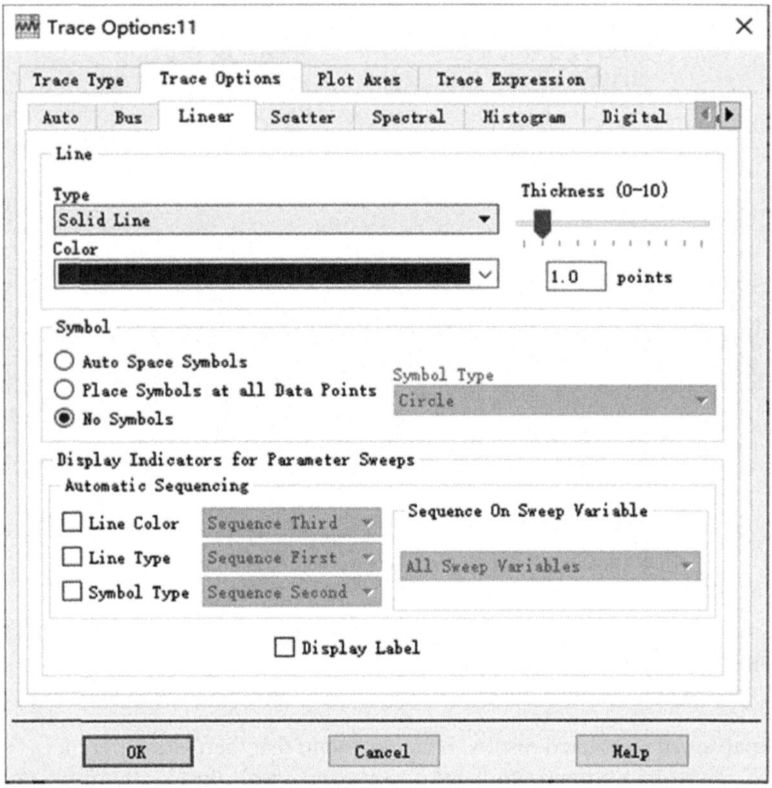

Fig. 2.36 Modification of the dB(S(1,1)) trace

Parameters] dialog box to close it, the variable component at this time is shown in Fig. 2.63.

(2) Tune. Execute menu commands [Simulate] → [Tuning] (Or click the [Tuning] icon in the toolbar) and the [Tune Parameters] dialog box shown in Fig. 2.64 will pop up. The set parameter values can be adjusted by sliding the slider or clicking the up/down button. Meanwhile, S-parameter traces in the data display window will also change accordingly as the tuning of the parameters.

(3) Select the optimal parameter values. Compare the output S-parameter traces with the expected performance of the millimeter-wave microstrip bandpass filter to select the optimal parameter values. The values shown in Fig. 2.65 are selected in the proposed filter, and the corresponding traces dB(S(1,1)) and dB(S(2,1) are shown in Fig. 2.66. Click the [Update Schematic] button at the left side of the [Tune Parameters] dialog box to update the parameter values to the schematic, then close the dialog box.

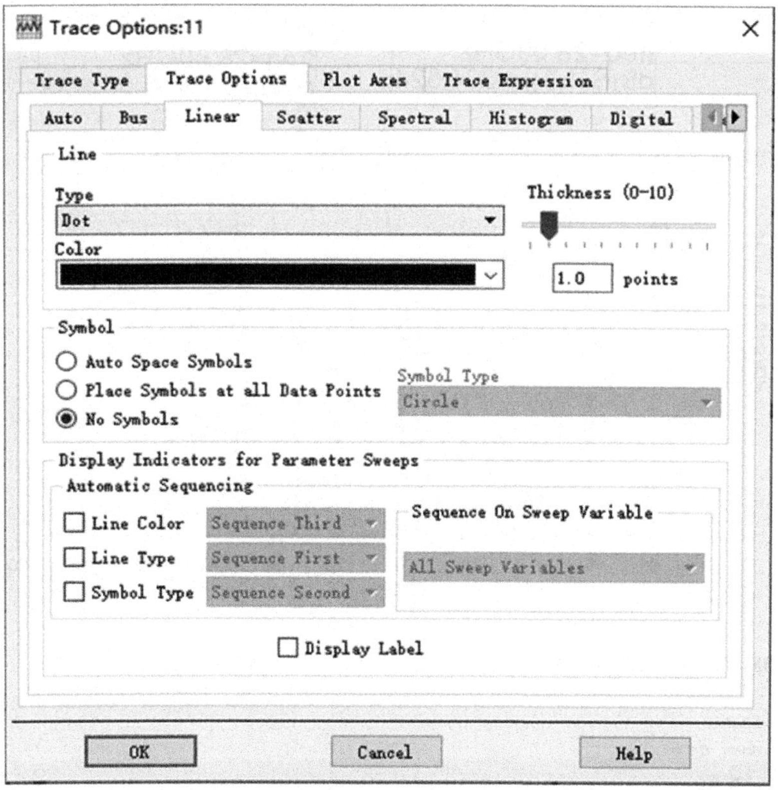

Fig. 2.37 Modification of the dB(S(2,1)) trace

2.3 Layout of Millimeter-Wave Microstrip Bandpass Filter

The performance of the actual circuit often has some differences comparing with the theoretical results, so the influence of the interference, coupling, and some other factors should be taken into consideration. Therefore, it is necessary to use the ADS software for the layout simulation.

2.3.1 New Substrate File

Since the bandpass filter in this chapter is built by microstrip lines which must be grounded, so this chapter adopts the new TFIPD technology that can achieve the process of punching holes. All circuit elements are constructed on the gallium arsenide (GaAs) substrate with the thickness of 100 μm, the relative dielectric

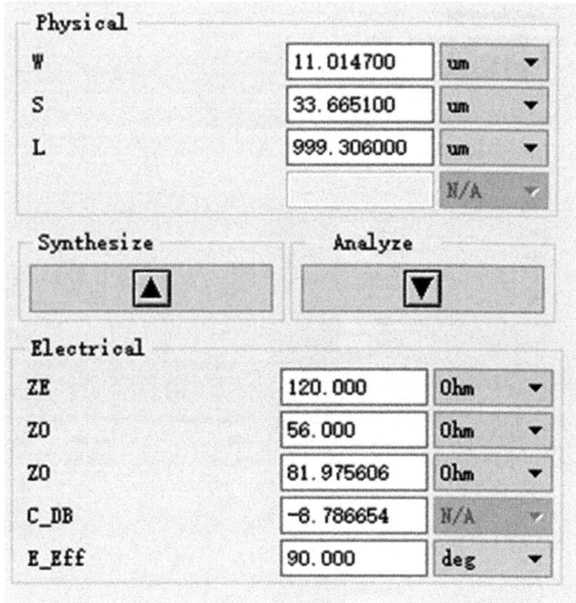

Fig. 2.42 Physical size results of TL1

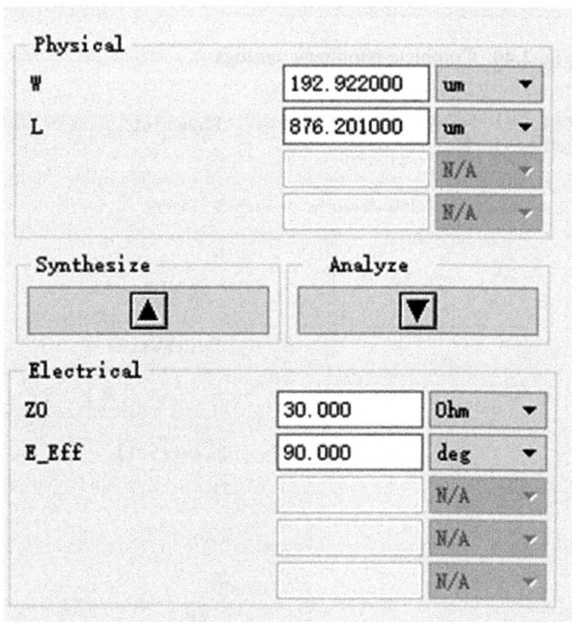

Fig. 2.43 Physical size results of TL3

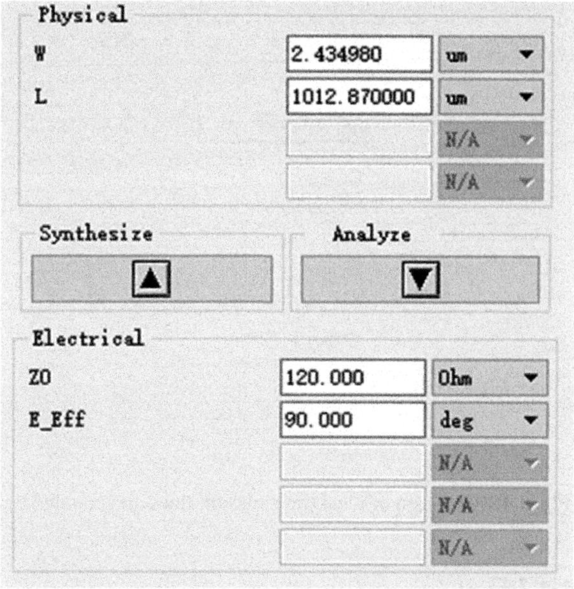

Fig. 2.44 Physical size results of TL4

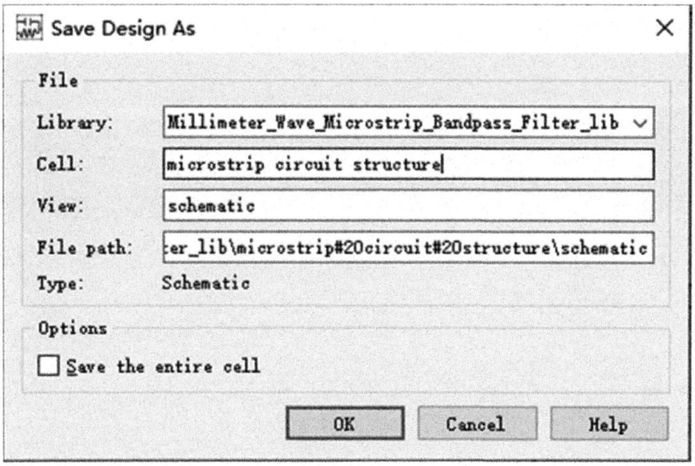

Fig. 2.45 [Save Design As] dialog box

constant of 12.9, and the loss tangent of 0.001. The metal layer with the conductivity of 4.1e7 Siemens/m is used to realize microstrip lines.

Before drawing the layout, the substrate file needs to be set according to the used TFIPD technology.

1. New Layout

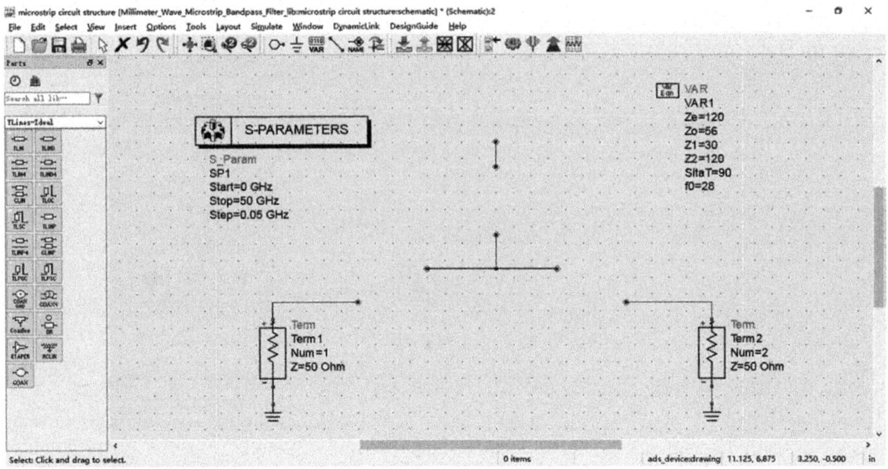

Fig. 2.46 Deletion of ideal transmission lines and coupled transmission lines

Fig. 2.47
[TLines-Microstrip] list

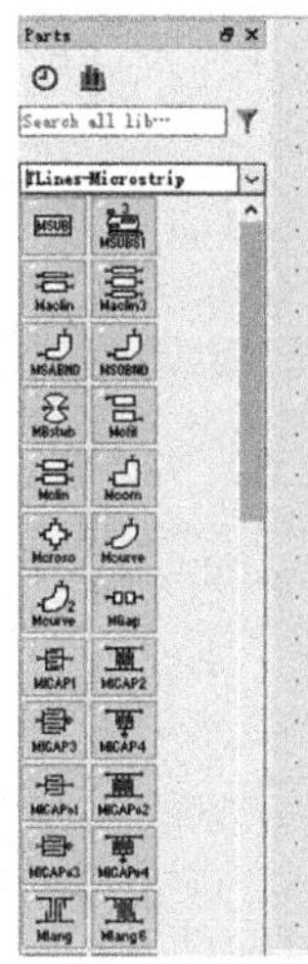

Return to the "Millimeter_Wave_Microstrip_Bandpass_Filter_wrk" main interface of the workspace. Execute menu commands [File] → [New] → [Layout…] to open the [New Layout] dialog box in Fig. 2.67. Modify the "Cell" name to "millimeter-wave microstrip bandpass filter", then click the [Create Layout] button. Since the layout accuracy has been set in the previous microstrip circuit model simulation, the [Choose Layout Technology] dialog box will not pop up here, while the new layout window will pop up directly, as shown in Fig. 2.68.

2. New Substrate

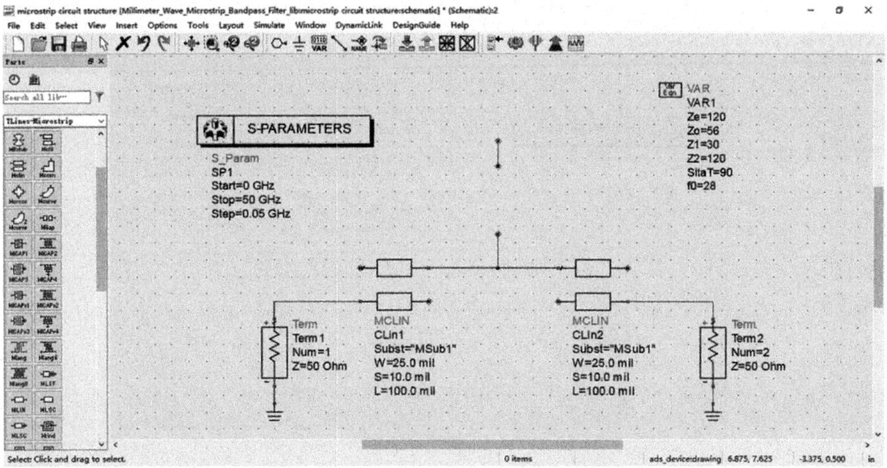

Fig. 2.48 Insertion of coupled lines with physical size parameters

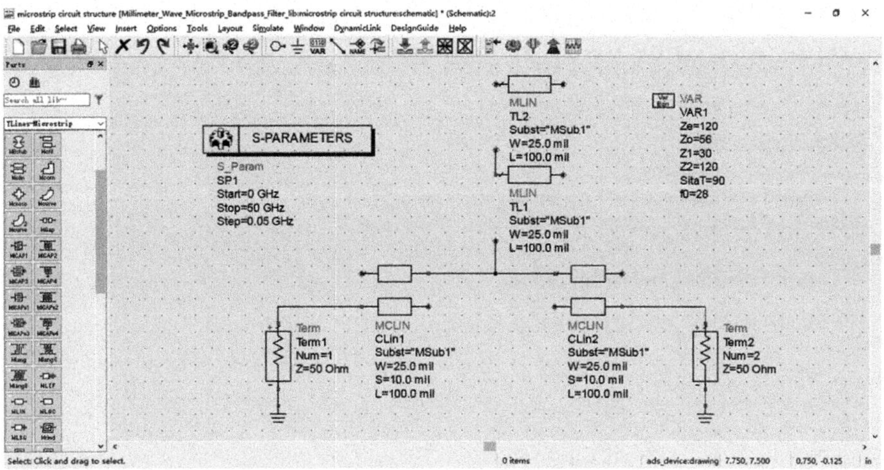

Fig. 2.49 Insertion of microstrip lines with physical size parameters

(1) Create a new substrate. Execute menu commands [EM] → [Substrate…] in the layout window, then click [OK] in the pop-up dialog box. The [New Substrate] dialog box in Fig. 2.69 will be displayed, in which "File name" and "Template" can be modified. Keep the default file name here. Because the used TFIPD technology is based on the GaAs substrate, select "100µmGaAs" in the "Template" column and click the [Create Substrate] button. Then execute menu commands [View] → [View All] in the pop-up window to view all substrate items, as shown in Fig. 2.70.

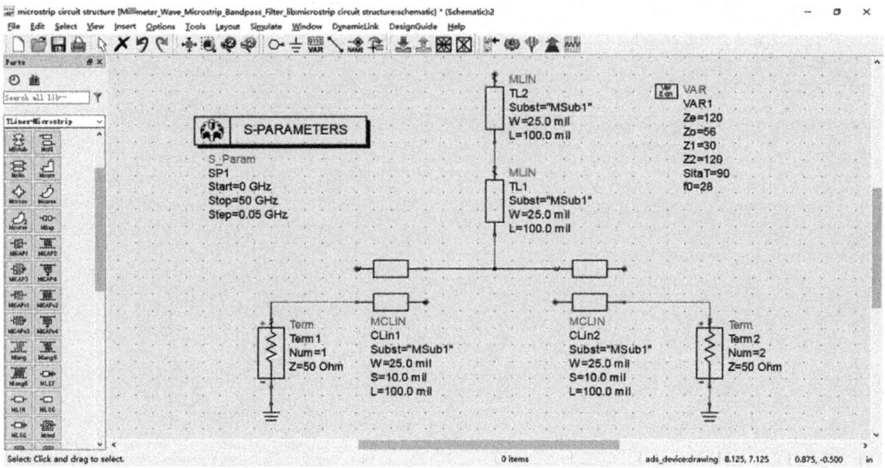

Fig. 2.50 Rotation of microstrip lines

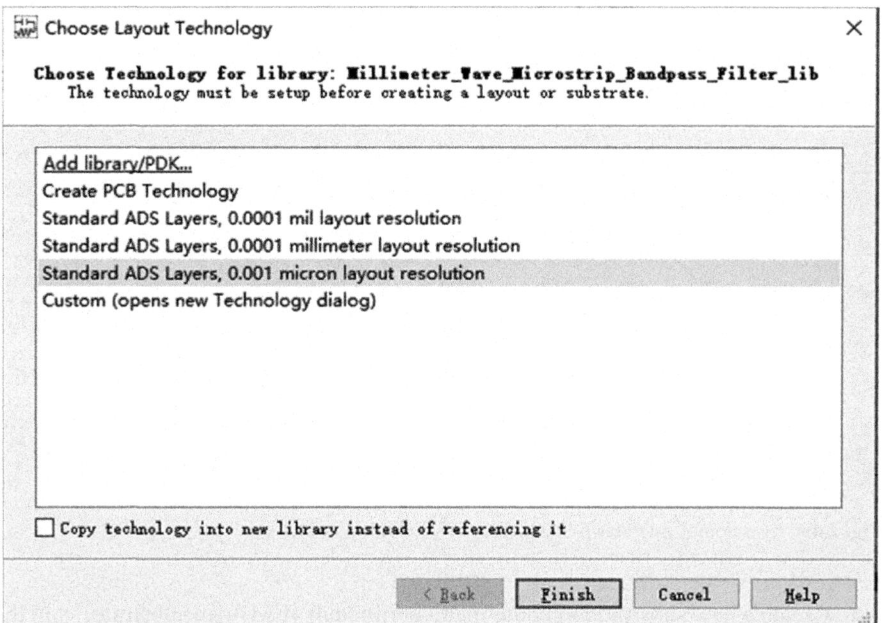

Fig. 2.51 [Choose Layout Technology] dialog box

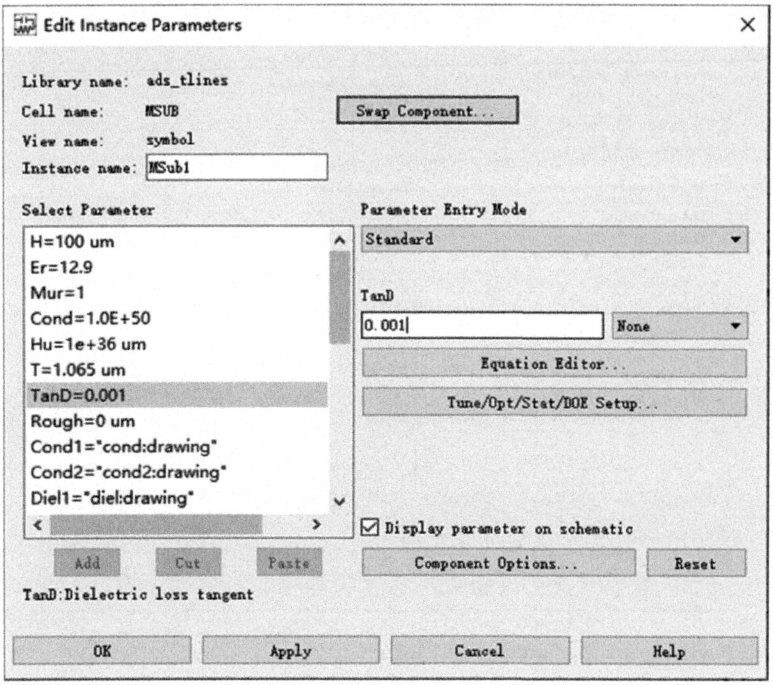

Fig. 2.52 Substrate and conductor parameter setting

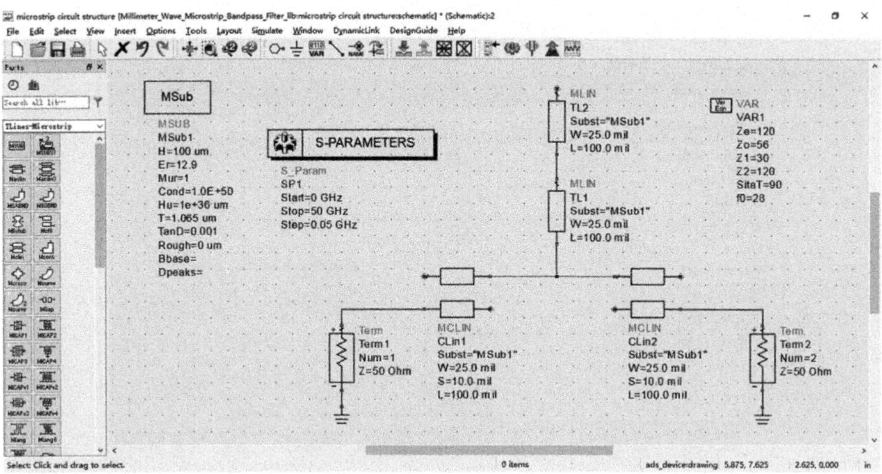

Fig. 2.53 Schematic window after the parameter setting

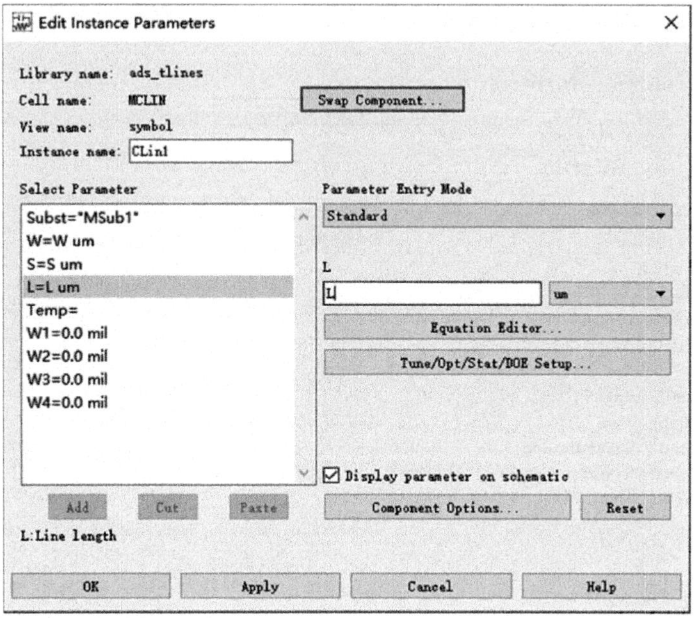

Fig. 2.54 Physical parameter settings of CLin1

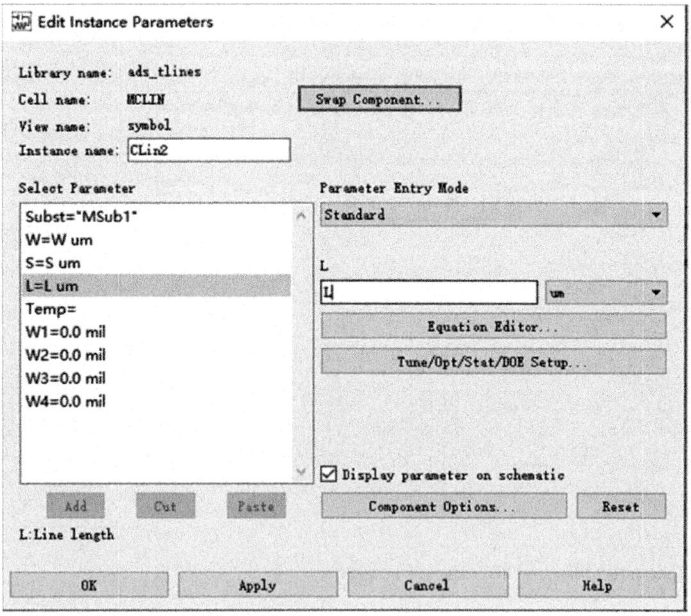

Fig. 2.55 Physical parameter settings of CLin2

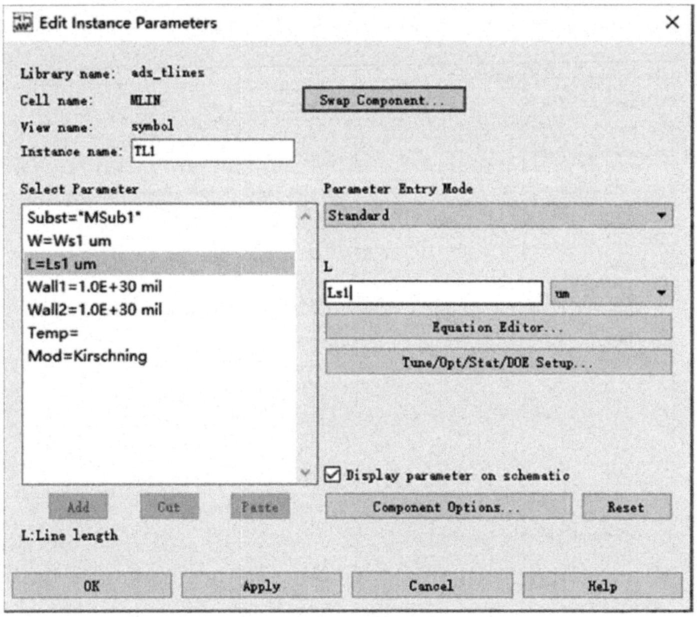

Fig. 2.56 Physical parameter settings of TL1

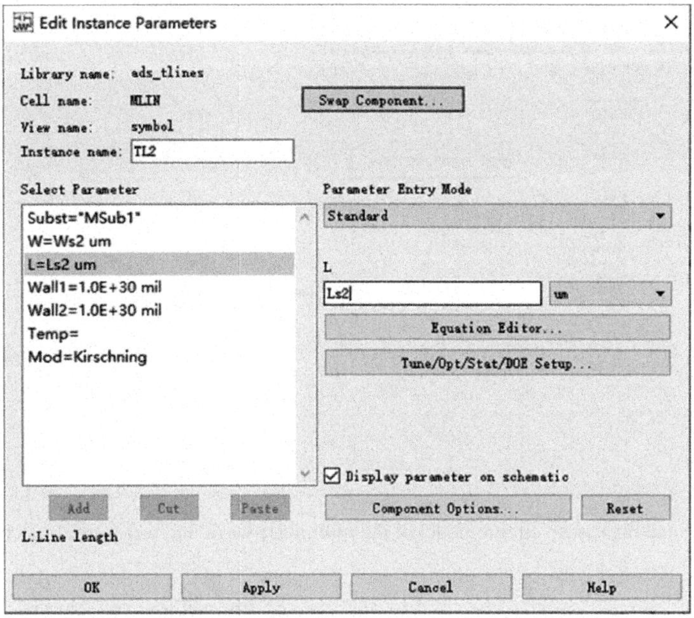

Fig. 2.57 Physical parameter settings of TL2

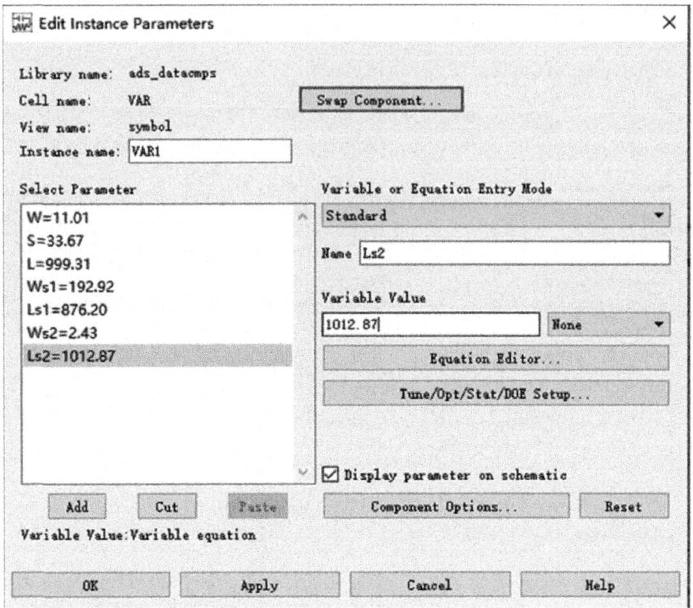

Fig. 2.58 Definitions of all variables in the Physical circuit

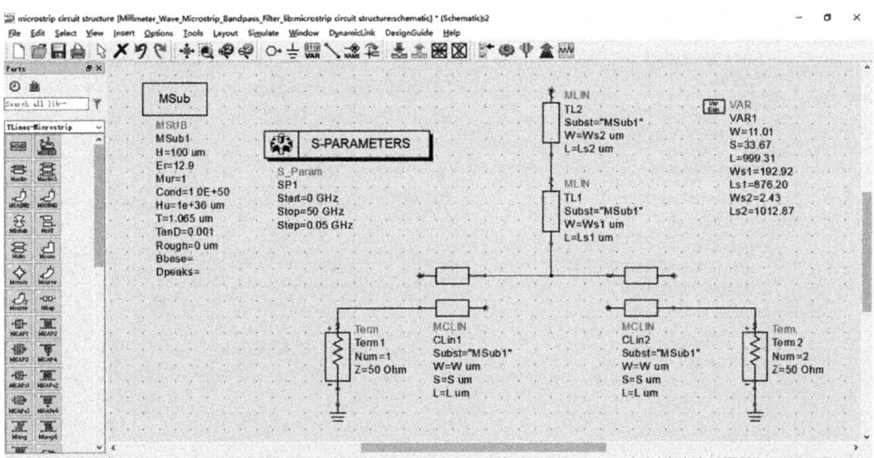

Fig. 2.59 Final microstrip circuit model of the millimeter-wave microstrip bandpass filter

(2) Add conductors. Execute menu commands [Technology] → [Material Definitions…] to open the [Material Definitions] dialog box shown in Fig. 2.71. Select the [Conductors] tab and define the relevant conductors in Fig. 2.72. The specific steps are listed as follows: Click the [Add From Database…] button in

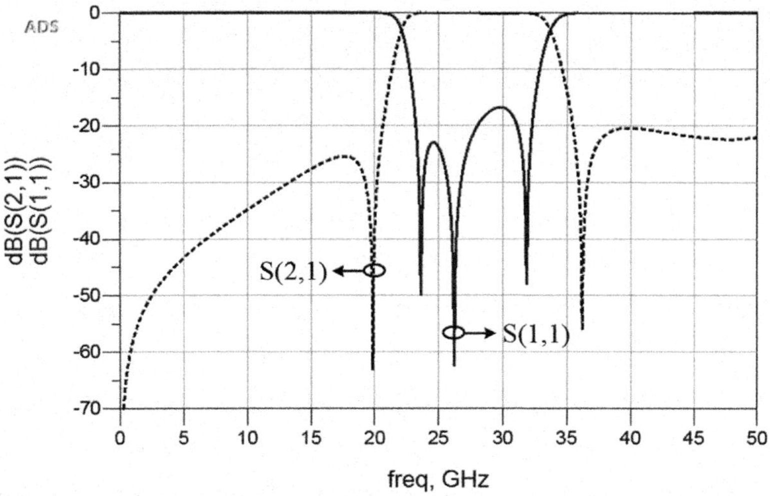

Fig. 2.60 dB(Scc11) and dB(Scc12) traces of microstrip circuit simulation

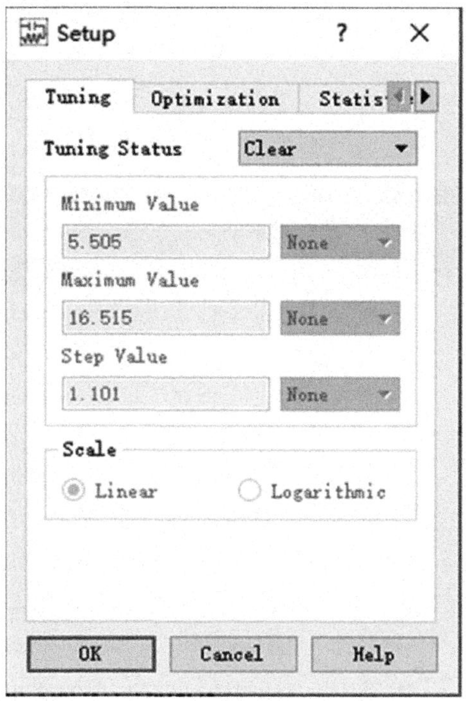

Fig. 2.61 [Setup] dialog box (Before tuning settings)

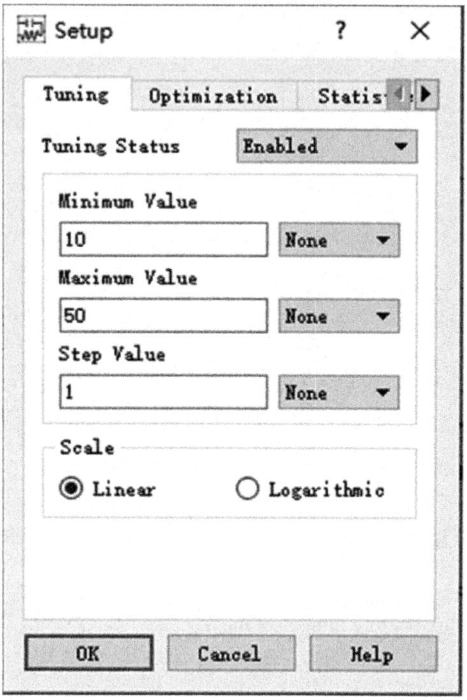

Fig. 2.62 [Setup] dialog box (After tuning settings)

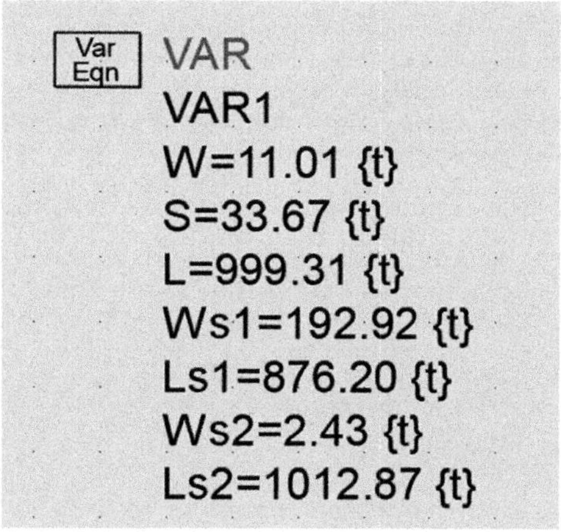

Fig. 2.63 Variable components after complete tuning settings

the lower right corner of Fig. 2.72. If the conductor to be added exists in the pop-up [Add Materials From Database] dialog box shown in Fig. 2.73, select the conductor and click [OK] to add it. If the conductor to be added does not exist in the [Add Materials From Database] dialog box, return to the [Material Definitions] dialog box and click the [Add Conductor] button to add a conductor

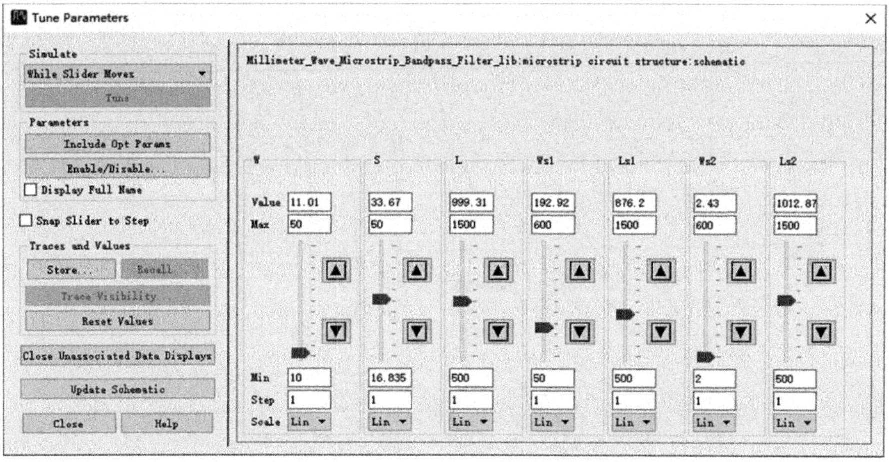

Fig. 2.64 [Tune Parameters] dialog box

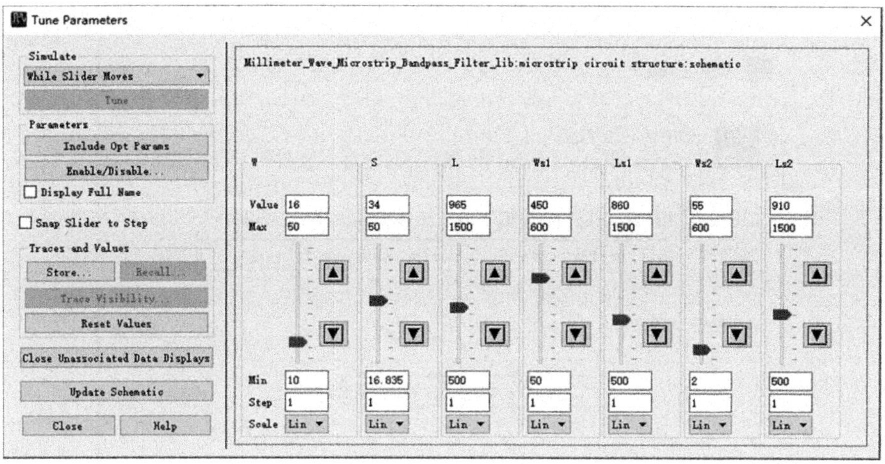

Fig. 2.65 Select of the optimal parameter values

and modify its properties. In addition, you can remove unwanted conductors by clicking the [Remove Conductor] button in the [Material Definitions] dialog box. After finishing all the steps, click the [Apply] button.

(3) Add dielectrics. Select the [Dielectrics] tab and define the relevant dielectrics in Fig. 2.74. The specific steps are listed as follows: Click the [Add From Database…] button in the lower right corner of Fig. 2.74. If the dielectric to be added exists in the pop-up [Add Materials From Database] dialog box shown in Fig. 2.75, select the dielectric and click [OK] to add it. If the dielectric to be added does not exist in the [Add Materials From Database] dialog box,

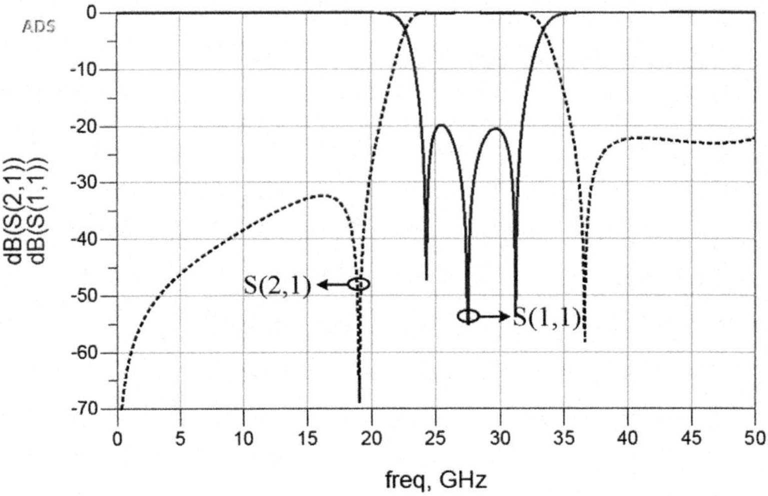

Fig. 2.66 Traces dB(S(1,1)) and dB(S(2,1)) after the tuning

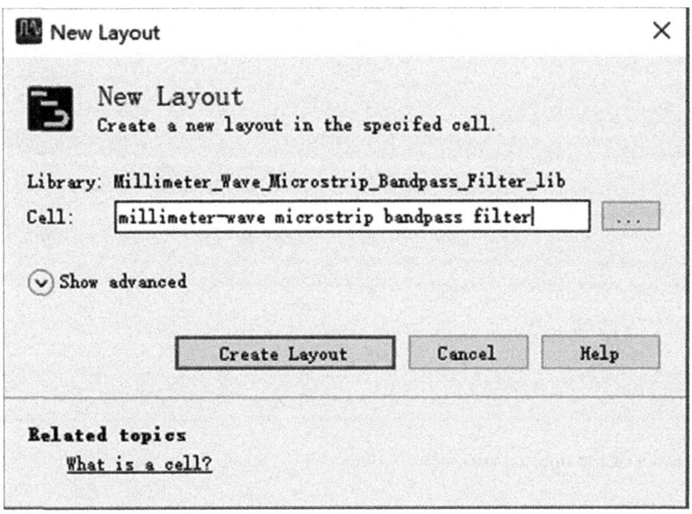

Fig. 2.67 [New Layout] dialog box

return to the [Material Definitions] dialog box and click the [Add Dielectric] button to add a dielectric and modify its properties. In addition, you can remove unwanted dielectrics by clicking the [Remove Dielectric] button in the [Material Definitions] dialog box. After finishing all the steps, click the [OK] button to close the [Material Definitions] dialog box.

3. Define layers

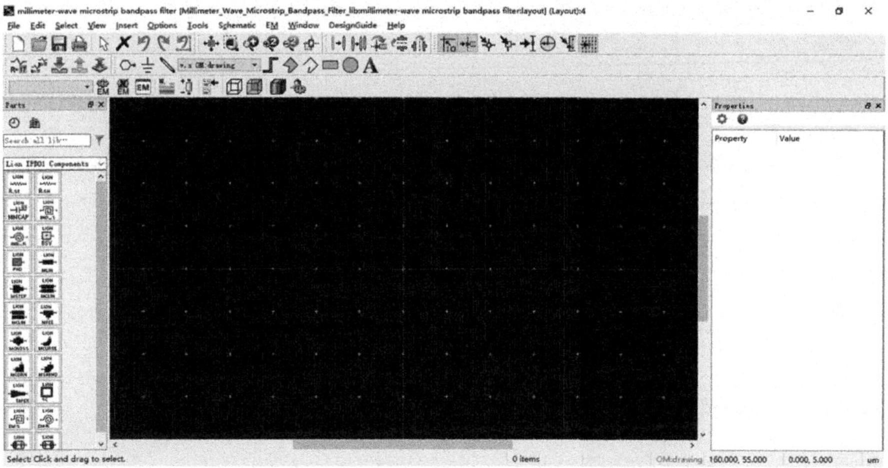

Fig. 2.68 New layout window

Fig. 2.69 [New Substrate] dialog box

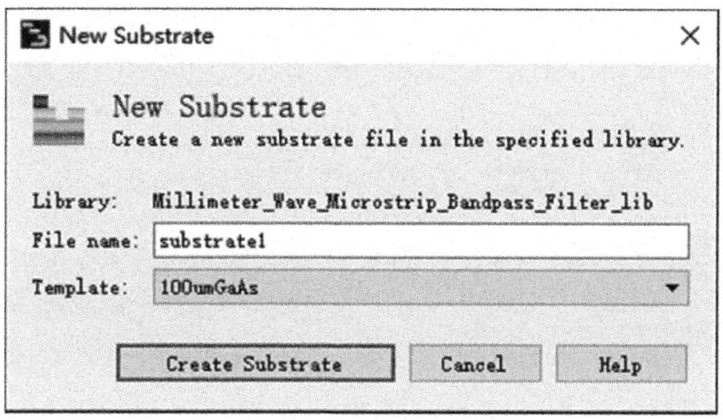

Execute menu commands [Technology] → [Layer Definitions…], then select the [Layers] tab to open the [Layer Definitions] dialog box shown in Fig. 2.76. Define the correlation layers according to Fig. 2.77. Click the [Add Layer] button to add a layer and modify its properties.

4. Edit Substrate Items

(1) Select the "cond" layer and click the right mouse button to pop up Unmap . Click the "Unmap" to delete this layer (Or select the "cond" layer, then press "Delete" on the keyboard to delete it). Delete the "cond2" layer in the same way.
(2) Edit substrate layers. Select an existing substrate layer, right-click the mouse, and choose the [Insert Substrate Layer] from the pop-up menu to insert a new substrate layer. Select the substrate layer to be modified to view and modify its properties under the [Substrate Layer] at the right side of the window.

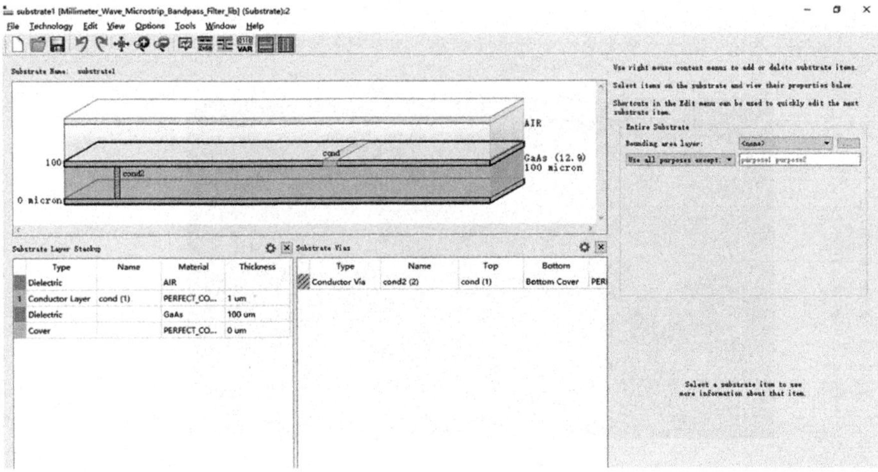

Fig. 2.70 View of all substrate items

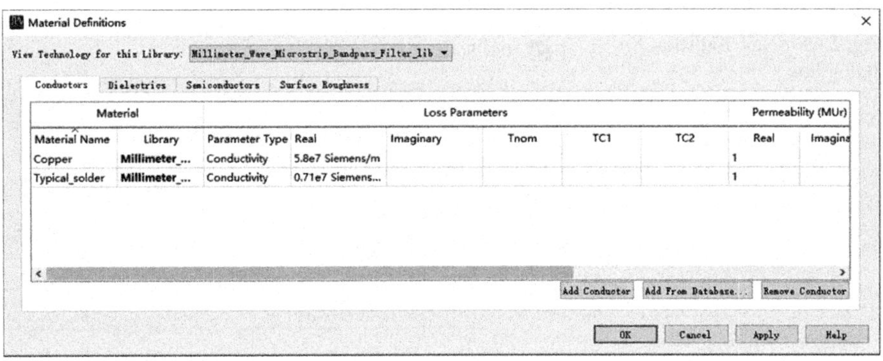

Fig. 2.71 [Material Definitions] dialog box

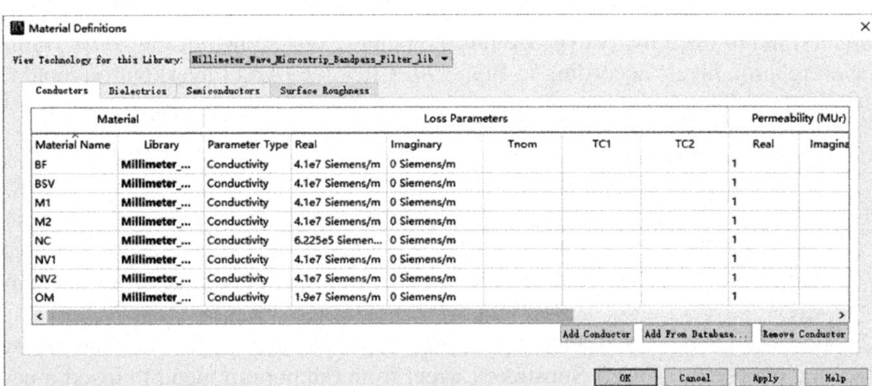

Fig. 2.72 Add of conductors and modification of their properties

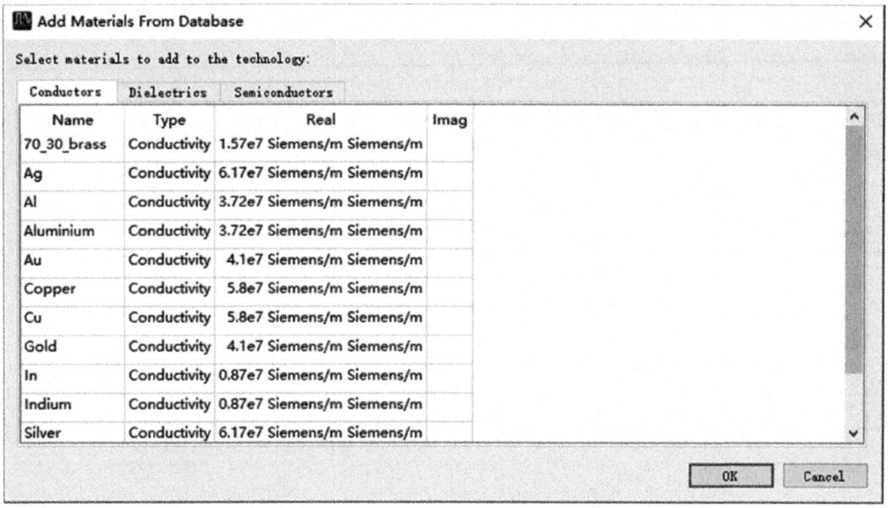

Fig. 2.73 [Conductors] tab of the [Add Materials From Database] dialog box

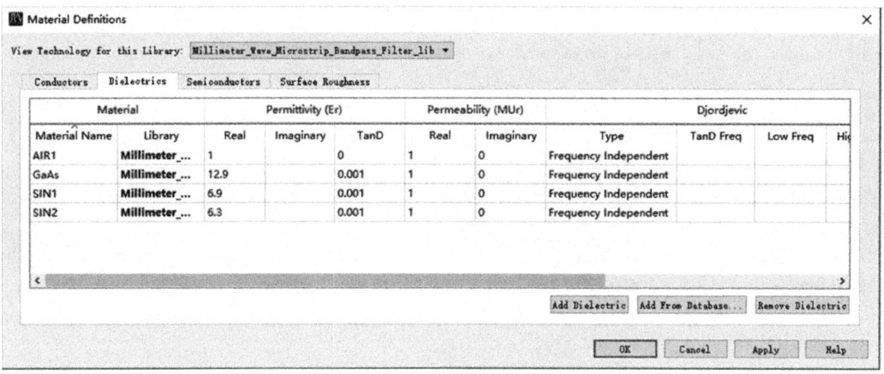

Fig. 2.74 Add of dielectrics and modification of their properties

(3) Edit conductor layers. Select the surface of the substrate layer on which you want to insert a conductor layer, right-click the mouse, and choose the [Map Conductor Layer] from the pop-up menu to insert a new conductor layer. Select the conductor layer to be modified to view and modify its properties under the [Conductor Layer] at the right side of the window.

(4) Edit conductor vias. Select the substrate layer into which you want to insert a conductor via, right-click the mouse, and choose the [Map Conductor Via] from the pop-up menu to insert a new conductor via. Select the conductor via to be modified to view and modify its properties under the [Conductor Via] at the right side of the window.

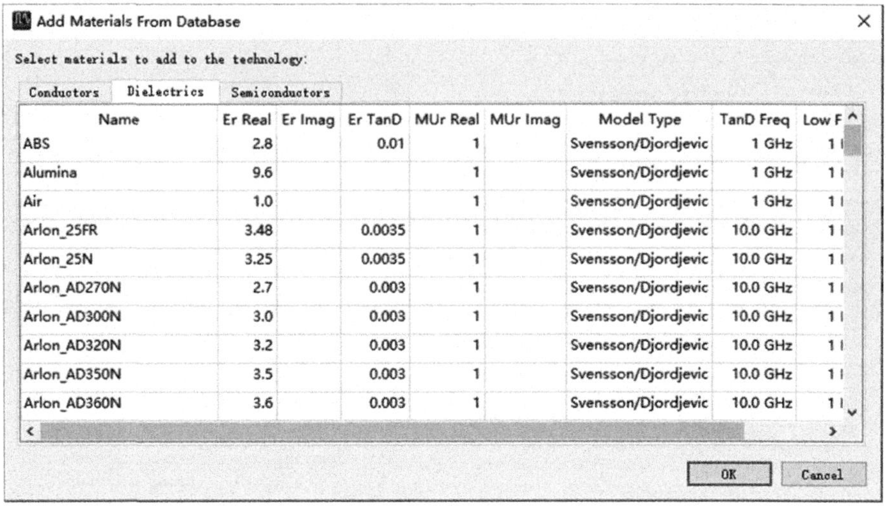

Fig. 2.75 [Dielectrics] tab of the [Add Materials From Database] dialog box

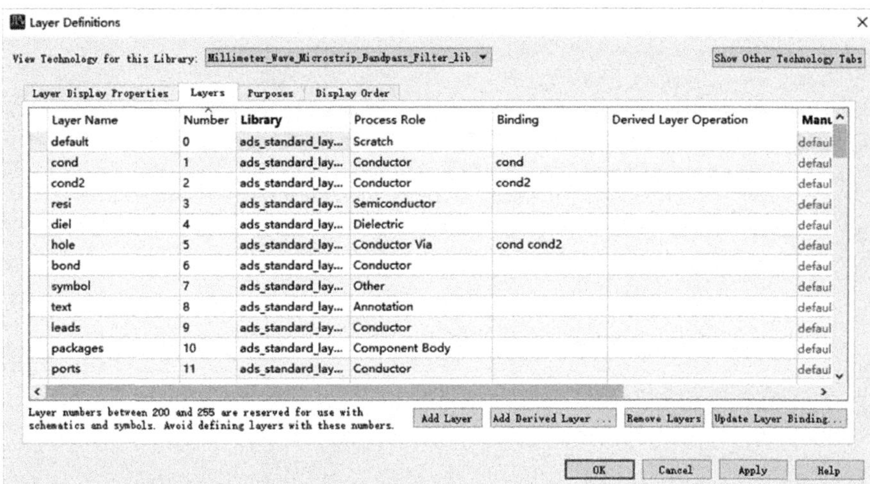

Fig. 2.76 [Layer Definitions] dialog box

(5) All substrate items are shown in Fig. 1.59. The bottom layer is "Cover". The thickness of the GaAs substrate layer is 100 μm. For the OM layer, the "Process Role" is selected as "Conductor", the "Material" is selected as "OM", the "Operation" is selected as "Expand the substrate", the "Position" is selected as "Above interface", and the thickness is 0.276 μm. The thickness of the first SIN1 layer is 0.1 μm. For the NC layer, the "Process Role" is selected as "Conductor", the "Material" is selected as "NC", the "Operation" is selected as "Expand the

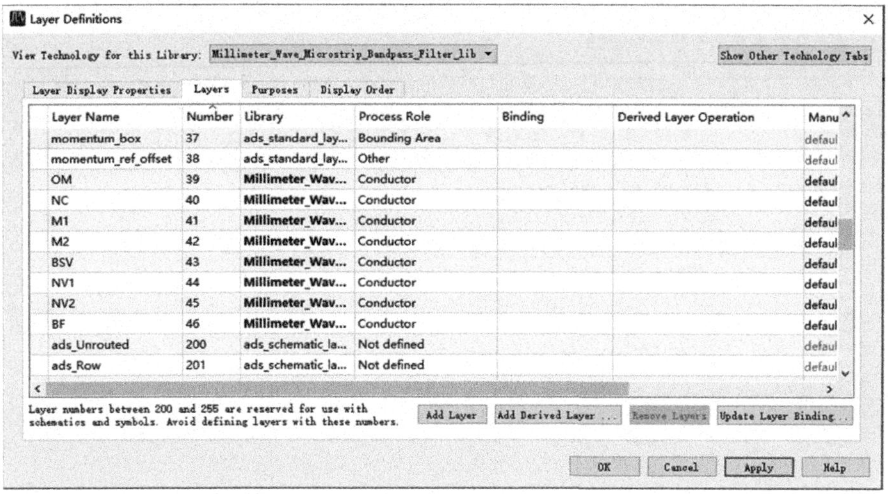

Fig. 2.77 Definitions of all layers

substrate", the "Position" is selected as "Above interface", and the thickness is 0.035 μm. The thickness of the second SIN1 layer is 0.0515 μm. For the M1 layer, the "Process Role" is selected as "Conductor", the "Material" is selected as "M1", the "Operation" is selected as "Expand the substrate", the "Position" is selected as "Above interface", and the thickness is 1.065 μm. The thickness of the SIN2 layer is 0.37 μm. The thickness of the AIR1 layer is 2.19 μm. For the M2 layer, the "Process Role" is selected as "Conductor", the "Material" is selected as "M2", the "Operation" is selected as "Expand the substrate", the "Position" is selected as "Above interface", and the thickness is 4 μm. The thickness of the third SIN1 layer is 0.1 μm. The top layer is the "AIR". For the BSV layer, the "Process Role" is selected as "Conductor Via" and the "Material" is selected as "BSV". For the NV1 layer, the "Process Role" is selected as "Conductor Via" and the "Material" is selected as "NV1". For the NV2 layer, the "Process Role" is selected as "Conductor Via" and the "Material" is selected as "NV2". For the BF layer, the "Process Role" is selected as "Conductor Via" and the "Material" is selected as "BF" (Fig. 2.78).

(6) Moreover, execute menu commands [Technology] → [Layer Definitions…] to open the [Layer Definitions] dialog box shown in Fig. 2.79, where you can modify the display properties of each layer such as color and pattern, etc. Here, keep all the settings in default.

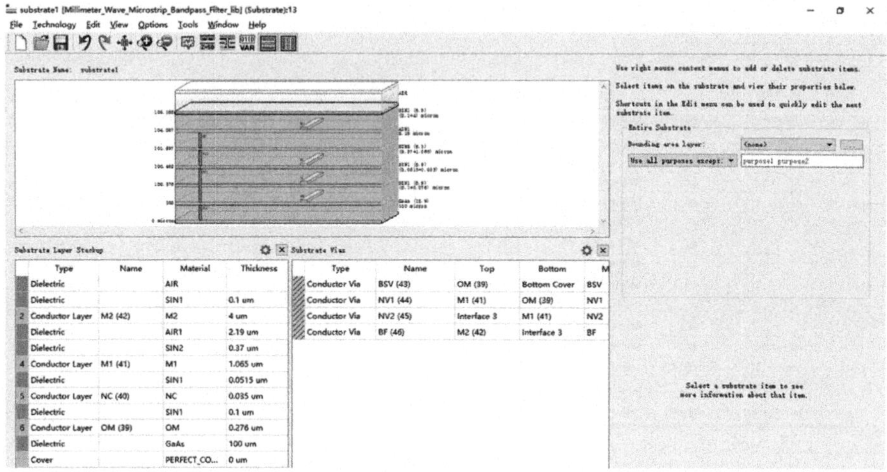

Fig. 2.78 Substrate layer stackup and substrate vias

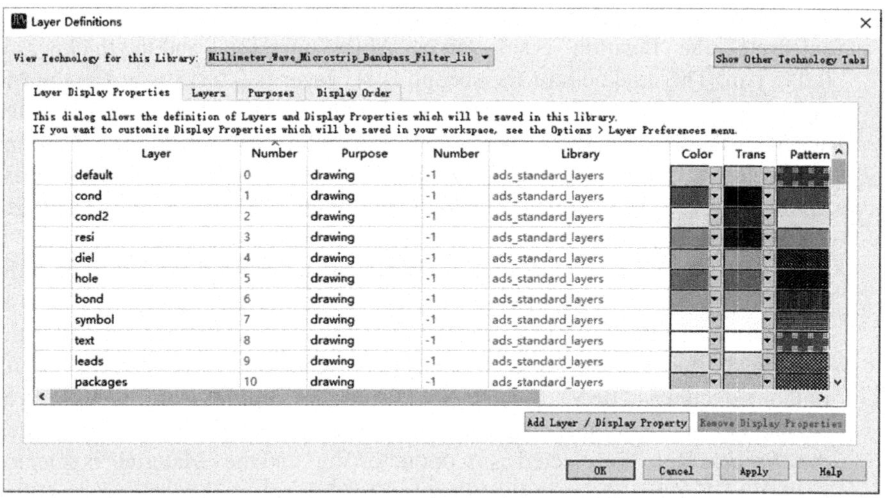

Fig. 2.79 [Layer Display Properties] tab of the [Layer Definitions] dialog box

2.3.2 *Layout of Millimeter-Wave Microstrip Bandpass Filter*

(1) To facilitate the layout drawing, select the function keys in the box of Fig. 2.80 in the "millimeter-wave microstrip bandpass filter" cell. Execute menu commands [Options] → [Preferences…], select the [Grid/Snap] tab in the pop-up [Preferences for Layout] dialog box to modify display grid in the layout. Set the "Snap Grid Distance (in layout units)", "Snap Grid Per Minor Display Grid", and "Minor Grid Per Major Display Grid" to appropriate values, as shown in

Fig. 2.80 Selection of the function key in the box

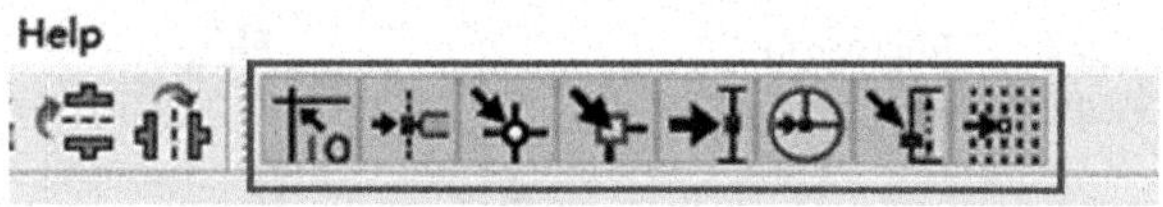

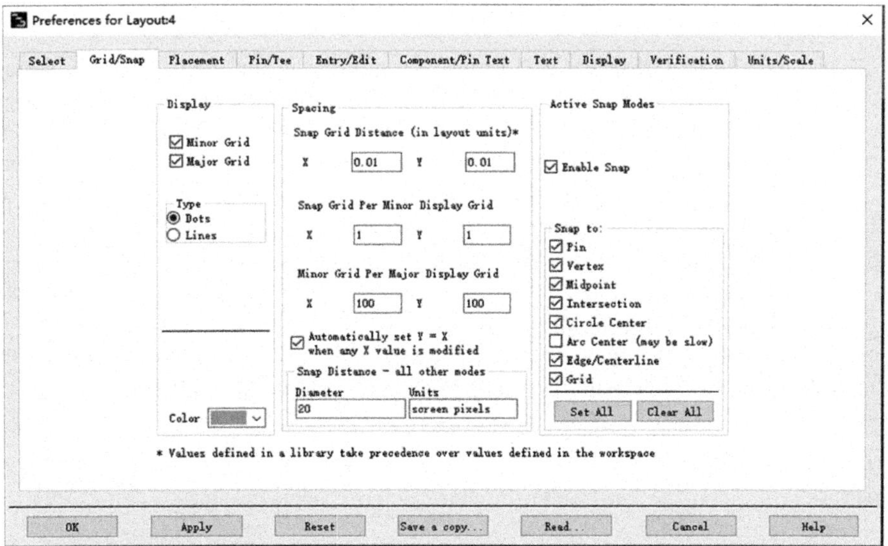

Fig. 2.81 Modification of the display grid in the layout

Fig. 2.81 (Or right-click in the layout drawing area and select " < 0.1–1–100 > " under [Grid Spacing…] in the pop-up menu; or use the shortcut key "Ctrl + Shift + 8").

(2) The layout drawing of microstrip coupled lines and microstrip lines. Execute menu commands [Insert] → [Rectangle] to insert a rectangle into the layout, then press "Esc" to exit. Select the inserted rectangle, select "M1: drawing" in the drop-down menu of the [All Shapes] → [Layer] under [Properties] at the right side of the window, and set the [Rectangles] → [Width] and [Height] to optimal values based on Fig. 2.65. For the drawing of the coupled microstrip line, two microstrip lines with the same size should be drawn according to the length and width of the coupled microstrip line (Or draw a microstrip line, select the microstrip line, and press the shortcut keys "Ctrl + C" and "Ctrl + V" to copy and paste). Then select one of the microstrip lines, long press the left mouse button and move its position to keep a certain coupling gap with the other microstrip line, as shown in Fig. 2.82.

(3) "Sharp angle" of the microstrip line. To reduce the influence of the microstrip line connection in the circuit, the microstrip line is processed with the sharp angle. The specific steps are listed as follows: Execute menu commands [Insert] → [Polygon] (or click the [Create A New Polygon] icon  in the toolbar; or

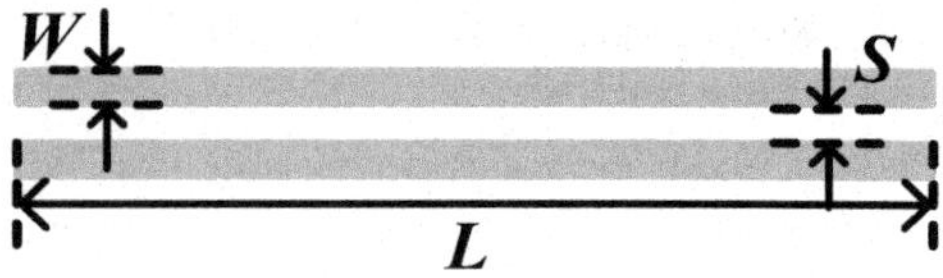

Fig. 2.82 Microstrip coupled lines

use the shortcut "Shift + P") to insert a triangle in the microstrip line which needs to be sharpened, then press "Esc" to exit. Select the inserted triangle, then select "OM:drawing" in the drop-down menu of the [All Shapes] → [Layer] under [Properties] at the right side of the window. Similarly, insert a symmetrical triangle in the opposite position to ensure the symmetry of the sharp angle, as shown in Fig. 2.83. Long press the "Ctrl" key, select the microstrip line to be sharpened and the two inserted triangles. Execute menu commands [Edit] → [Boolean Logical…] and set the Boolean operation between the M1 layer and the OM layer in the pop-up [Boolean Logical Operation Between Layers] dialog box, as shown in Fig. 2.84. Click the [OK] button to complete the operation. The microstrip line with the sharp angle is shown in Fig. 2.85.

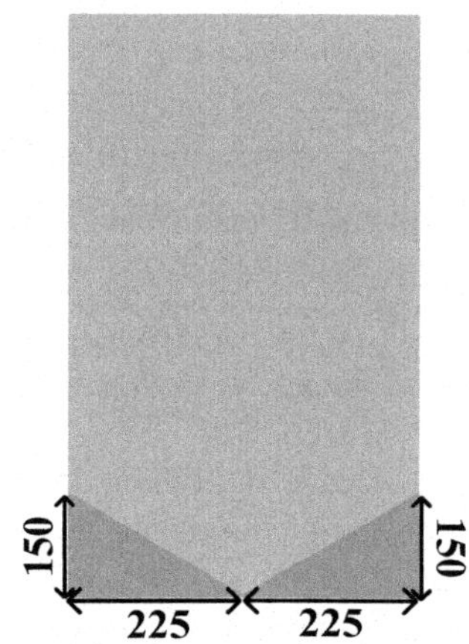

Fig. 2.83 Insertion of triangles

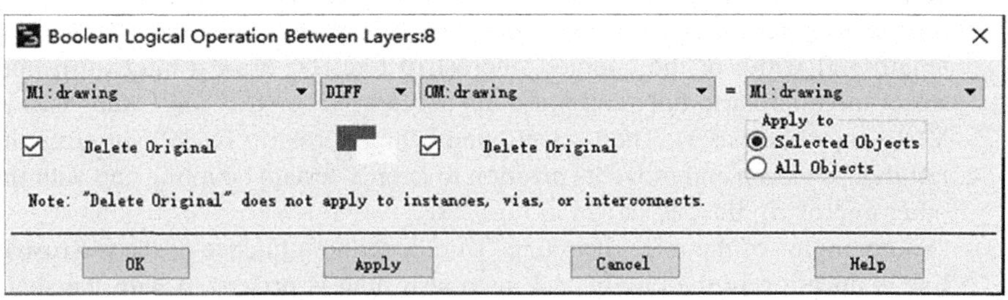

Fig. 2.84 [Boolean Logical Operation Between Lalyers] dialog box

Fig. 2.85 Microstrip line with the sharp angle

(4) Draw signal pads. Signal pads should be added to the I/O ports in the layout because "ground-single-ground" ("GSG") probes are used in the measurement. The overall size of the signal pad is 100 μm × 100 μm.

- Note

  Considering that the probe needs to get down in a certain depth during the measurement, multi-layer structure is used to draw pads to avoid being penetrated by the probe. Holes ensure that the top metal layer is connected to the bottom ground layer.

The specific steps of drawing pads are listed as follows: Execute menu commands [Insert] → [Rectangle] to insert a rectangle in the layout, then press "Esc" to exit. Select the inserted rectangle, then select "OM:drawing" in the drop-down menu of the [All Shapes] → [Layer] under [Properties] at the right side of the window. Set the [Rectangles] → [Width] and [Height] to appropriate values (Here, both are set to 100 μm). Execute menu commands [Edit] → [Copy/Paste] → [Copy To Layer…], select [M1:drawing] in the pop-up [Copy to Layer] dialog box, and click the [Apply] button to copy an M1 layer in the original position. Similarly, copy an M2 layer, a BF layer, an NV1 layer, and an NV2 layer in the original position, respectively. After copying all the needed layers, click the [Cancel] button to close this dialog box. Then indent the layers. Select the M1 layer and execute menu commands [Edit] → [Scale/Oversize] → [Oversize…]. Since the M1 layer is indented by 0.5 μm compared with the OM layer, enter -0.5 in the "Oversize(+)/Undersize(−)" column of the pop-up [Oversize] dialog box and click [Apply] button. Similarly, indent the M2 layer, BF layer, NV1 layer, and NV2 layer by 1, 1.5, 2, and 3 μm relative to the OM layer.

(5) Draw grounding pads. The grounding pad is constructed by two pads in parallel, and one of the pads has a circular grounding hole at the center (The radius of the

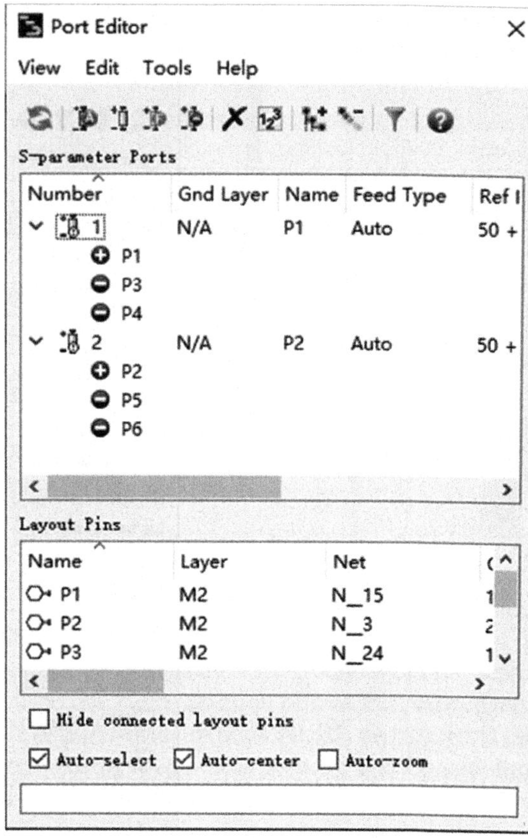

Fig. 2.91 [Port Editor] dialog box after the editing

2.3.4 Parameter Optimization

(1) Parameter optimization. Comparing Fig. 2.66 with Fig. 2.94, it can be observed that there are differences between the layout simulated results and the corresponding microstrip circuit simulated results of the millimeter-wave microstrip bandpass filter. Therefore, the layout parameters should be optimized. By adjusting physical sizes, the optimized layout of the millimeter-wave microstrip bandpass filter is shown in Fig. 2.95 (unit: μm).
(2) Layout simulation. Refer to the steps in Sect. 2.3.3 to simulate the optimized layout. The simulated traces dB(S(1,1)) and dB(S(2,1)) are shown in Fig. 2.96.
(3) Analysis of simulated results . By comparing Fig. 2.38 with Fig. 2.96, it can be

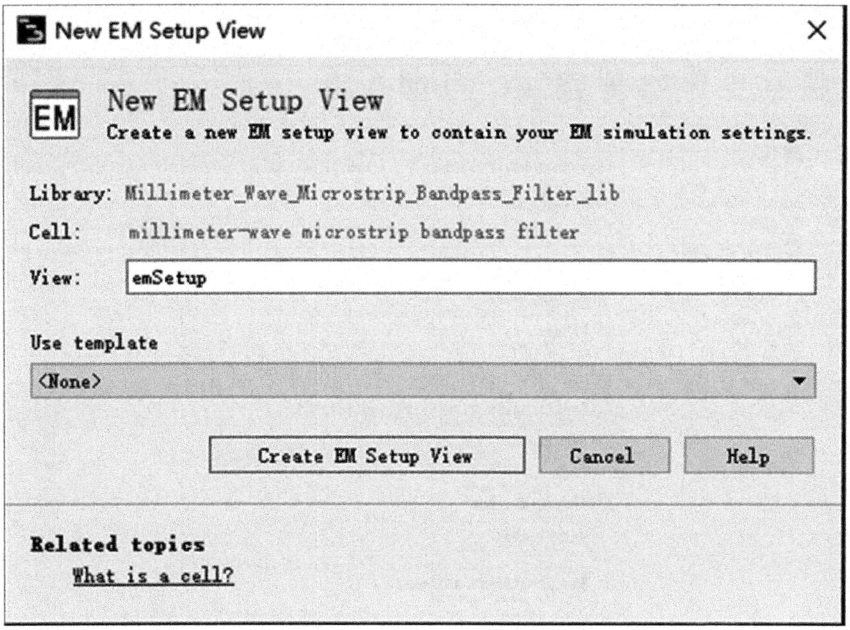

Fig. 2.92 [New EM Setup View] dialog box

seen that the theoretical simulated results are in good agreement with the EM simulated results of the millimeter-wave microstrip bandpass filter. The EM simulated results show that the 3-dB fractional bandwidth is 36.8% (from 22.84 to 33.15 GHz). The three transmission poles are located at 24, 27.08 GHz, and 31.54 GHz, respectively. Two transmission zeros at 19.56 and 36.74 GHz have enhanced the frequency selectivity of the filter. In addition, the passband return loss is more than 19 dB and the out-of-band suppression is over 24.5 dB.

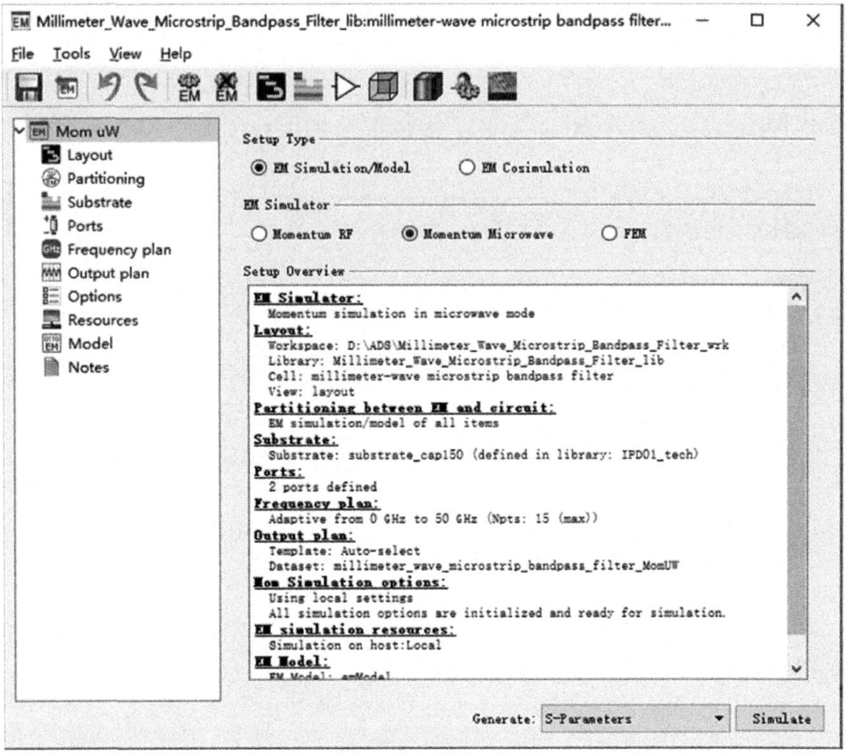

Fig. 2.93 [EM Setup for simulation] window

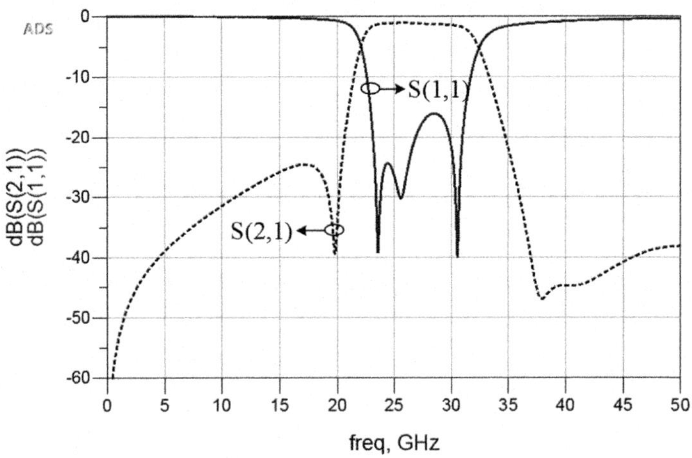

Fig. 2.94 Simulated results of the millimeter-wave microstrip bandpass filter

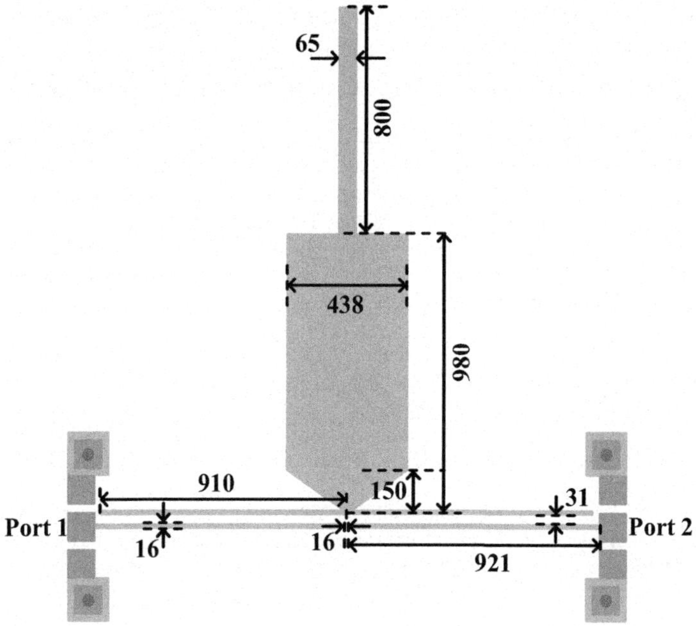

Fig. 2.95 Optimized layout

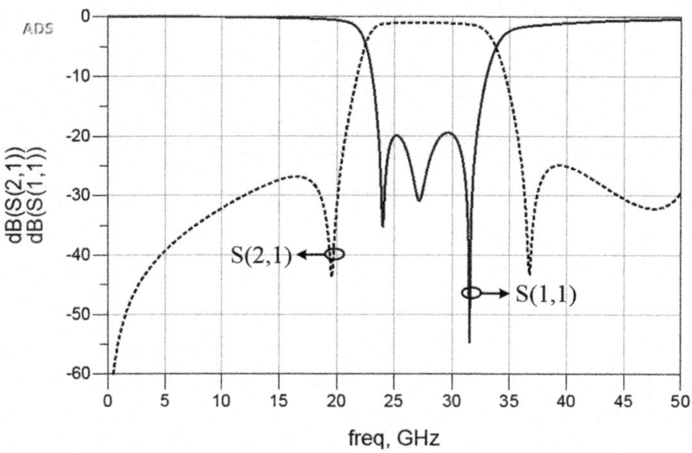

Fig. 2.96 Simulated results of the optimized layout of the millimeter-wave microstrip bandpass filter

References

1. X. Xie, Y. Xu, L. Xia, *Microwave Integrated Circuits* (Publishing House of Electronics Industry, Beijing, 2018)
2. Q. Gu, J. Xiang, X. Yuan, *Microwave Integrated Circuit Design* (People's Posts and Telecommunications Press, Beijing, 1978)
3. R. Li, *Key Issues in RF/RFIC Circuit Design* (Higher Education Press, Beijing, 2007)
4. B. Zhang, Y. Wu, Y. Liu, Wideband single-ended and differential bandpass filters based on terminated coupled line structures. IEEE Trans. Microw. Theory Techn. **65**(3), 761–774 (2017)
5. Y. Wu, L. Cui, Z. Zhuang, W. Wang, Y. Liu, A simple planar dual-band bandpass filter with multiple transmission poles and zeros. IEEE Trans. Circuits Syst. II, Exp. Briefs **65**(1), 56–60 (2018)
6. Y. Wu, L. Cui, W. Zhang, L. Jiao, Z. Zhuang, Y. Liu, High performance single-ended wideband and balanced bandpass filters loaded with stepped-impedance stubs. IEEE Access **5**, 5972–5981 (2017)
7. Z.-C. Guo, L. Zhu, S.-W. Wong, A quantitative approach for direct synthesis of bandpass filters composed of transversal resonators. IEEE Trans. Circuits Syst. II, Exp. Briefs **66**(4), 577–581 (2019)
8. Z. Zhuang, Y. Wu, Y. Liu, Z. Ghassemlooy, Wideband bandpass-to-all-stop reconfigurable filtering power divider with bandwidth control and all-passband isolation. IET Microw. Antennas Propag. **12**(11), 1852–1858 (2018)

Chapter 3
Design and Simulation of Input-Absorptive Bandstop Filter

As a two-port frequency-selective device in the RF front-end, filters have been widely investigated in the bandpass, bandstop, and reconfigurable types, etc. Among them, the absorptive bandstop filter can perform better stopband absorption in the case of strong interferences, and reduce the impact of the RF signal power reflection on the performance of adjacent components. In this chapter, we are going to introduce the basic theory of an input-absorptive bandstop filter based on the TFIPD technology [1]. This chapter is going to introduce the basic theories of the input-absorptive bandstop filter and how to use the ADS software to build, simulate, and optimize its ideal simulated model and full-wave electromagnetic simulated model. Finally, the manufactured input-absorptive bandstop filter chip is measured, and the measured results are also analyzed.

3.1 Overview of Absorptive Bandstop Filter

3.1.1 Theoretical Basis

Bandstop filters are widely used in RF/microwave systems. Most traditional bandstop filters are reflective, namely, the unwanted signals can be reflected by the filter. In this case, the unwanted reflected power can cause the performance degradations of the adjacent circuit components. Therefore, it is necessary to design the absorptive bandstop filters to absorb unwanted signals.

Y. Wu and W. Wang, *Microwave and Millimeter-Wave Chips Based on Thin-Film Integrated Passive Device Technology*,
https://doi.org/10.1007/978-981-99-1455-5_3

3.1.2 Propaedeutics

1. One-TZ Input-Absorptive Bandstop Filter

Figure 3.1 shows the circuit schematic of the one-TZ input-absorptive bandstop filter based on a basic bandstop block and one resistor connected with the input port. The basic bandstop block is composed of the conventional 3rd-order bandstop filter to realize a broad stopband. Thus, according to the Chebyshev reflective prototype values (g-parameters), the lumped elements of the basic bandstop block can be calculated as follows:

$$\begin{cases} L_1 = L_3 = \frac{FBW}{2\pi f_0}\frac{Z_0}{\gamma} \\ C_1 = C_3 = \frac{1}{2\pi f_0 FBW}\frac{\gamma}{Z_0} \\ L_2 = \frac{3+4\gamma^2}{16\pi f_0 FBW}\frac{Z_0}{\gamma} \\ C_2 = \frac{4FBW}{\pi f_0 (3+4\gamma^2)}\frac{\gamma}{Z_0} \end{cases}, \tag{3.1a}$$

in which,

$$\gamma = \sinh[\frac{\ln\left[-\coth\left(\frac{10\log_{10}(1-10^{-RL_{\min}/10})}{17.37}\right)\right]}{6}], \tag{3.1b}$$

where f_0 and FBW are the center frequency and fractional bandwidth of the stopband filter, respectively. $RL_{\min}$ represents the allowable minimum return loss in the passband ($RL_{\min} > 0$). To absorb the input signal power in Port 1, the resistor loaded between the input port and the conventional 3rd-order bandstop filter in Fig. 3.1 should satisfy the following condition:

$$R = Z_0, \tag{3.2}$$

where Z_0 is the terminal impedance, which is set as 50 Ω in general.

2. Three-TZs Input-Absorptive Bandstop Filter

To further improve the frequency selectivity and expand the stopband bandwidth, the series-inductor-capacitor resonators (SICRs) are loaded in the basic bandstop

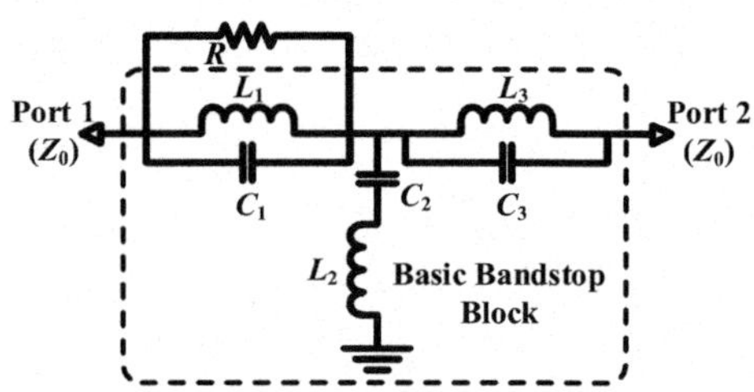

Fig. 3.1 Lumped-element circuit schematic of the one-TZ input-absorptive bandstop filter

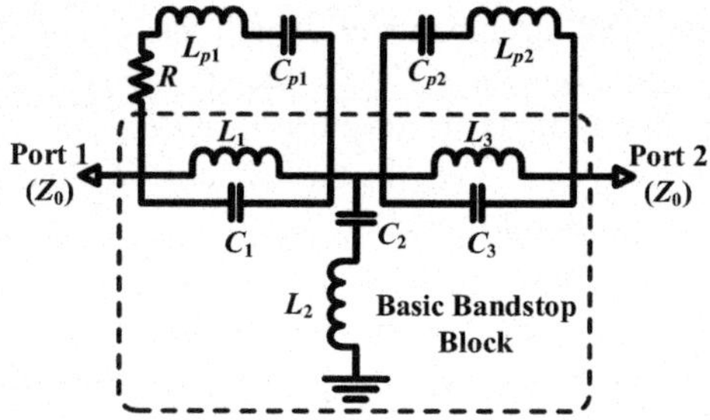

Fig. 3.2 Lumped-element circuit schematic of the three-TZs input-absorptive bandstop filter

block, as shown in Fig. 3.2. The lumped elements (L_1, C_1, L_2, C_2, L_3, and C_3) of the basic bandstop block and the resistor (R) can still be calculated by Eqs. (3.1) and (3.2), while the lumped elements of the SICR (L_{p1}, C_{p1}, L_{p2}, C_{p2}) are taken as free variables. The values of L_{p1} and C_{p1} can affect the stopband center frequency f_0 and the return loss, while the value changes of L_{p2} and C_{p2} can generate two extra transmission zeros. According to the impedance matching conditions of $S_{11} = 0$, L_{p1} and C_{p1} should meet the following condition:

$$L_{p1} = \frac{1}{C_{p1}\omega_0^2}, \tag{3.3a}$$

in which $\omega_0 = 2\pi f_0$. For simplification, assume that L_{p2} and C_{p2} also meet the condition:

$$L_{p2} = \frac{1}{C_{p2}\omega_0^2}. \tag{3.3b}$$

According to (3.1a), (3.3b), and $S_{21} = 0$, the extra TZs are derived as:

$$\begin{cases} f_{z1} = \dfrac{\sqrt{2\omega_0^2 + \frac{1}{C_3 L_{p2}} - \omega_0^3\sqrt{(4 + C_{p2}L_3\omega_0^2)C_{p2}L_3}}}{2\sqrt{2}\pi} \\ f_{z2} = \dfrac{\sqrt{2\omega_0^2 + \frac{1}{C_3 L_{p2}} + \omega_0^3\sqrt{(4 + C_{p2}L_3\omega_0^2)C_{p2}L_3}}}{2\sqrt{2}\pi} \end{cases}. \tag{3.4}$$

It is obvious that when the lumped elements ($L_1, C_1, L_2, C_2, L_3, C_3$) and the resistor ($R$) are determined by (3.1) and (3.2), L_{p2} and C_{p2} can determine the frequencies of the two extra TZs according to (3.4).

3.2 Schematic of Input-Absorptive Bandstop Filter

ADS can realize the parameterized model simulation. Take the one-TZ input-absorptive bandstop filter as an example to introduce how to build and simulate its ideal circuit model, which has the center frequency $f_0 = 5$ GHz, the terminal impedance $Z_0 = 50\ \Omega$, the fractional bandwidth $FBW = 1.6$, and the allowable minimum passband return loss $RL_{\min} = 30$ dB. The lumped elements (L_1, C_1, L_2, C_2, L_3, and C_3) of the basic bandstop block and the resistor (R) can be parameterized in the circuit model.

3.2.1 New Workspace and Circuit Model

1. New Workspace

(1) Double-click the ADS shortcut icon, and click the [OK] in the displayed dialog box to start the ADS software. After ADS runs, the [Get Started] dialog box will pop up automatically. Click the [Close] button in the lower right corner to enter the main interface [Advanced Design System 2020 (Main)], as shown in Fig. 3.3.
(2) Create a new workspace to store all files of the design simulation. Execute menu commands [File] → [New] → [Workspace…] to open the new workspace dialog box shown in Fig. 3.4, where you can set the workspace name [Name] and the working path [Create in]. Here, change the workspace name to "Input_Absorptive_Bandstop_Filter_wrk" and keep the working path in the default setting. Click the [Create Workspace] button to complete the creation of the new workspace.
(3) The [Folder View] in the ADS main interface can display the name and working path of the created workspace, as shown in Fig. 3.5. The workspace name is "Input_Absorptive_Bandstop_Filter_wrk", the corresponding path

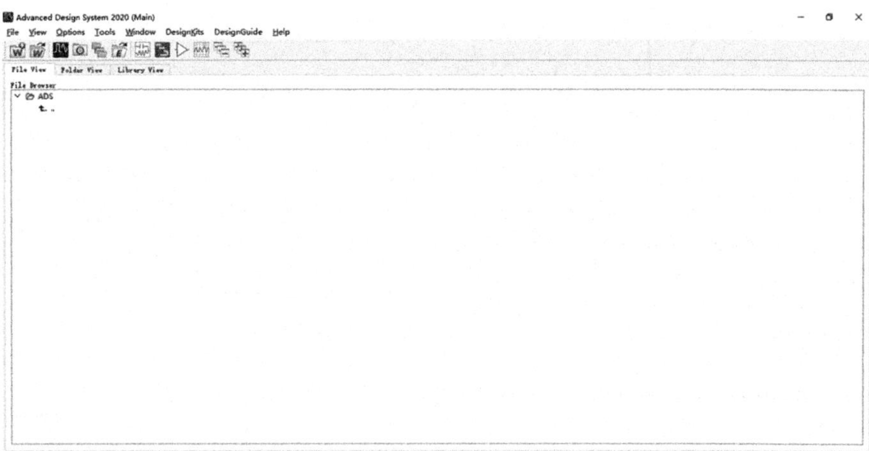

Fig. 3.3 Main interface of ADS

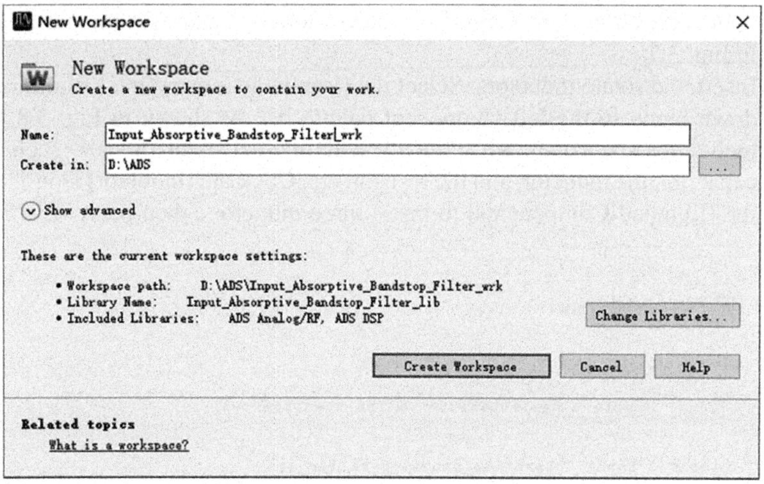

Fig. 3.4 [New workspace] dialog box

is "D:\ADS\Input_Absorptive_Bandstop_Filter_wrk", and a subfolder named "Input_Absorptive_Bandstop_Filter_wrk" in the ADS folder of disk D can be found in your computer.

2. New Circuit Model

(1) Create a new schematic. Execute menu commands [File] → [New] → [Schematic…] to open the [New Schematic] dialog box shown in Fig. 3.6. Modify the name of "Cell" to "circuit structure" and click the [Create

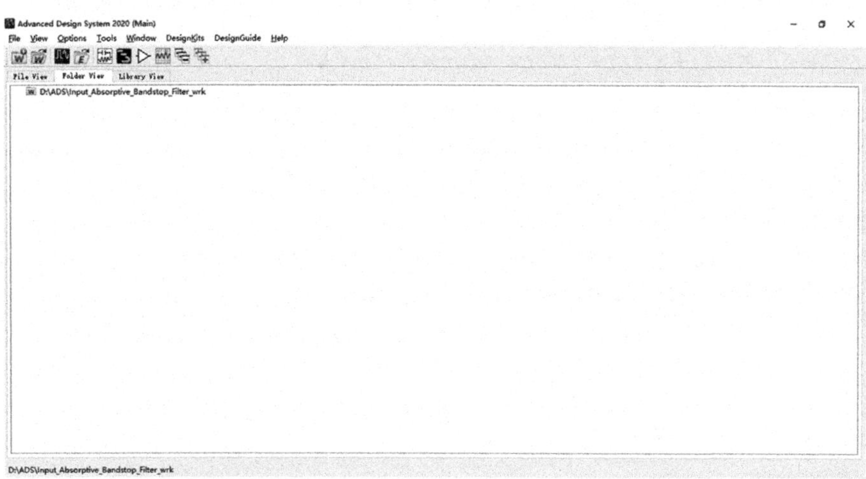

Fig. 3.5 Folder view of the new workspace

Schematic] button to complete the creation of a new schematic, as shown in Fig. 3.7.

(2) Insert and rotate inductors. Select the [Lumped-Components] from the drop-down menu in the left component palette list, as shown in Fig. 3.8, which includes some commonly used ideal lumped-element models, such as the capacitor, the inductor, and the resistor, etc. Click the [Inductor] icon under the [Lumped-Components] to insert three inductors, then press "Esc" on the

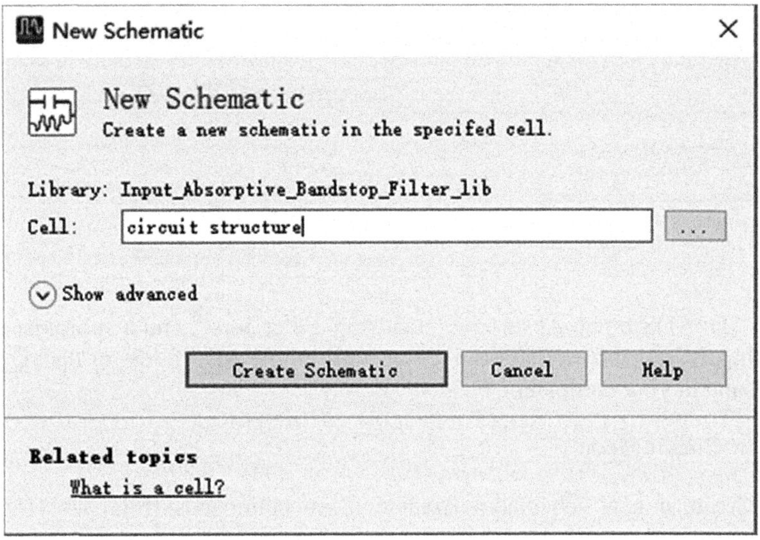

Fig. 3.6 [New schematic] dialog box

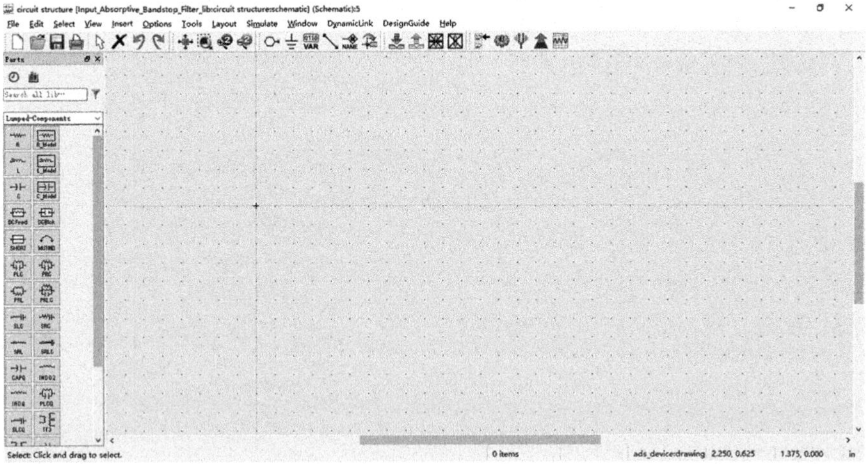

Fig. 3.7 New schematic window

keyboard to exit, as shown in Fig. 3.9. Right-click the inductors to be rotated, select [Rotate] from the pop-up menu to rotate them in 90° clockwise. Inductors have been inserted and rotated, as shown in Fig. 3.10.

(3) Insert and rotate capacitors. Click the [Capacitor] icon under the [Lumped-Components] to insert three capacitors, then press "Esc" on the keyboard to exit, as shown in Fig. 3.11. Right-click the capacitors to be rotated, select

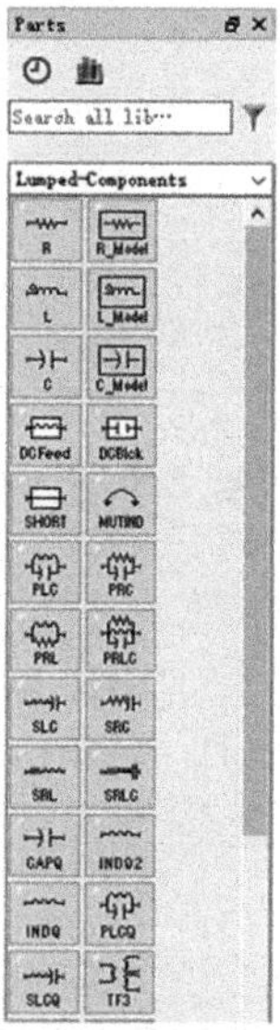

Fig. 3.8 [Lumped-Components] list

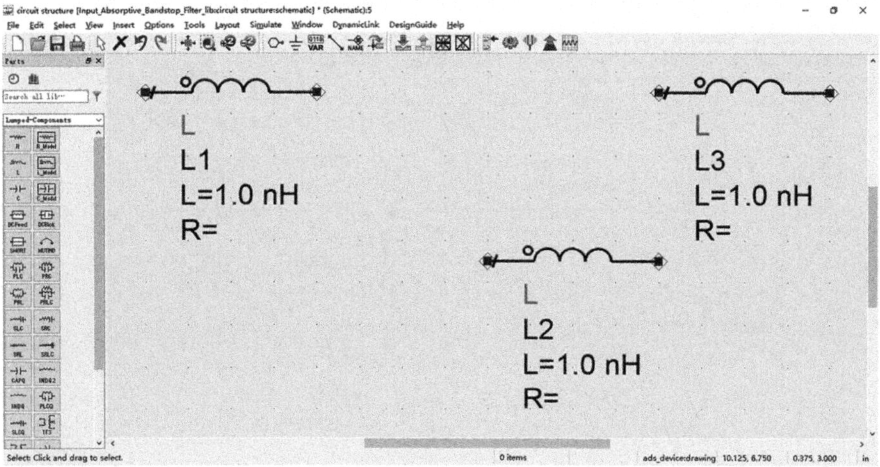

Fig. 3.9 Insertion of inductors

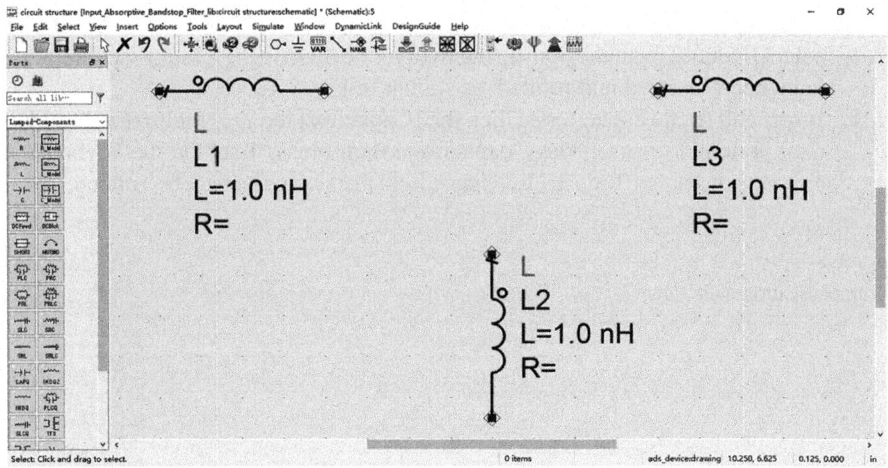

Fig. 3.10 Rotation of inductors

[Rotate] from the pop-up menu to rotate them in 90° clockwise. Capacitors have been inserted and rotated, as shown in Fig. 3.12.

(4) Insert resistor. Click the [Resistor] icon under the [Lumped-Components] to insert one resistor, then press "Esc" on the keyboard to exit. The resistor has been inserted, as shown in Fig. 3.13.

(5) Place ground. Execute menu commands [Insert] → [GROUND] to place a ground, then press "Esc" on the keyboard to exit. The ground symbol has been inserted, as shown in Fig. 3.14.

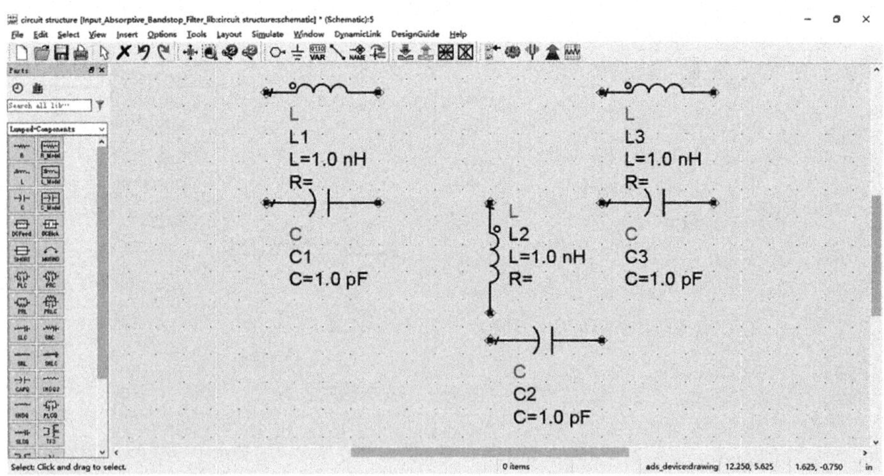

Fig. 3.11 Insertion of capacitors

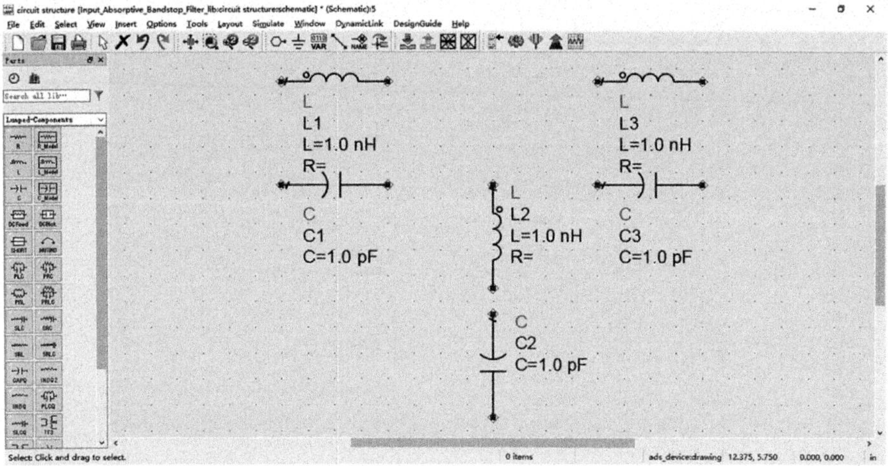

Fig. 3.12 Rotation of capacitors

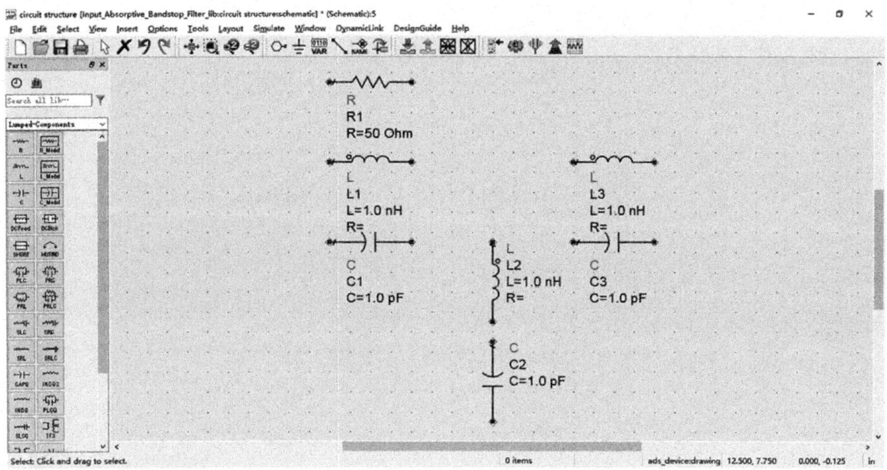

Fig. 3.13 Insertion of the resistor

(6) Connect components. Execute menu commands [Insert] → [Wire] to connect each component according to the circuit structure shown in Fig. 3.1. Figure 3.15 shows the circuit structure after the components are connected.

(7) Modify circuit model parameters. Double-click the inductor L1, then modify the parameter value to L1 (nH) in the pop-up [Edit Instance parameters] dialog box, as shown in Fig. 3.16 (Check whether the unit settings are consistent). Click the [OK] button to save the parameter and close the dialog box. Similarly, modify the parameter value of L2 to L2 (nH), the parameter values of L3 to L3 (nH), the parameter value of C1 to C1 (pF), the parameter value of C2 to

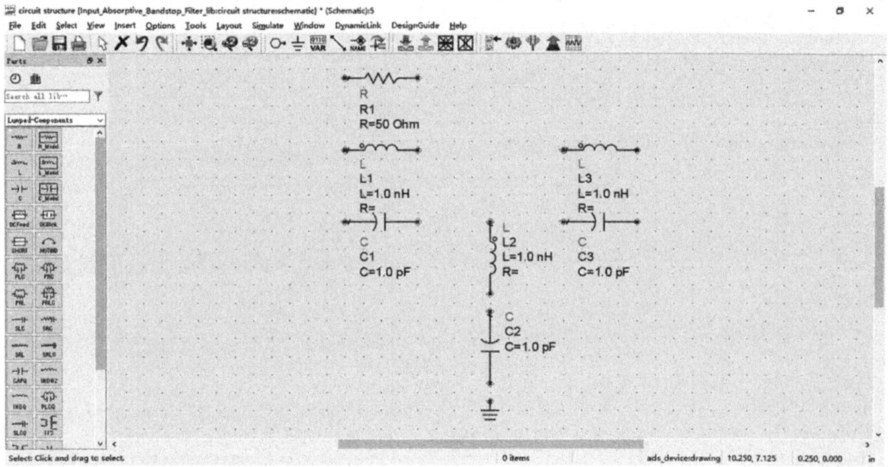

Fig. 3.14 Placement of the ground

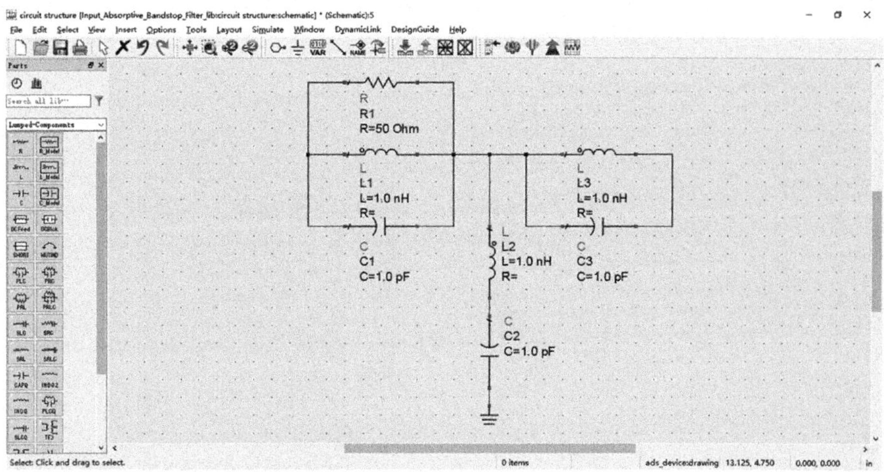

Fig. 3.15 Connection of components

C2 (pF), the parameter value of C3 to C3 (pF), and the parameter value of R1 to R (Ohm), as shown in Figs. 3.17, 3.18, 3.19, 3.20, 3.21 and 3.22.

(8) Define parameter values of variables. Click the [Insert VAR: Variable Equations] icon in the toolbar to insert a VAR item in the schematic. Double-click the component to open the [Edit Instance Parameter] dialog box. The "Variable or Equation Entry Mode" is "Standard" by default. Input the variable name "L1" in the "Name" column and the variable value "1.36" in the "Variable Value" column. Click the [Apply] button, and the dialog box after the setting is shown in Fig. 3.23. If you click [OK], the dialog box can be

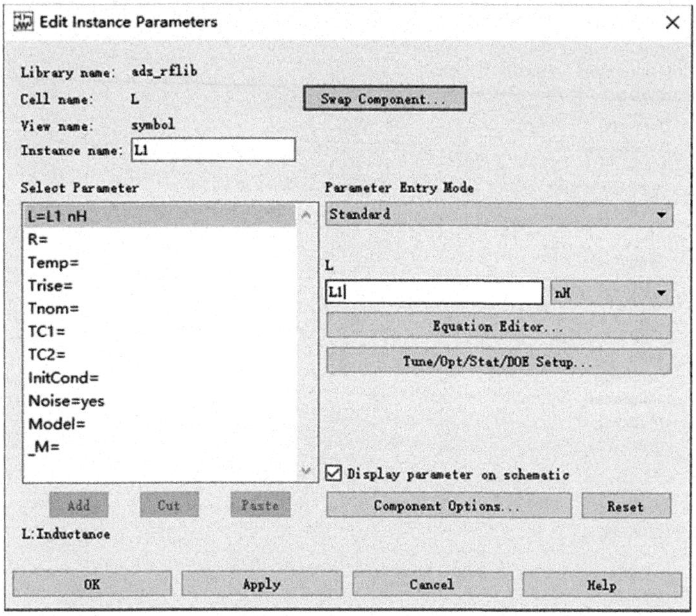

Fig. 3.16 Parameter setting of the inductor L1

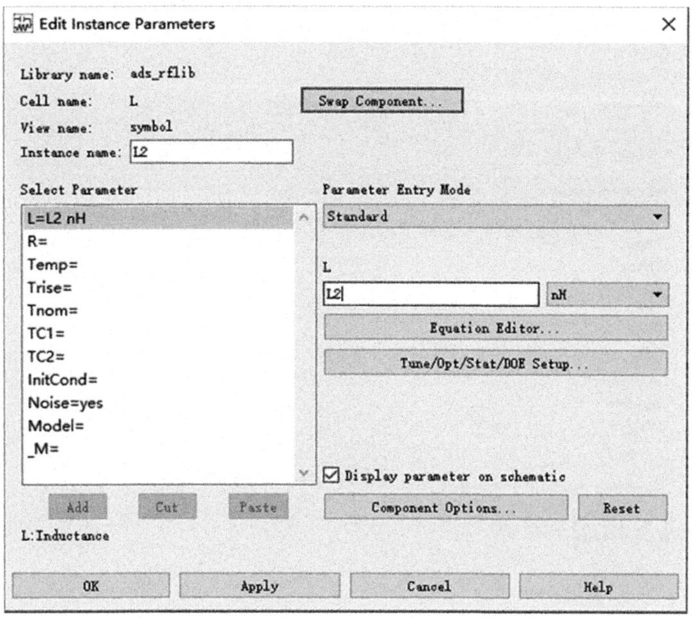

Fig. 3.17 Parameter setting of the inductor L2

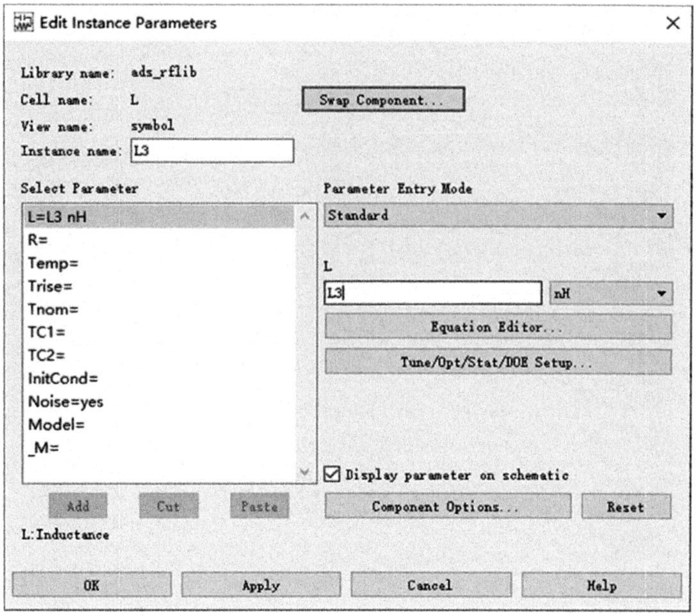

Fig. 3.18 Parameter setting of the inductor L3

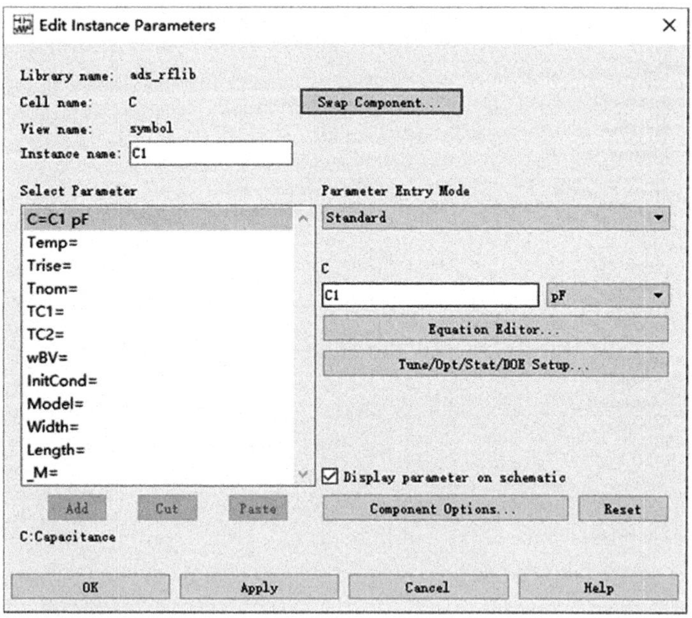

Fig. 3.19 Parameter setting of the capacitor C1

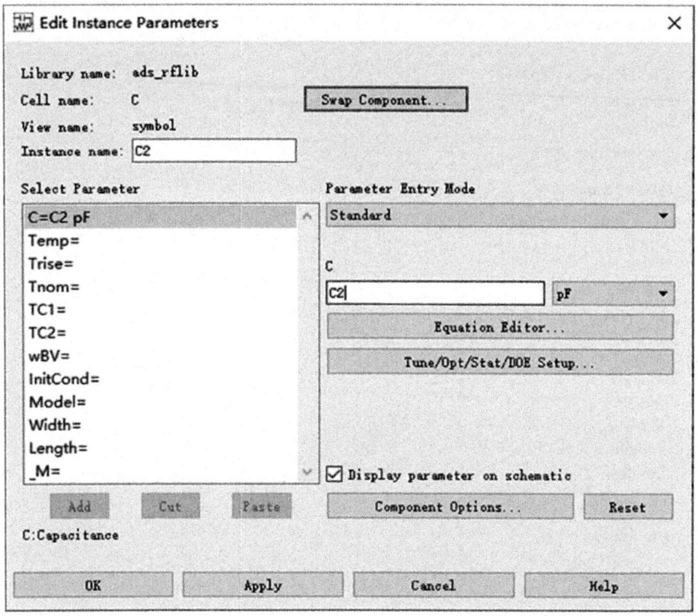

Fig. 3.20 Parameter setting of the capacitor C2

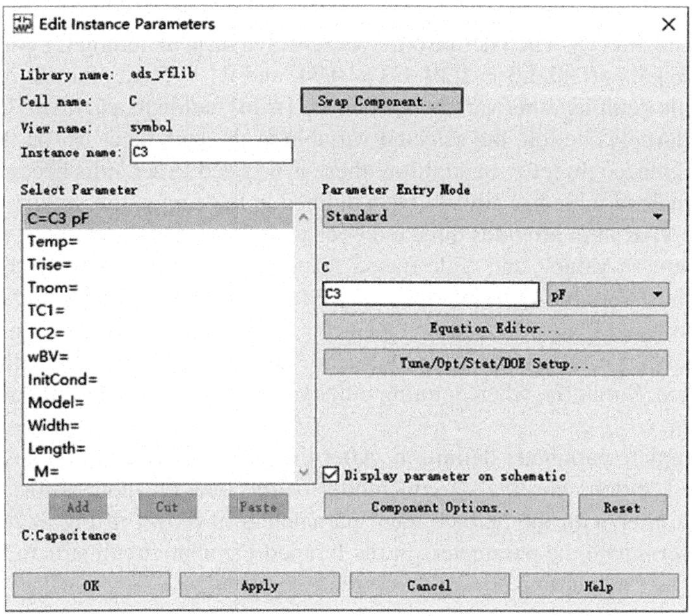

Fig. 3.21 Parameter setting of the capacitor C3

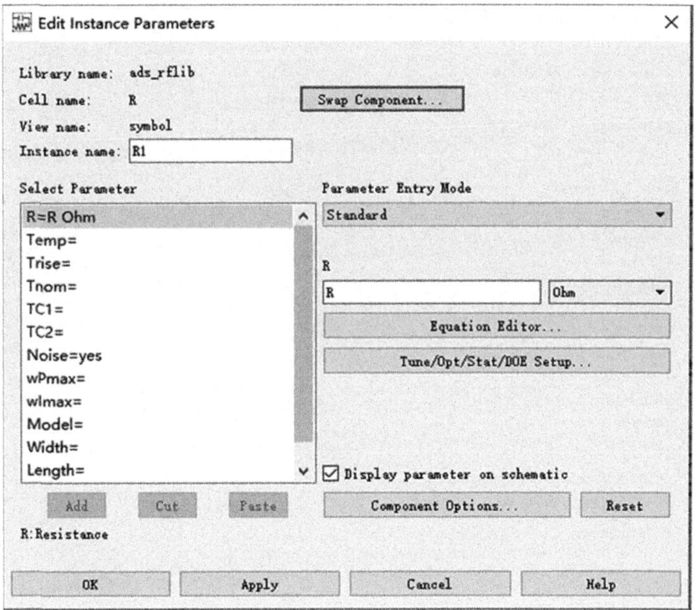

Fig. 3.22 Parameter setting of the resistor R1

closed directly. Then define other variables in turn, including C1 = 0.74, L2 = 1.13, C2 = 0.90, L3 = 1.39, C3 = 0.74, and R1 = 50, as shown in Fig. 3.24. While defining other variables, click the [Add] button to add them. If you click the [Apply] button, the selected variable in the parameter list on the left can be replaced directly. In addition, there is no need to set units because the unit of each variable has already been defined in the component model.

(9) The VAR item provides three methods to add variables, including "Standard", "Name = Value", and "File Based". You can select the most suitable method in the "Variable or Equation Entry Mode" column. The second method is introduced below. Select the "Name = Value" mode, as shown in Fig. 3.25. Input "L1 = 1.36" under the "Variable Value" column, then click the [Apply] button. Similarly, when defining other variables, click the [Add] button to add them.

(10) Complete parameter definition. After the definition of all variables, click the [OK] button, the final circuit model of the one-TZ input-absorptive bandstop filter with the defined ideal parameters is shown in Fig. 3.26. Readers can compare the parameters in the lumped-element circuit structure to check whether all parameters are set correctly in detail.

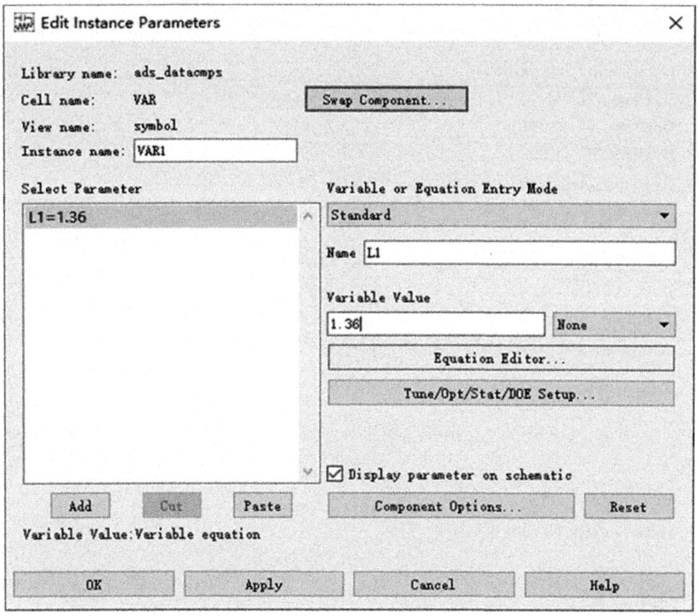

Fig. 3.23 Definition of the variable L1

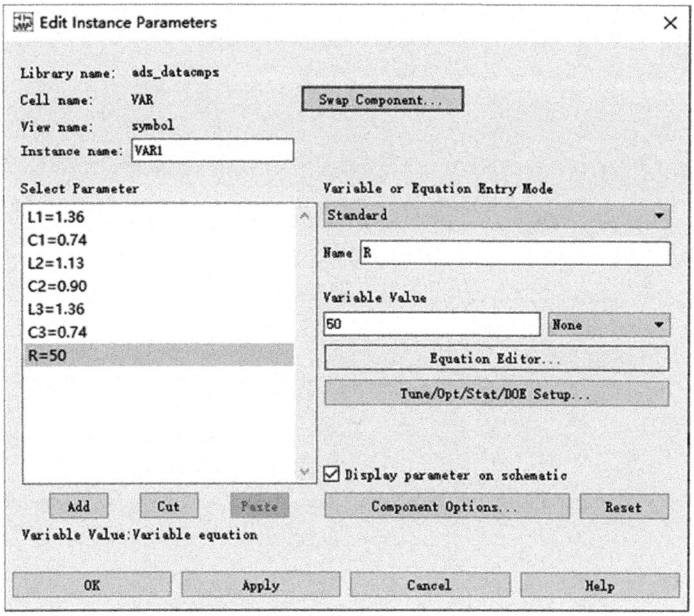

Fig. 3.24 Definitions of all variables in the circuit

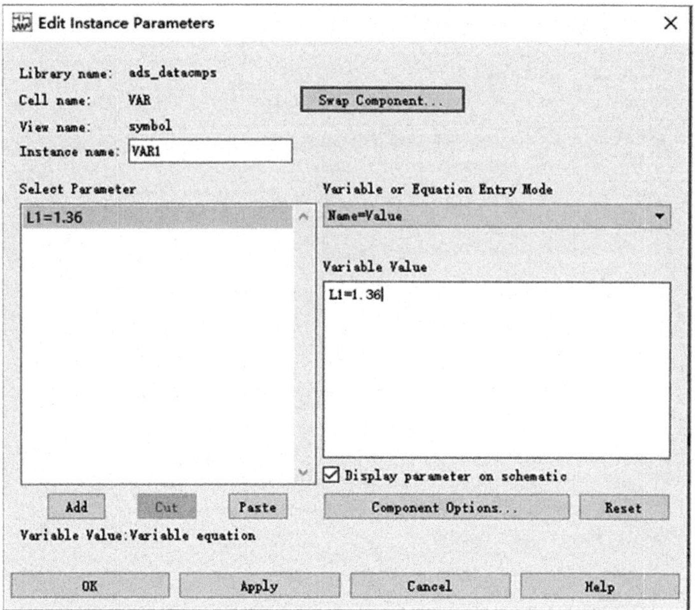

Fig. 3.25 The second method to define the variable L1

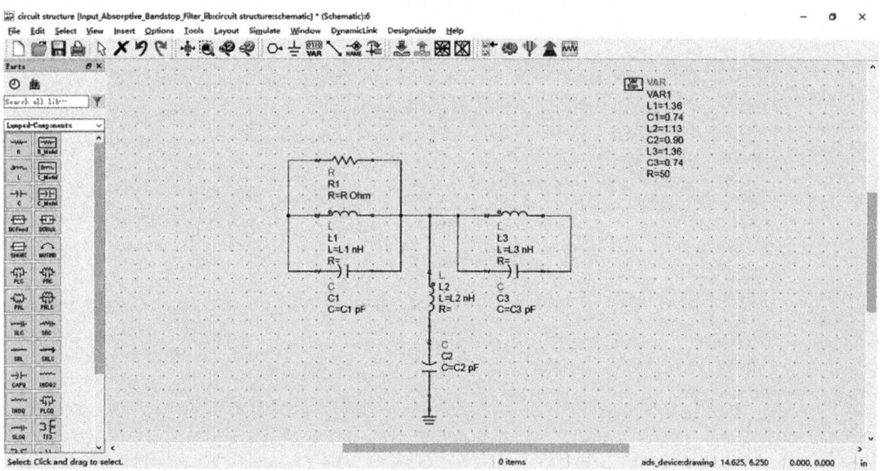

Fig. 3.26 One-TZ input-absorptive bandstop filter circuit schematic with the defined ideal parameters

3.2.2 Schematic Simulation

1. Set Simulation Parameters

(1) Insert the S-parameter simulator, term ports, and grounds. As shown in Fig. 3.27, select the [Simulation-S_Param] from the drop-down menu in the left component palette list and click the [S-parameter Simulator] icon to insert an S-parameter simulator. Click the [Port Impedance Termination for S-Parameters] icon to insert two term ports, then press "Esc" on the keyboard to exit. After that, execute menu commands [Insert] → [GROUND] to place two grounds (Or click the [TermG Port Impedance Termination for S-Parameters with Grounded Reference] icon to insert two term ports with the grounded reference). Execute menu commands [Insert] → [Wire] to connect all the inserted components. After finishing all the steps, press "Esc" to exit.

(2) Set simulation frequency. Double-click the S-parameter simulator [S-PARAMETERS]. Complete the settings shown in Fig. 3.28, in which the start frequency is 0 GHz, the stop frequency is 16 GHz, and the step-size is 0.01 GHz, then click the [OK] button. The final ideal parameter simulated circuit model of the one-TZ input-absorptive bandstop filter is obtained, as shown in Fig. 3.29.

2. View Simulated Results

(1) Execute menu commands [Simulate] → [Simulate]. After the simulation completes, the data display window will be opened, as shown in Fig. 3.30.

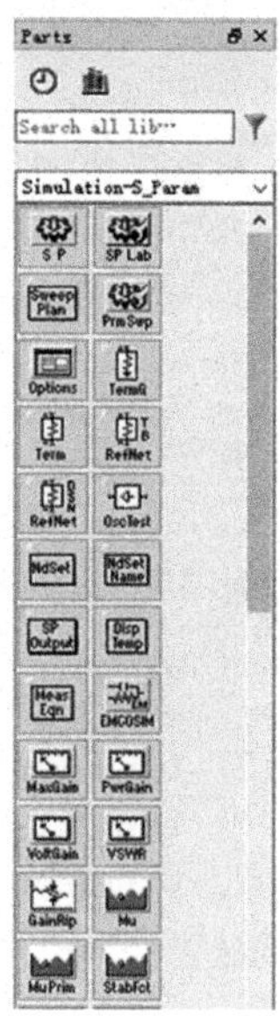

Fig. 3.27 [Simulation-S_Param] list

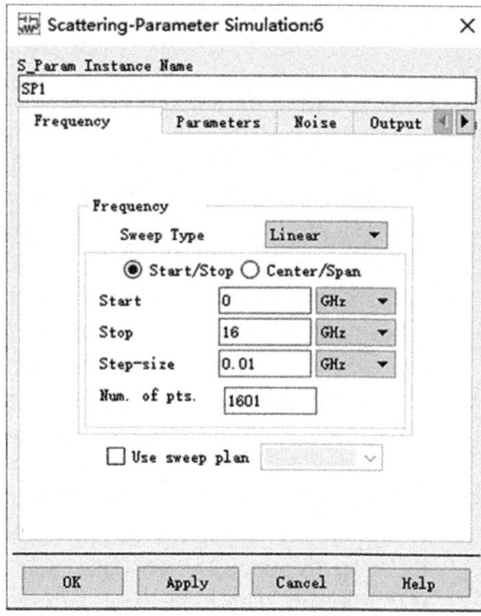

Fig. 3.28 Frequency settings of the S-parameter simulation

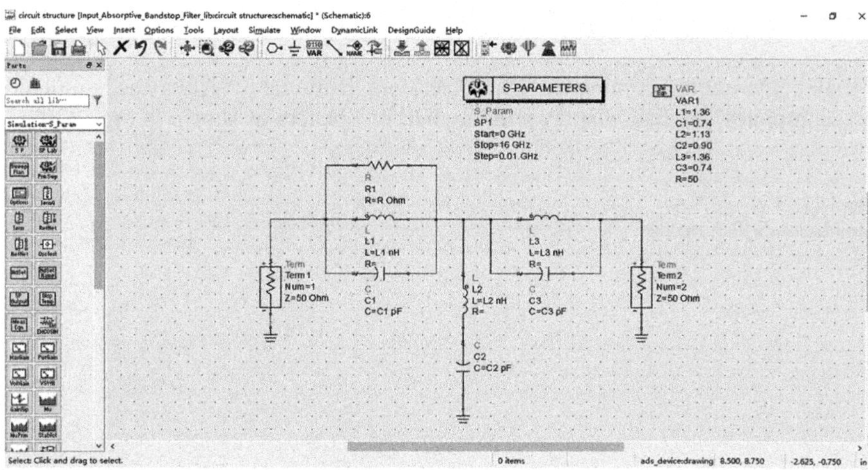

Fig. 3.29 Final ideal parameter simulated circuit model of the one-TZ input-absorptive bandstop filter

(2) Click the [Rectangular Plot] icon in the [Palette], then click the left mouse button at the blank data display area to open the [Plot Traces and Attributes] dialog box shown in Fig. 3.31, where the parameter traces can be set and plotted.

(3) Long press the "Ctrl" key to select S(1,1), S(2,1), and S(2,2), then click the [>> Add >>] button. Choose [dB] in the pop-up dialog box shown in Fig. 3.32,

Fig. 3.30 Data display window

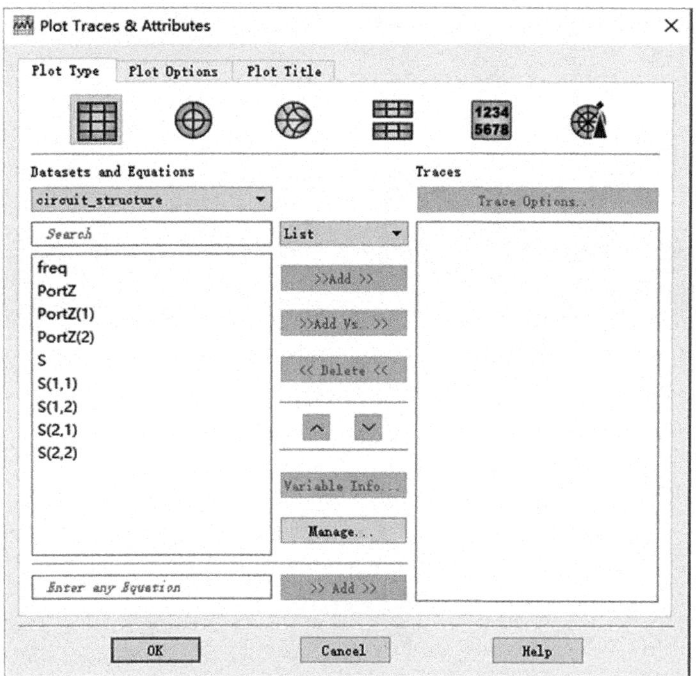

Fig. 3.31 [Plot Traces & Attributes] dialog box

then click the [OK] button. It can be seen that traces dB(S(1,1)), dB(S(2,1)), and dB(S(2,2)) have been added to the [Traces] list box, as shown in Fig. 3.33.

(4) Click the [OK] button in Fig. 3.33, the traces of dB(S(1,1)), dB(S(2,1)), and dB(S(2,2)) can be displayed in the data display area, the ordinate is the dB value, as shown in Fig. 3.34.

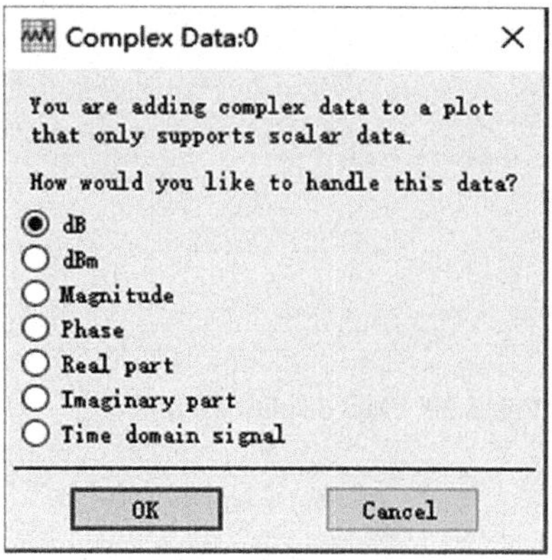

Fig. 3.32 Handling of the data

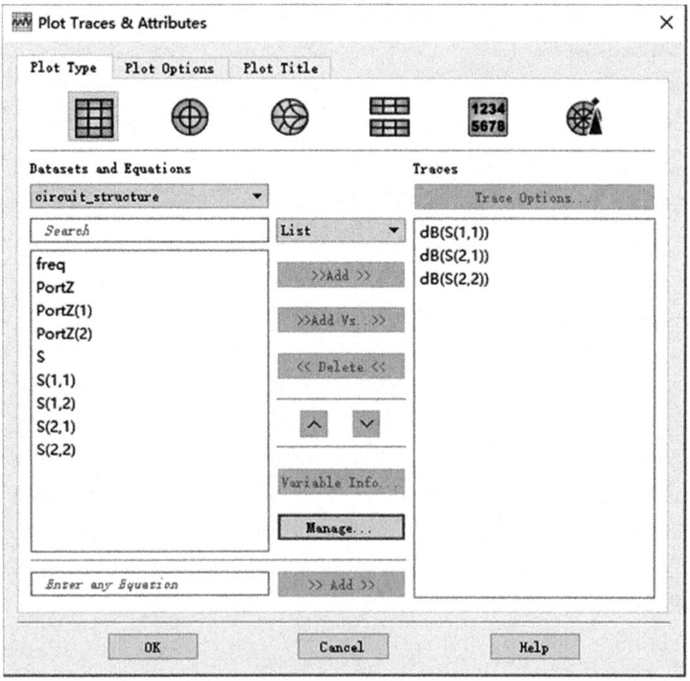

Fig. 3.33 Add of S(1,1), S(2,1), and S(2,2)

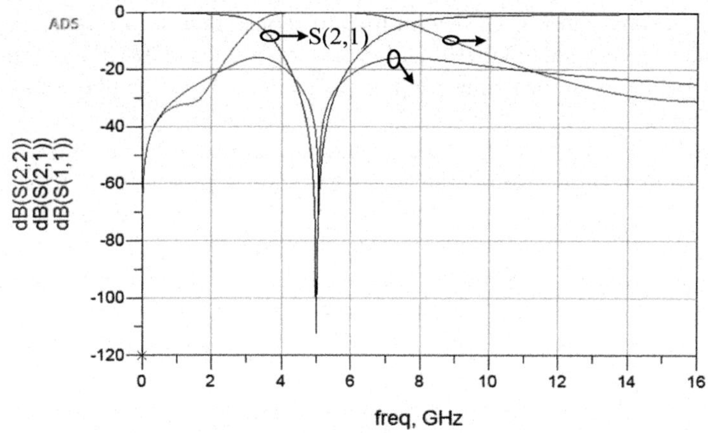

Fig. 3.34 Traces of dB(S(1,1)), dB(S(2,1)), and dB(S(2,2))

3. Handle Traces

Next, traces of dB(S(1,1)), dB(S(2,1)), and dB(S(2,2)) are taken as examples to introduce the curves handling. Similar steps can also be applied in other traces.

(1) By changing the frequency position of the marker, the values of any point on the trace can be read. Execute menu commands [Marker] → [New…] to open the dialog box shown in Fig. 3.35. Select a trace and click the left mouse button to insert a marker on it, as shown in Fig. 3.36. Similarly, insert markers for traces of dB(S(2,1)) and dB(S(2,2)). In addition, you can long press the left mouse button to move the position of the data frame.
(2) After selecting the inserted marker, the left and right arrow keys on the keyboard can be used to adjust the position of the abscissa (freq). You can also click the

Fig. 3.35 [Insert Marker] dialog box

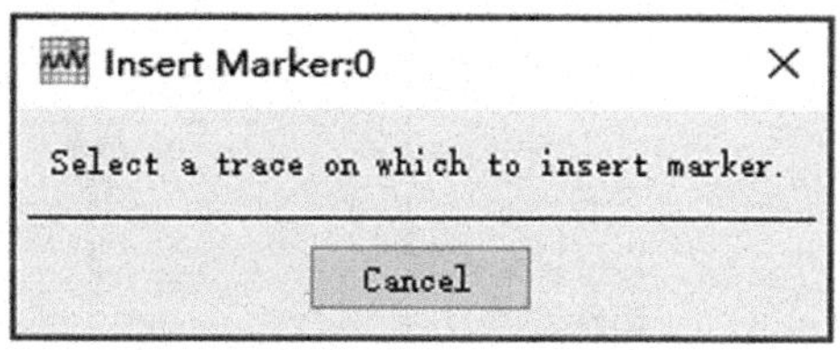

Fig. 3.36 Insertion of the marker on the trace

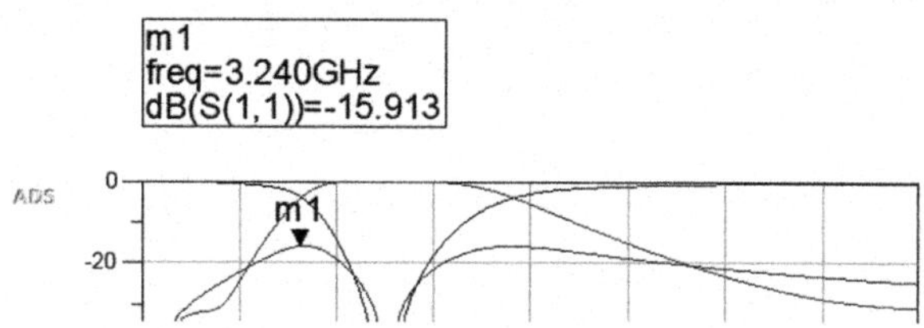

position shown in Fig. 3.37 with the left mouse button to modify the specific frequency you want to view directly. Here, the values of dB(S(1,1)), dB(S(2,1)), and dB(S(2,2)) at the center frequency $f_0 = 5$ GHz are displayed, as shown in Fig. 3.38.

(3) The editing function of the data display is introduced below. Take the modification and beautification of the Y axis as examples. Double-click the simulated parameters figure, then the [Plot Traces and Attributes] dialog box pops up. Click the [Plot Options] tab to deselect the "Auto Scale" (The automatic adjustment scale of the ADS software is not used). Adjust the Y axis scale shown in Fig. 3.39, then the adjusted S-parameter results in Fig. 3.40 can be obtained.

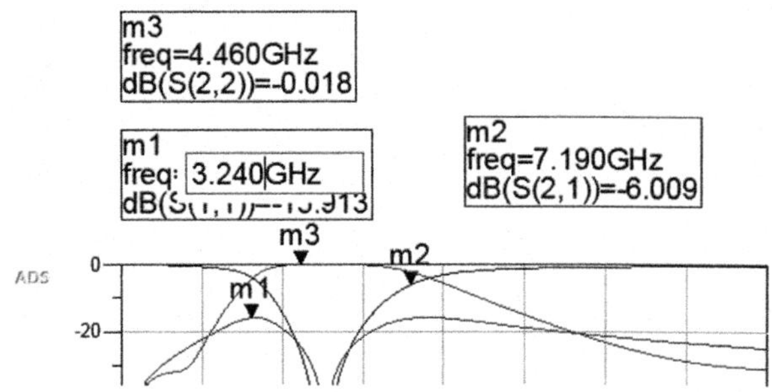

Fig. 3.37 Modification of the abscissa (freq) value

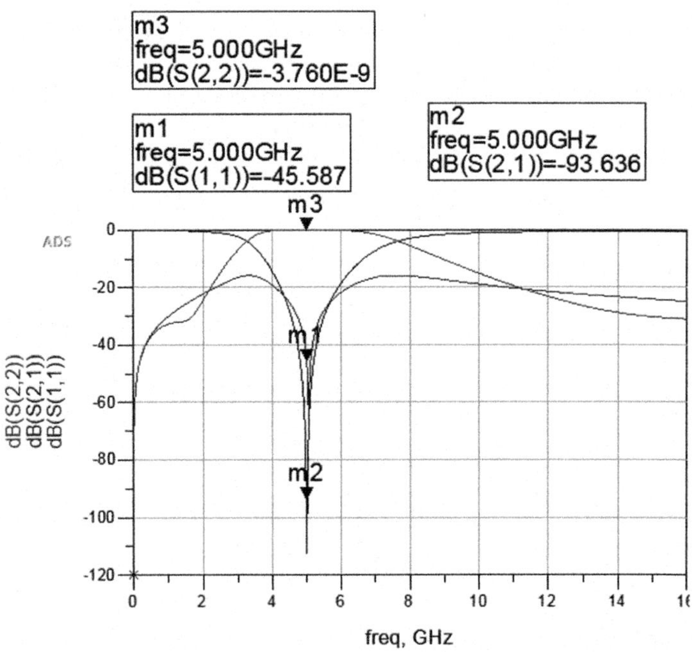

Fig. 3.38 Values of dB(S(1,1)), dB(S(2,1)), and dB(S(2,2)) at the center frequency

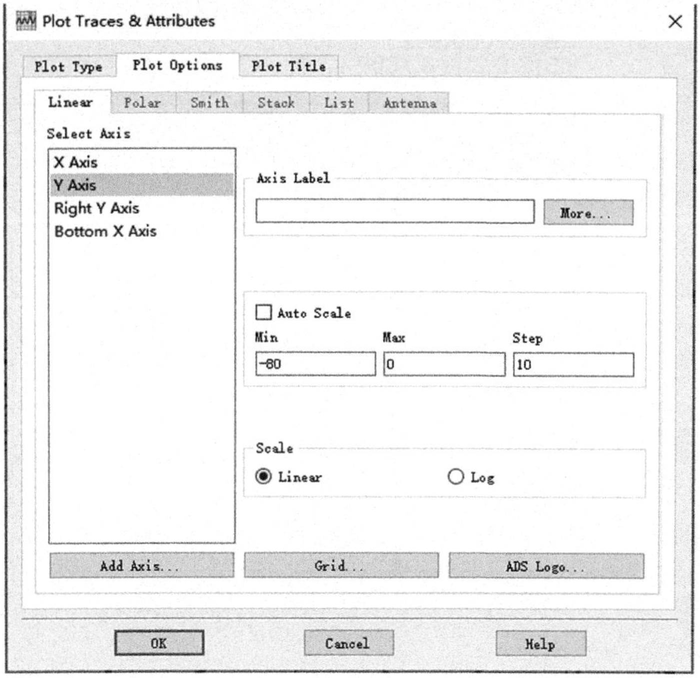

Fig. 3.39 Adjustment of the Y axis scale

(4) In addition, the type, color, thickness, and symbol of the trace can also be modified. Double-click the dB(S(1,1)) to open the [Trace Options] dialog box and modify the trace, as shown in Fig. 3.41 (Keep the trace color as default). Similarly, modify the traces of dB(S(2,1)) and dB(S(2,2)) shown in Figs. 3.42 and 3.43, respectively. The final traces of dB(S(1,1)), dB(S(2,1)), and dB(S(2,2)) are shown in Fig. 3.44.

3.3 Layout of Input-Absorptive Bandstop Filter

The performance of the actual circuit often has some differences from the theoretical results, so the influence of the interference, coupling, and some other factors should be taken into consideration. Therefore, it is necessary to use the ADS software for the layout simulation.

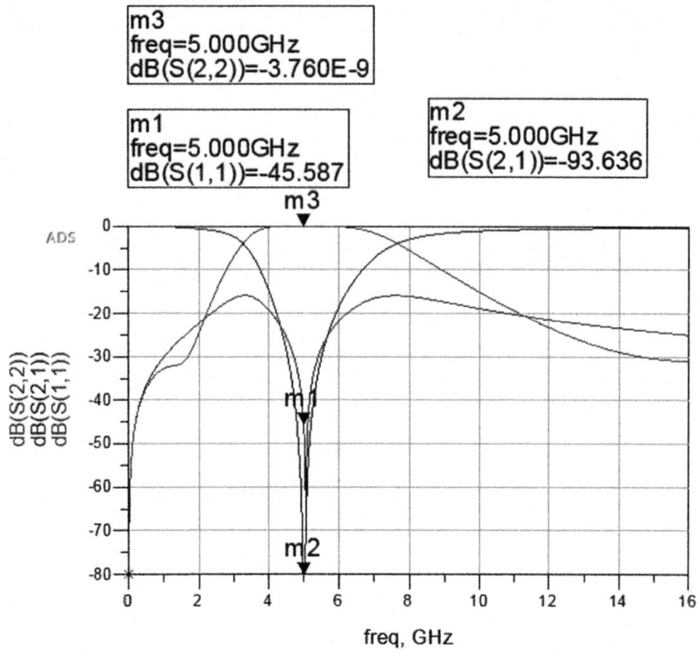

Fig. 3.40 The adjusted S-parameter results

3.3.1 New Substrate File

All circuit elements are constructed on the gallium arsenide (GaAs) substrate with the thickness of 200 μm, relative dielectric constant of 12.85, and the loss tangent of 0.006. The 75-nm-thick nickel chromium alloy (NiCr) layer with a block resistance of about 25 Ω/sq is used to realize the thin-film resistance, while the top and bottom 5-μm-thick metals and the 0.2-μm-thick intermediate Si_3N_4 layer construct the MIM capacitor. In addition, an air bridge can be built between the two copper layers to connect the circuits inside the spiral inductor and the peripheral circuits, which can make the layout design more flexible.

Before drawing the layout, the substrate file needs to be set according to the used TFIPD technology.

1. New Layout

Return to the "Input_Absorptive_Bandstop_Filter_wrk" main interface of the workspace. Execute menu commands [File] → [New] → [Layout…] to open the [New Layout] dialog box in Fig. 3.45. Modify the "Cell" name to "thin film resistor", then click the [Create Layout] button. After that, the [Choose Layout Technology] dialog box shown in Fig. 3.46 will pop up. Here, select "Standard ADS Layers, 0.001 micron layout resolution", which is 0.001 μm (note that the unit is μm in

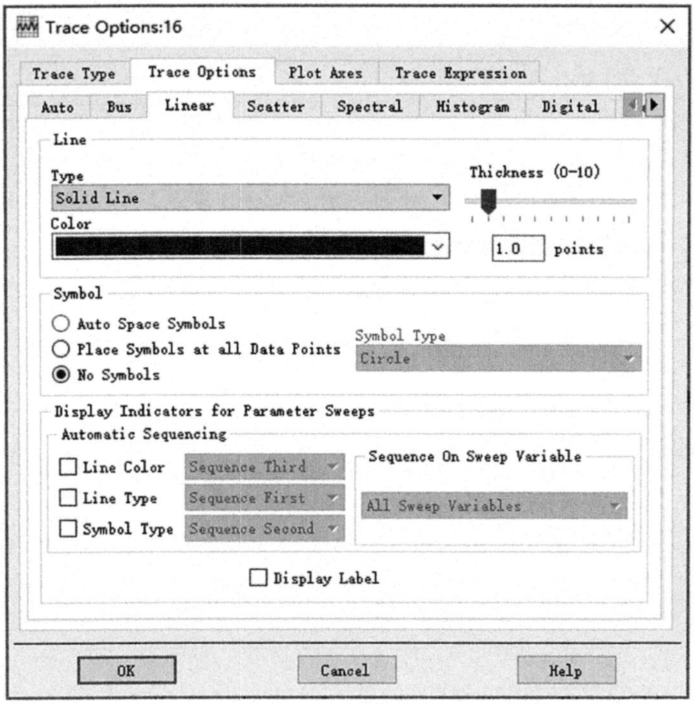

Fig. 3.41 Modification of the dB(S(1,1)) trace

this chapter), click [Finish], and the new layout window will pop up, as shown in Fig. 3.47.

2. New Substrate

(1) Create a new substrate. Execute menu commands [EM] → [Substrate…] in the layout window, then click [OK] in the dialog box which pops up. The [New Substrate] dialog box in Fig. 3.48 will be displayed, in which "File name" and "Template" can be modified. Keep the default file name here. Because the used TFIPD technology is based on the GaAs substrate, select "100μmGaAs" in the "Template" column and click the [Create Substrate] button. Then execute menu commands [View] → [View All] in the pop-up window to view all substrate items, as shown in Fig. 3.49.

(2) Add conductors. Execute menu commands [Technology] → [Material Definitions…] to open the [Material Definitions] dialog box shown in Fig. 3.50. Select the [Conductors] tab and define the relevant conductors in Fig. 3.51. The specific steps are listed as follows: Click the [Add From Database…] button in the lower right corner of Fig. 3.51. If the conductor to be added exists in the pop-up [Add Materials From Database] dialog box shown in Fig. 3.52, select the conductor and click [OK] to add it. If the conductor to be added does not

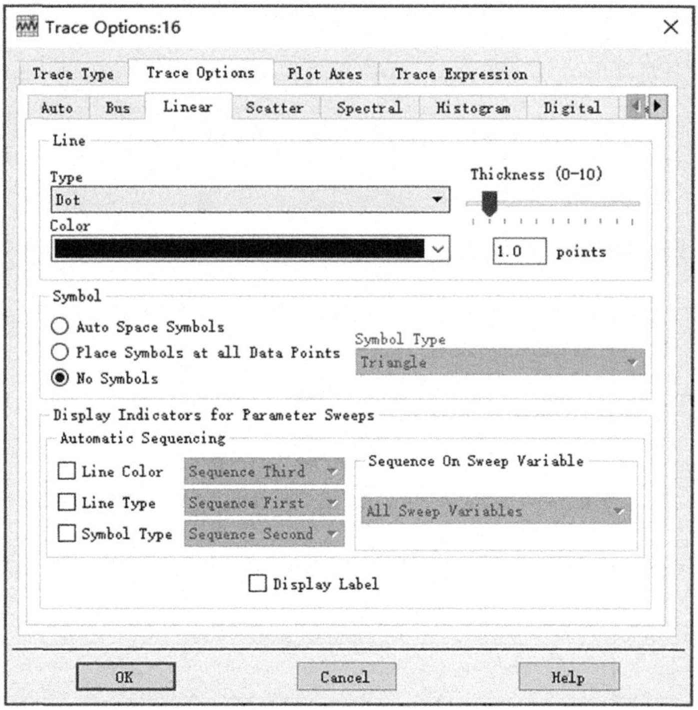

Fig. 3.42 Modification of the dB(S(2,1)) trace

exist in the [Add Materials From Database] dialog box, return to the [Material Definitions] dialog box and click the [Add Conductor] button to add a conductor and modify its properties. In addition, you can remove unwanted conductors by clicking the [Remove Conductor] button in the [Material Definitions] dialog box. After finishing all the steps, click the [Apply] button.

(3) Add dielectrics. Select the [Dielectrics] tab and define the relevant dielectrics in Fig. 3.53. The specific steps are listed as follows: Click the [Add From Database...] button in the lower right corner of Fig. 3.53. If the dielectric to be added exists in the pop-up [Add Materials From Database] dialog box shown in Fig. 3.54, select the dielectric and click [OK] to add it. If the dielectric to be added does not exist in the [Add Materials From Database] dialog box, return to the [Material Definitions] dialog box and click the [Add Dielectric] button to add a dielectric and modify its properties. In addition, you can remove unwanted dielectrics by clicking the [Remove Dielectric] button in the [Material Definitions] dialog box. After finishing all the steps, click the [OK] button to close the [Material Definitions] dialog box.

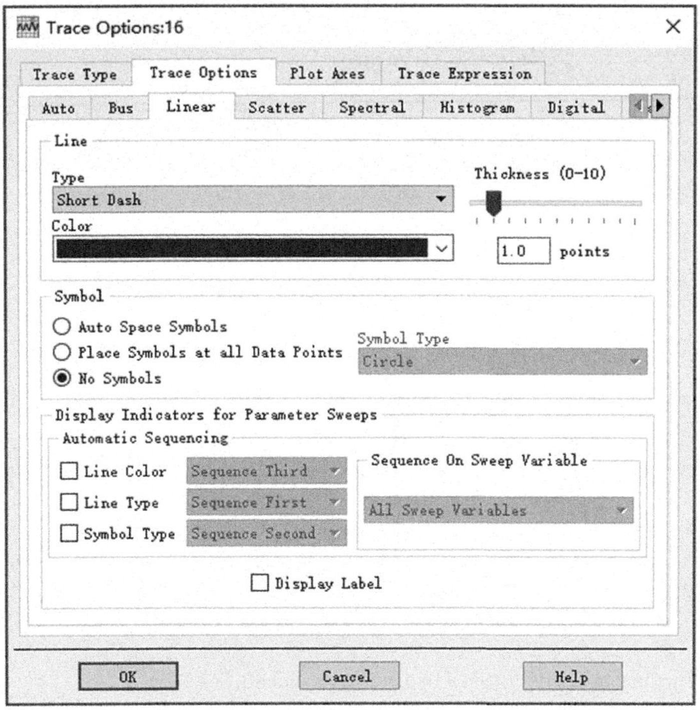

Fig. 3.43 Modification of the dB(S(2,2)) trace

3. Edit Substrate Items

(1) Select the "cond" layer and click the right mouse button to pop up Unmap. Click the "Unmap" to delete this layer (Or select the "cond" layer, then press "Delete" on the keyboard to delete it). Delete the "cond2" layer in the same way.
(2) Edit substrate layers. Select an existing substrate layer, right-click the mouse, and choose the [Insert Substrate Layer] from the pop-up menu to insert a new substrate layer. Select the substrate layer to be modified to view and modify its properties under the [Substrate Layer] at the right side of the window.
(3) Edit conductor layers. Select the surface of the substrate layer on which you want to insert a conductor layer, right-click the mouse, and choose the [Map Conductor Layer] from the pop-up menu to insert a new conductor layer. Select the conductor layer to be modified to view and modify its properties under the [Conductor Layer] at the right side of the window.
(4) Edit conductor vias. Select the substrate layer into which you want to insert a conductor via, right-click the mouse, and choose the [Map Conductor Via] from the pop-up menu to insert a new conductor via. Select the conductor via to be modified to view and modify its properties under the [Conductor Via] at the right side of the window.

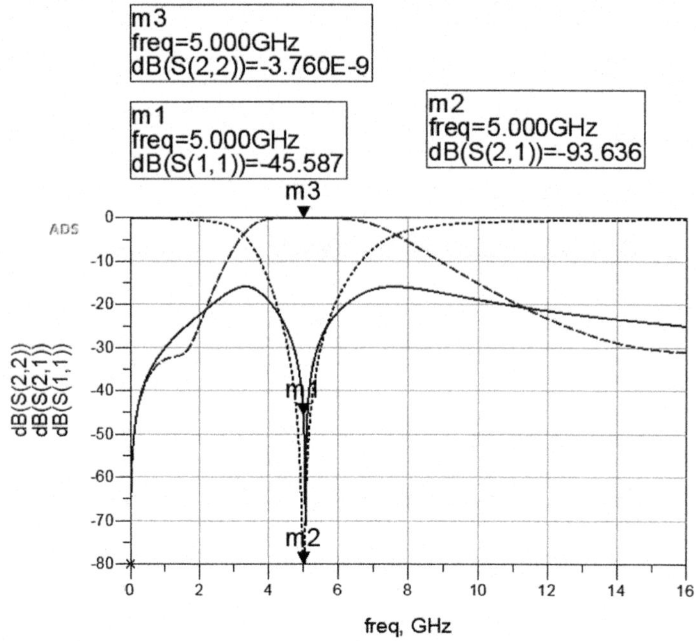

Fig. 3.44 The final traces of dB(S(1,1)), dB(S(2,1)), and dB(S(2,2))

Fig. 3.45 [New Layout] dialog box

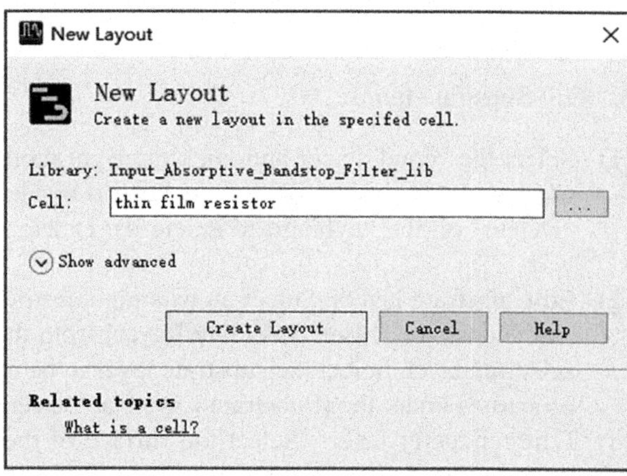

- Note

  Since both the top and bottom of the TFIPD technology are in the open air, a 0-μm-thick FreeSpace should be inserted under the GaAs layer. Then select the cover surface, click the right mouse button, and click [Delete Cover] to delete it.

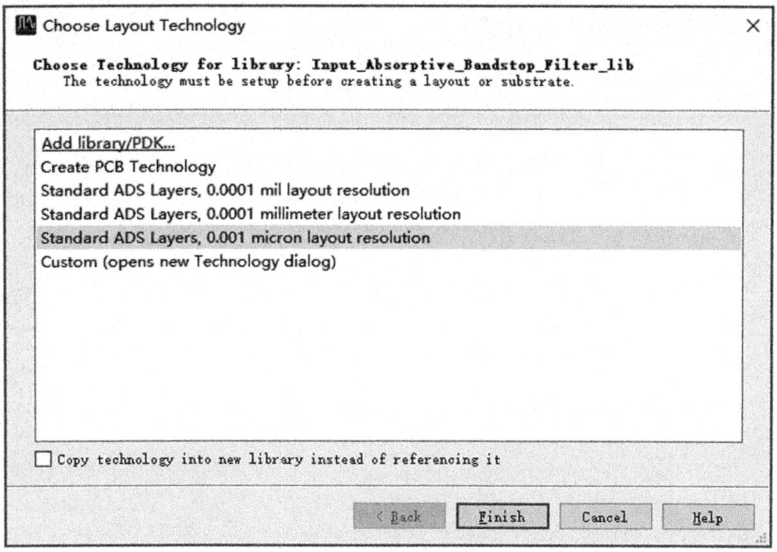

Fig. 3.46 [Choose Layout Technology] dialog box

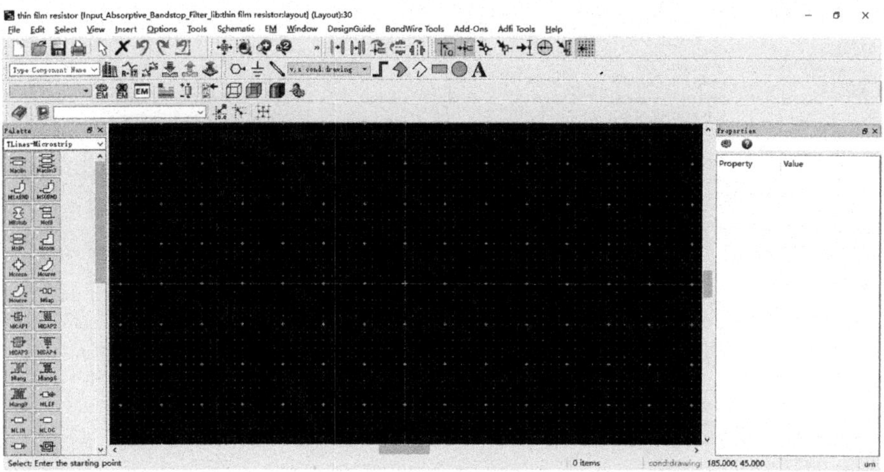

Fig. 3.47 New layout window

(5) All substrate items are shown in Fig. 3.55. The bottom layer is "FreeSpace", the thickness of the GaAs substrate layer is 200 µm, and the thickness of the first SiNx layer is 0.1 µm. For the diel layer, the "Process Role" is selected as "Conductor", the "Material" is selected as "NiCr", the "Operation" is selected as "Sheet", and the thickness is 75 nm. For the bond layer, the "Process Role"

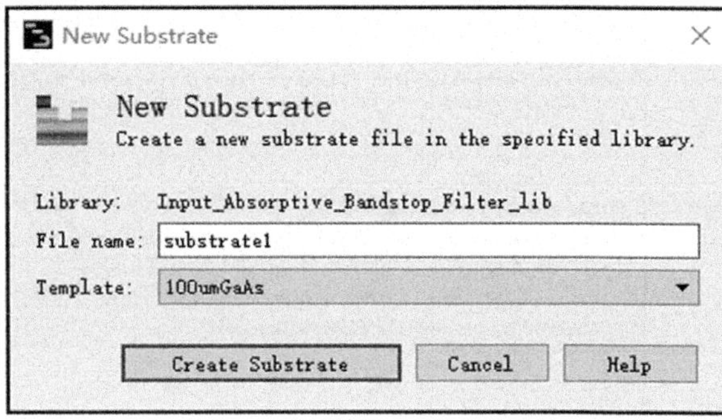

Fig. 3.48 [New Substrate] dialog box

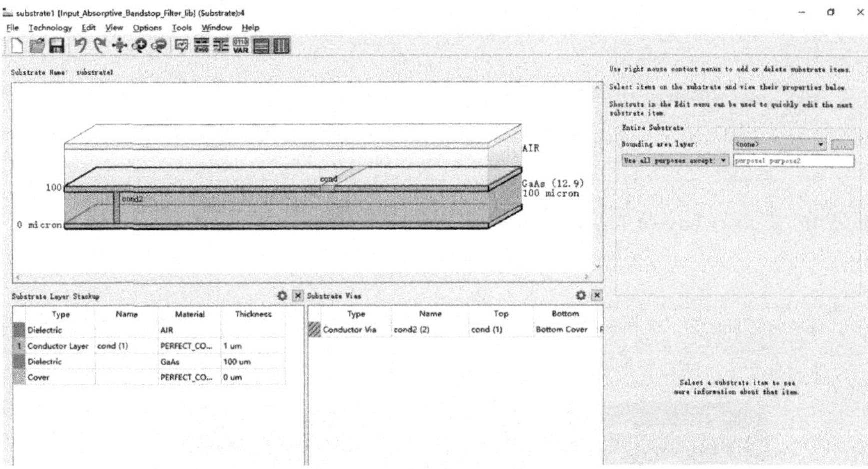

Fig. 3.49 View of all substrate items

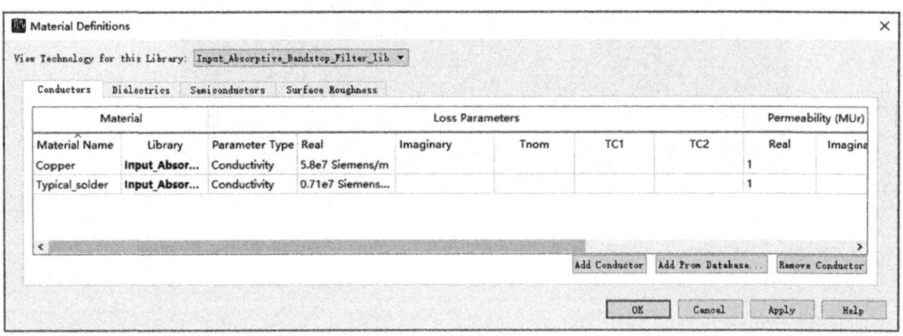

Fig. 3.50 [Material Definitions] dialog box

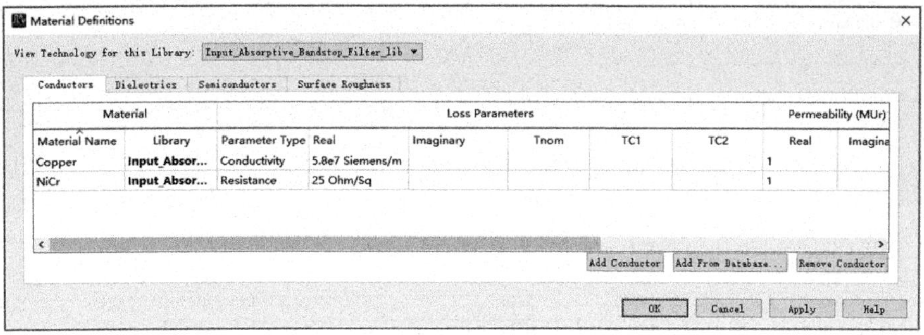

Fig. 3.51 Add of conductors and modification of their properties

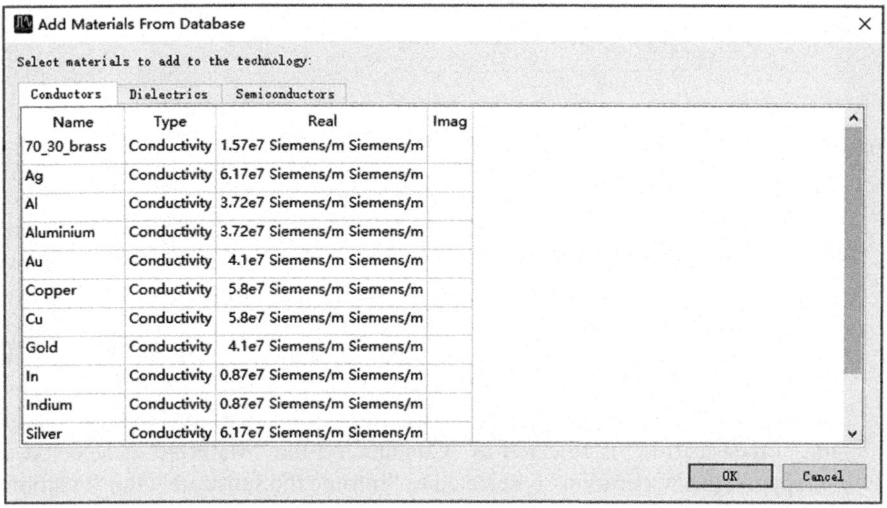

Fig. 3.52 [Conductors] tab of the [Add Materials From Database] dialog box

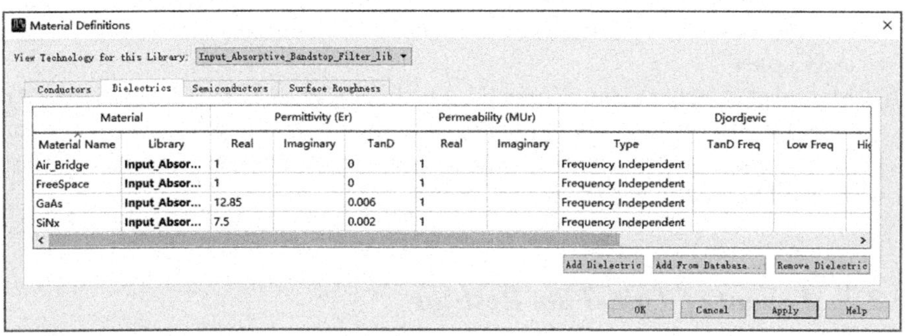

Fig. 3.53 Add of dielectrics and modification of their properties

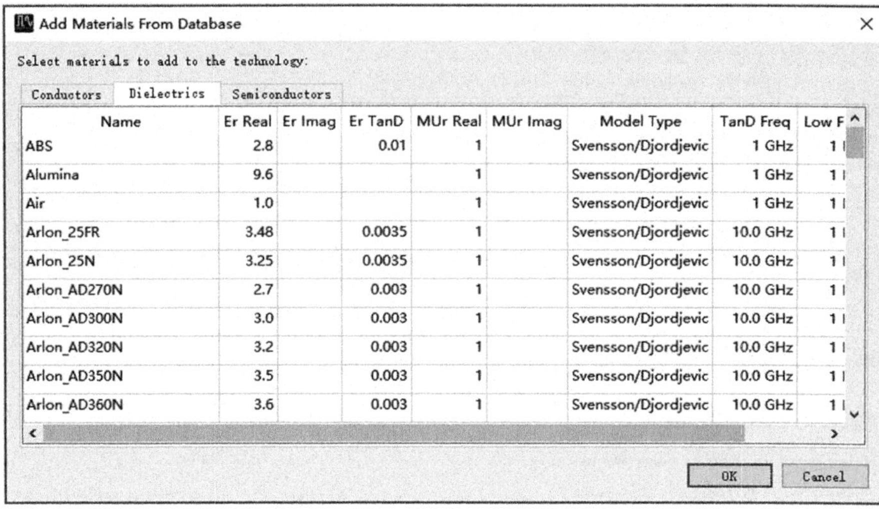

Name	Er Real	Er Imag	Er TanD	MUr Real	MUr Imag	Model Type	TanD Freq	Low F
ABS	2.8		0.01	1		Svensson/Djordjevic	1 GHz	1 I
Alumina	9.6			1		Svensson/Djordjevic	1 GHz	1 I
Air	1.0			1		Svensson/Djordjevic	1 GHz	1 I
Arlon_25FR	3.48		0.0035	1		Svensson/Djordjevic	10.0 GHz	1 I
Arlon_25N	3.25		0.0035	1		Svensson/Djordjevic	10.0 GHz	1 I
Arlon_AD270N	2.7		0.003	1		Svensson/Djordjevic	10.0 GHz	1 I
Arlon_AD300N	3.0		0.003	1		Svensson/Djordjevic	10.0 GHz	1 I
Arlon_AD320N	3.2		0.003	1		Svensson/Djordjevic	10.0 GHz	1 I
Arlon_AD350N	3.5		0.003	1		Svensson/Djordjevic	10.0 GHz	1 I
Arlon_AD360N	3.6		0.003	1		Svensson/Djordjevic	10.0 GHz	1 I

Fig. 3.54 [Dielectrics] tab of the [Add Materials From Database] dialog box

is selected as "Conductor", the "Material" is selected as "Copper", the "Operation" is selected as "Expand the substrate", the "Position" is selected as "Above interface", and the thickness is 5 μm. The thickness of the second SiNx layer is 0.2 μm. For the text layer, the "Process Role" is selected as "Conductor", the "Material" is selected as "Copper", the "Operation" is selected as "Expand the substrate", the "Position" is selected as "Above interface", and the thickness is 0.5 μm. The thickness of the Air_Bridge layer is 3 μm. For the leads layer, the "Process Role" is selected as "Conductor", the "Material" is selected as "Copper", the "Operation" is selected as "Intrude the substrate", the "Position" is selected as "Above interface", and the thickness is 5 μm. The top layer is the "FreeSpace". For the symbol layer, the "Process Role" is selected as "Conductor Via" and the "Material" is selected as "Copper". For the packages layer, the "Process Role" is selected as "Conductor Via" and the "Material" is selected as "Copper".

(6) Moreover, execute menu commands [Technology] → [Layer Definitions…] to open the [Layer Definitions] dialog box shown in Fig. 3.56, where you can modify the display properties of each layer such as color and pattern, etc. Here, keep all the settings in default.

3.3.2 Layout of Thin-Film Resistor

To facilitate the layout drawing, select the function key in the box of Fig. 3.57 in the "thin film resistor" cell. Execute menu commands [Options] → [Preferences…], select the [Grid/Snap] tab in the pop-up [Preferences for Layout] dialog box to modify the display grid in the layout. Set the "Snap Grid Distance (in layout units)",

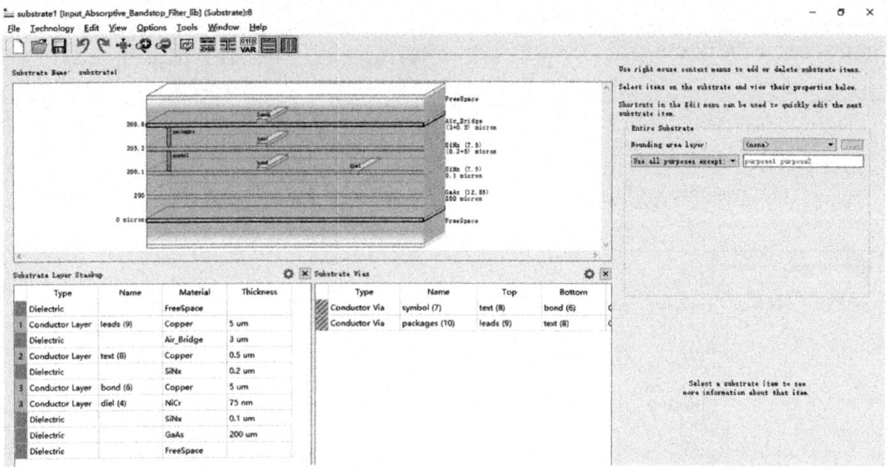

Fig. 3.55 Substrate layer stackup and substrate vias

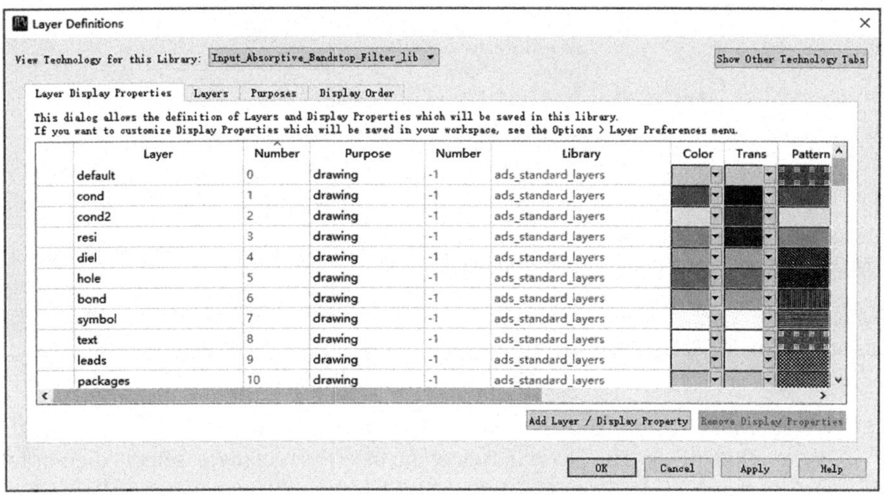

Fig. 3.56 [Layer Display Properties] tab of the [Layer Definitions] dialog box

"Snap Grid Per Minor Display Grid", and "Minor Grid Per Major Display Grid" to appropriate values, as shown in Fig. 3.58 (Or right-click in the layout drawing area and select " < 0.1–1–100 > " under [Grid Spacing…]; Or use the shortcut key "Ctrl + Shift + 8").

1. Layout Drawing of Thin-Film Resistor

The thin-film resistance is realized by the 75-nm-thick nickel chromium alloy (NiCr) layer with a block resistance of about 25 Ω/sq. The thin-film resistor with the

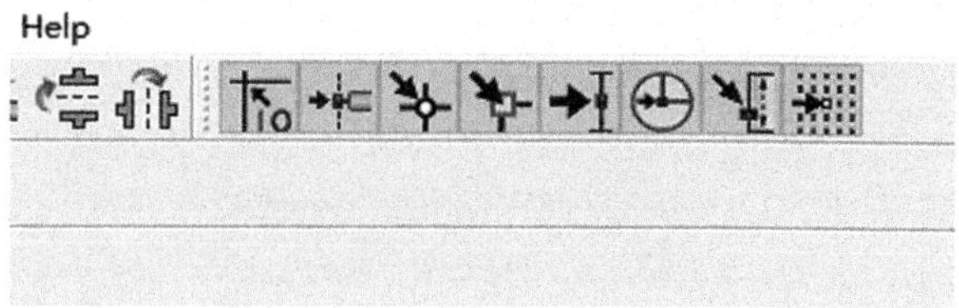

Fig. 3.57 Selection of the function key in the box

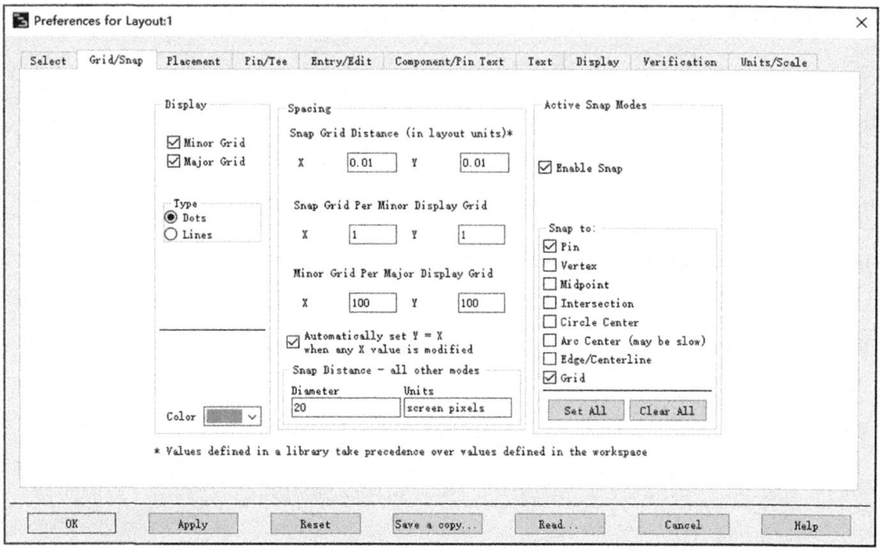

Fig. 3.58 Modification of the display grid in the layout

resistance of 50 Ω is taken as an example to describe its drawing steps in detail below.

(1) Execute menu commands [Insert] → [Rectangle] to insert a rectangle into the layout, then press "Esc" to exit. Select the inserted rectangle, select "diel: drawing" in the drop-down menu of the [All Shapes] → [Layer] under [Properties] at the right side of the window, and set the [Rectangles] → [Width] to 70 μm, [Height] to 15 μm.
(2) Execute menu commands [Insert] → [Rectangle] to insert a rectangle into the layout, then press "Esc" to exit. Select the inserted rectangle, select "bond: drawing" in the drop-down menu of the [All Shapes] → [Layer] under [Properties] at the right side of the window, and set the [Rectangles] → [Width] to 50 μm, [Height] to 20 μm. Select and long press the modified rectangle to move it to the middle position of a 15 μm wide overlap with the diel layer rectangle. Similarly, insert a bond layer rectangle at the opposite position. The layout of the thin-film resistor has been drawn, as shown in Fig. 3.59.

Fig. 3.59 Final layout of the thin-film resistor

Fig. 3.60 Insert pins on thin-film resistor

2. Layout Simulation of Thin-Film Resistor

(1) Insert ports. Execute menu commands [Insert] → [Pin] and click the left mouse button to insert two pins into the I/O ports of the thin-film resistor, as shown in Fig. 3.60.
(2) Modify simulation settings. Execute menu commands [EM] → [Simulation Settings…] and the [New EM Setup View] dialog box shown in Fig. 3.61 will pop up. Click the [Create EM Setup View] button and the [EM Setup for simulation] window will pop up, as shown in Fig. 3.62. Select the EM simulator at first. The second method "Momentum Microwave" is used generally, because this method runs fast and its high accuracy can also meet the application requirement (The first method "Momentum RF" is the fastest with the least accuracy. The third method "FEM" has the highest accuracy, but it shows the slowest simulation speed, so it is mainly used for some complex three-dimensional structure simulations). Select the [Frequency Plan] tab to set the simulation frequency. Select "Adaptive" in the "Type" column, then set the "Fstart" to 0 GHz, the "Fstop" to 16 GHz, and the "Npts" to 50. Finally, select the [Options] tab. Click "Preprocessor", select the "User specified snap distance" option under the "Heal the layout", and set the distance to 2.5 μm. Click "Mesh" and select the "Edge Mesh" option. Keep other settings in default. After finishing all the steps, close the window and click the [OK] button to save the settings.
(3) Layout simulation. Execute menu commands [EM] → [Simulate] to simulate the layout of the thin-film resistor. During the simulation, a status window will pop up to display the current item progress. After the finishing of the simulation, the data display window will be opened automatically. Refer to the steps in Sect. 3.2.2 to view and handle traces dB(S(1,1)) and dB(S(2,1)), the final results are shown in Fig. 3.63.

3. Co-simulation of Thin-Film Resistor

To verify whether the resistance of the drawn thin-film resistor is 50 Ω, the co-simulation of the resistor schematic and layout is required.

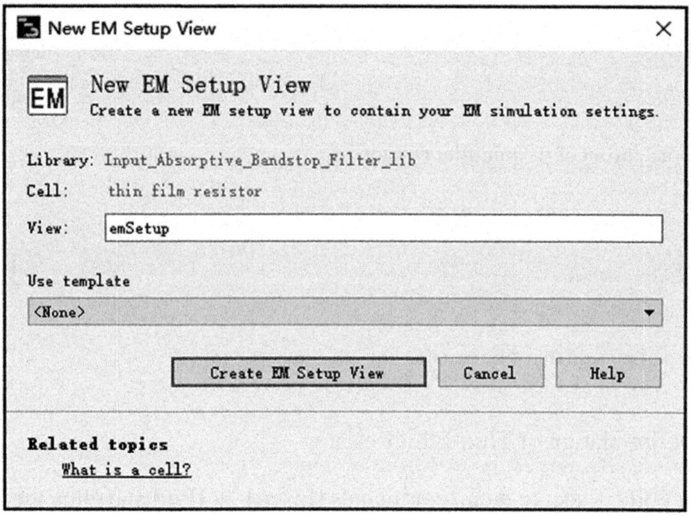

Fig. 3.61 [New EM Setup View] dialog box

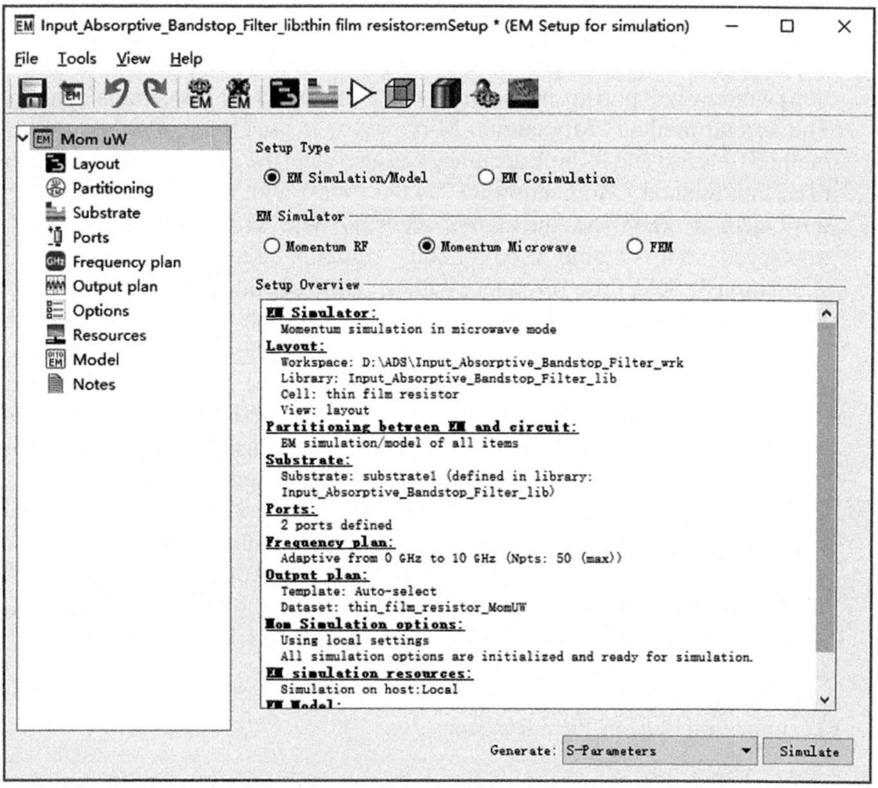

Fig. 3.62 [EM Setup for simulation] window

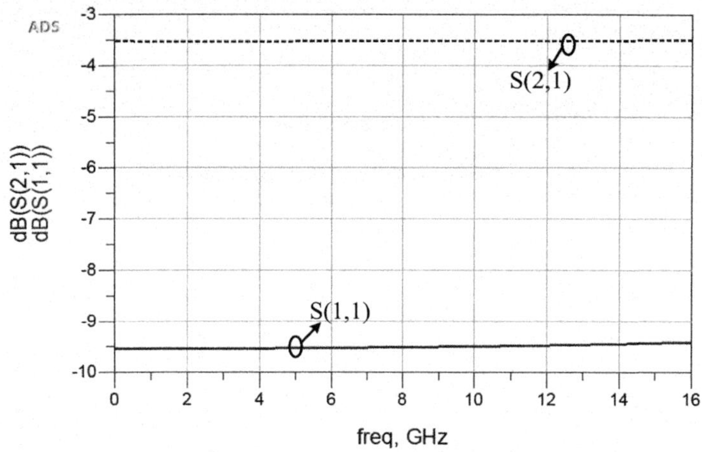

Fig. 3.63 Final simulated results of the drawn thin-film resistor

(1) Create EM model and symbol of the thin-film resistor. In the layout window, execute menu commands [EM] → [Component] → [Create EM Model And Symbol…], then click the [OK] button in the pop-up dialog boxes. Execute menu commands [Edit] → [Component] → [Update Component Definitions…], then click the [OK] button in the pop-up dialog box to complete the creation of the EM model and the symbol in the thin-film resistor.

(2) Create a new schematic and insert the thin-film resistor component. Return to the "Input_Absorptive_Bandstop_Filter_wrk" main interface of the workspace. Execute menu commands [File] → [New] → [Schematic…], modify the "Cell" name to "thin film resistor-cosimulation" in the opened [New Schematic] dialog box, and click the [Create Schematic] button to create a new schematic. Click the [Open the Library Browser] icon at the left side of the schematic window, then select the "thin film resistor" component under the [Workspace Libraries] in the displayed [Component Library] window, as shown in the Fig. 3.64. Click the right mouse button, then click the [Place Component] in the pop-up menu to insert a thin-film resistor component in the schematic. Press "Esc" to exit.

(3) Insert ideal resistor. Select the [Lumped-Components] from the drop-down menu in the left component palette list and click the [Resistor] icon to insert a resistor. Press "Esc" on the keyboard to exit. Double-click the resistor and modify R = 50 Ω in the [Edit Instance Parameter] pop-up dialog box (check whether the unit setting is consistent), then click the [OK] button to save the parameter.

(4) Insert the S-parameter simulator, term ports, and grounds. Select the [Simulation-S_Param] from the drop-down menu in the left component palette list and click the [S-parameter Simulator] icon to insert an S-parameter simulator. Click the [Port Impedance Termination for S-Parameters] icon to insert four term ports, then press "Esc" to exit. Execute menu commands [Insert] →

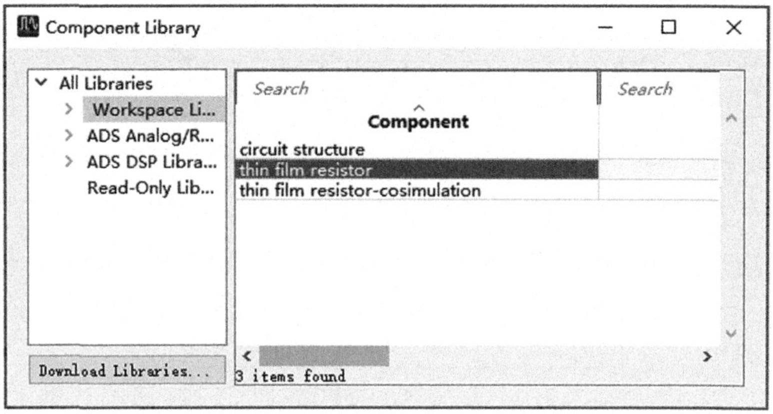

Fig. 3.64 Thin-film resistor component in the [Component Library] window

[GROUND] to place four grounds (Or click the [TermG Port Impedance Termination for S-Parameters with Grounded Reference] icon to insert four term ports with the grounded reference). Execute menu commands [Insert] → [Wire] to connect the inserted components. After finishing all the steps, press "Esc" to exit.

(5) Set simulation frequency. Double-click the S-parameter simulator S-PARAMETERS. Set the "Start" to 0 GHz, the "Stop" to 16 GHz, and the "Step-size" to 0.01 GHz, then click the [OK] button. The final resistor co-simulation circuit is obtained, as shown in Fig. 3.65.

(6) Co-simulation. Execute menu commands [Simulate] → [Simulate]. After the simulation completes, the data display window will be opened automatically. Refer to the steps in Sect. 3.2.2 to view and handle traces of dB(S(1,1)), dB(S(2,1)), dB(S(3,3)), and dB(S(4,3)), the final results are shown in Fig. 3.66. It can be seen that traces dB(S(1,1)) and dB(S(3,3)), traces dB(S(2,1)) and dB(S(4,3)) are almost coincident, indicating that the value of the drawn thin-film resistor is approximately equal to 50 Ω. If the differences between these traces are large, go back to modify the thin-film resistor layout and repeat the steps above until the error of two traces is in the range of acceptance (within an acceptable range).

3.3.3 Layout of MIM Capacitor

1. Layout Drawing of MIM Capacitor

Two 5-μm-thick copper layers (bond and leads layers) and a 0.2-μm-thick intermediate silicon nitride (Si_3N_4) layer are used to construct the MIM capacitors. The area and thickness of the intermediate dielectric layer determine the value of the MIM

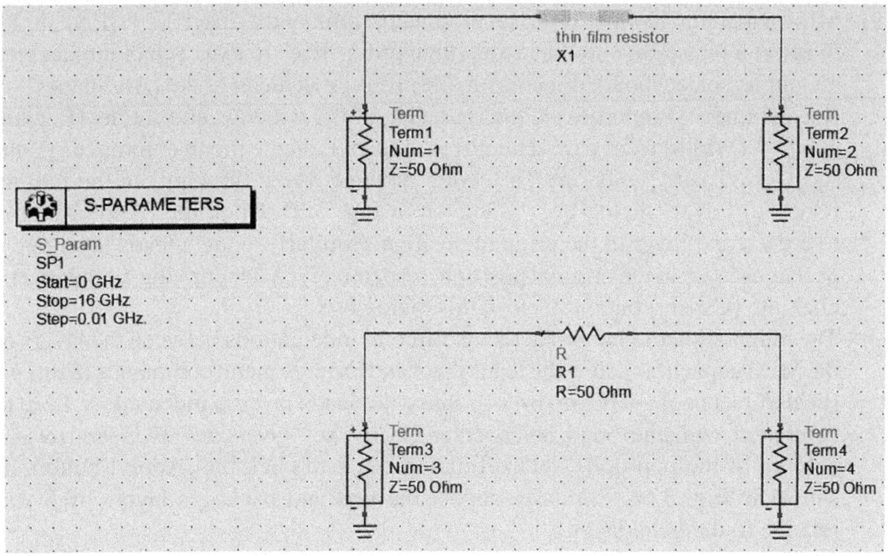

Fig. 3.65 Resistor co-simulation circuit

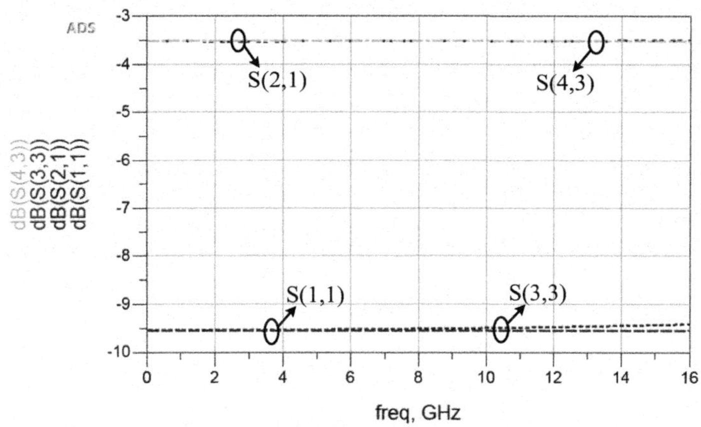

Fig. 3.66 S-parameter results of the resistor co-simulation

capacitor. The MIM capacitor with the capacitance of 0.74 pF is taken as an example to describe its drawing steps in detail below.

(1) Create a new layout. In the "Input_Absorptive_Bandstop_Filter_wrk" main interface of the workspace, execute menu commands [File] → [New] → [Layout…], and modify the "Cell" name to "MIM capacitor" in the opened [New Layout] dialog box. Click the [Create Layout] button, then a new layout window will pop up.

(2) MIM capacitor lamination. Execute menu commands [Insert] → [Rectangle] to insert a rectangle into the layout, then press "Esc" to exit. Select the inserted rectangle, select "bond: drawing" in the drop-down menu of the [All Shapes] → [Layer] under [Properties] at the right side of the window, and set the [Rectangles] → [Width] to 54 μm, [Height] to 49 μm. Execute menu commands [Edit] → [Copy/Paste] → [Copy To Layer…], select [text: drawing] in the pop-up [Copy to Layer] dialog box, as shown in Fig. 3.67. Click the [Apply] button to copy a text layer in the original position. Similarly, copy a leads layer and a packages layer in the original position, respectively. After copying all the layers, click the [Cancel] button to close this dialog box.

(3) The indentation of layers. There are different indentations between the layers of the MIM capacitor. Select the leads layer and execute menu commands [Edit] → [Scale/Oversize] → [Oversize…]. Since the leads layer is indented by 1.5 μm compared with the bond layer, enter-1.5 in the "Oversize(+)/Undersize(−)" column of the pop-up [Oversize] dialog box and click the [Apply] button, as shown in Fig. 3.68. Similarly, indent the text and packages layers by 3 μm relative to the bond layer.

(4) Draw connection parts. Execute menu commands [Insert] → [Rectangle] to insert a rectangle in the layout, then press "Esc" to exit. Select the inserted rectangle, select "bond: drawing" in the drop-down menu of the [All Shapes] → [Layer] under [Properties] at the right side of the window, and set the [Rectangles] → [Width] to 50 μm, [Height] to 20 μm. Select and long press the modified rectangle by the left mouse button to move it to the middle position of the MIM capacitor and connect it with the original bond layer. Similarly, a leads layer rectangle is inserted in the opposite position and is connected to the original leads layer. The layout of a MIM capacitor has been drawn, as shown in Fig. 3.69.

2. Layout Simulation of MIM Capacitor

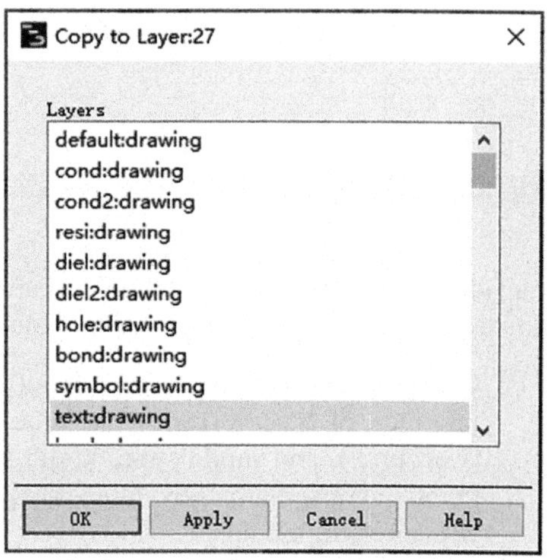

Fig. 3.67 [Copy to Layer] dialog box

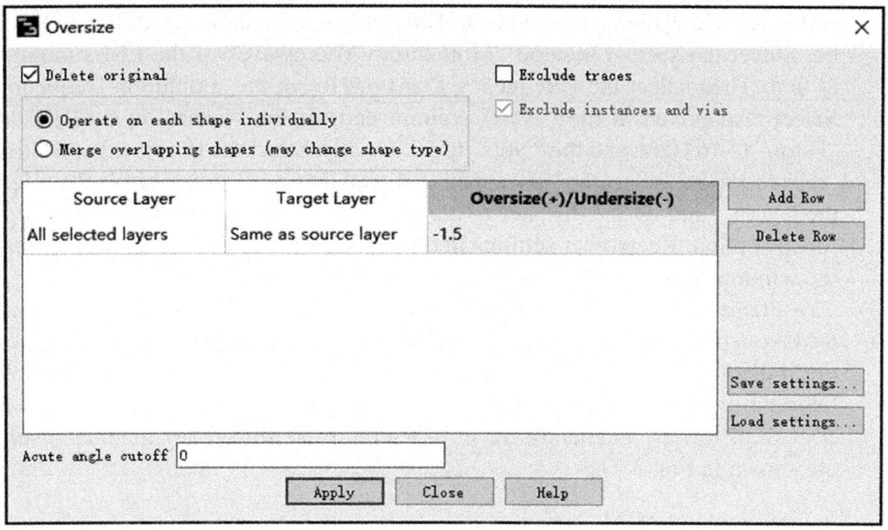

Fig. 3.68 [Oversize] dialog box

Fig. 3.69 Final layout of a MIM capacitor

(1) Insert ports. Execute menu commands [Insert] → [Pin] and click the left mouse button to insert two pins into the I/O ports of the MIM capacitor, as shown in Fig. 3.70.
(2) Modify simulation settings. Execute menu commands [EM] → [Simulation Settings…], click the [Create EM Setup View] button in the pop-up the [New

Fig. 3.70 Insert pins on MIM capacitor

EM Setup View] dialog box, and the [EM Setup for simulation] window will pop up. Select the second method "Momentum Microwave" of the EM simulator at first. Then select the [Frequency Plan] tab to set the simulation frequency. Select "Adaptive" in the "Type" column and set the "Fstart" to 0 GHz, the "Fstop" to 16 GHz, and the "Npts" to 50. Finally, select the [Options] tab. Click "Preprocessor", select the "User specified snap distance" option under the "Heal the layout", and set the distance to 2.5 μm. Click "Mesh" and select the "Edge Mesh" option. Keep other settings in default. After finishing all the steps, close the window and click the [OK] button to save the settings.

(3) Layout simulation. Execute menu commands [EM] → [Simulate] to simulate the layout of the MIM capacitor. During the simulation, a status window will pop up to display the current item progress. After the finishing of the simulation, the data display window will be opened automatically. Refer to the steps in Sect. 3.2.2 to view and handle traces dB(S(1,1)) and dB(S(2,1)), the final results are shown in Fig. 3.71.

3. Co-simulation of MIM Capacitor

To verify whether the capacitance of the drawn MIM capacitor is 0.74 pF, the co-simulation of capacitor schematic and layout is required.

(1) Create EM model and symbol of the MIM capacitor. In the layout window, execute menu commands [EM] → [Component] → [Create EM Model And Symbol...], then click the [OK] button in the pop-up dialog boxes. Execute menu commands [Edit] → [Component] → [Update Component Definitions...], then click the [OK] button in the pop-up dialog box to complete the creation of the EM model and the symbol in the MIM capacitor.
(2) Create a new schematic and insert the MIM capacitor component. Return to the "Input_Absorptive_Bandstop_Filter_wrk" main interface of the workspace.

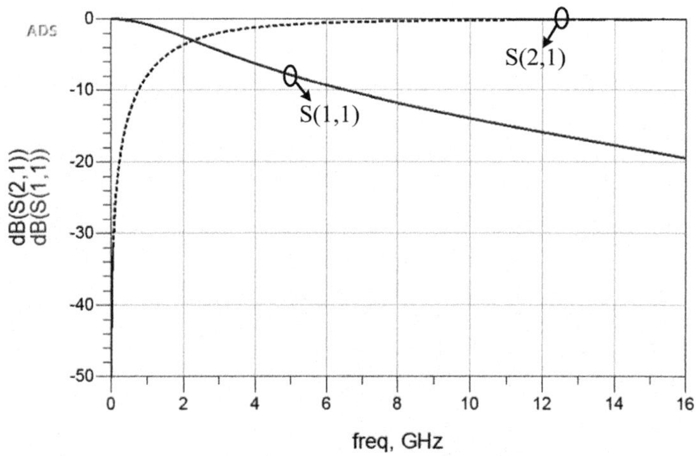

Fig. 3.71 Final simulated results of the drawn MIM capacitor

Execute menu commands [File] → [New] → [Schematic…], modify the "Cell" name to "MIM capacitor-cosimulation" in the opened [New Schematic] dialog box, and click the [Create Schematic] button to create a new schematic. Click the [Open the Library Browser] icon at the left side of the schematic window, then select the "MIM capacitor" component under the [Workspace Libraries] in the displayed [Component Library] window. Click the right mouse button, then click the [Place Component] in the pop-up menu to insert a MIM capacitor component in the schematic. Press "Esc" to exit.

(3) Insert ideal capacitor. Select the [Lumped-Components] from the drop-down menu in the left component palette list and click the [Capacitor] icon to insert a capacitor. Press "Esc" on the keyboard to exit. Double-click the capacitor and modify C = 0.74 pF in the [Edit Instance Parameter] pop-up dialog box (check whether the unit setting is consistent), then click the [OK] button to save the parameter.

(4) Insert the S-parameter simulator, term ports, and grounds. Select the [Simulation-S_Param] from the drop-down menu in the left component palette list and click the [S-parameter Simulator] icon to insert an S-parameter simulator. Click the [Port Impedance Termination for S-Parameters] icon to insert four term ports, then press "Esc" to exit. Execute menu commands [Insert] → [GROUND] to place four grounds (Or click the [TermG Port Impedance Termination for S-Parameters with Grounded Reference] icon to insert four term ports with the grounded reference). Execute menu commands [Insert] → [Wire] to connect the inserted components. After finishing all the steps, press "Esc" to exit.

(5) Set simulation frequency. Double-click the S-parameter simulator S-PARAMETERS. Set the "Start" to 0 GHz, the "Stop" to 8 GHz, and the "Step-size" to 0.01 GHz, and click the [OK] button. The final capacitor co-simulation circuit is obtained, as shown in Fig. 3.72.

(6) Co-simulation. Execute menu commands [Simulate] → [Simulate]. After the simulation completes, the data display window will be opened automatically. Refer to the steps in Sect. 3.2.2 to view and handle traces of dB(S(1,1)), dB(S(2,1)), dB(S(3,3)), and dB(S(4,3)), the final results are shown in Fig. 3.73. It can be seen that traces dB(S(1,1)) and dB(S(3,3)), traces dB(S(2,1)) and dB(S(4,3)) are almost coincident, indicating that the capacitance of the drawn MIM capacitor is approximately equal to 0.74 pF. If the differences between these traces are large, go back to modify the MIM capacitor layout and repeat the steps above until the error of two traces is in the range of acceptance (within an acceptable range).

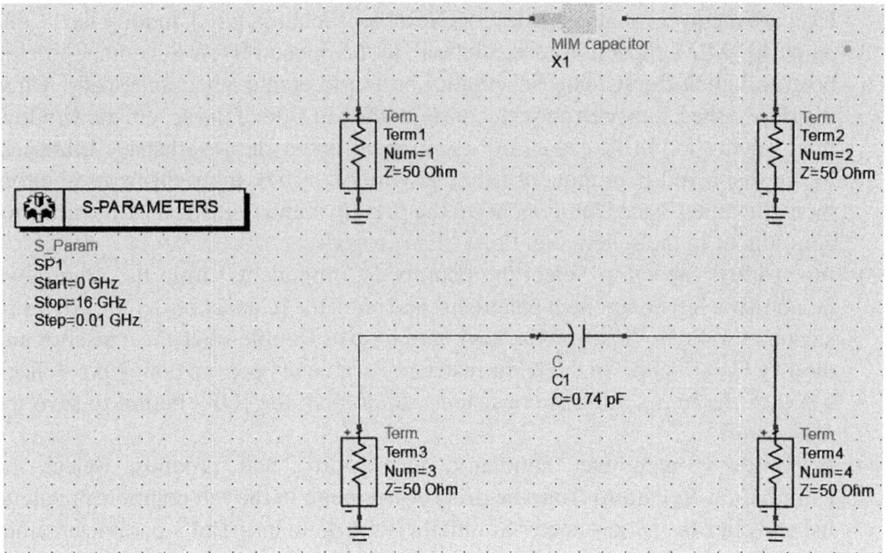

Fig. 3.72 Capacitor co-simulation circuit

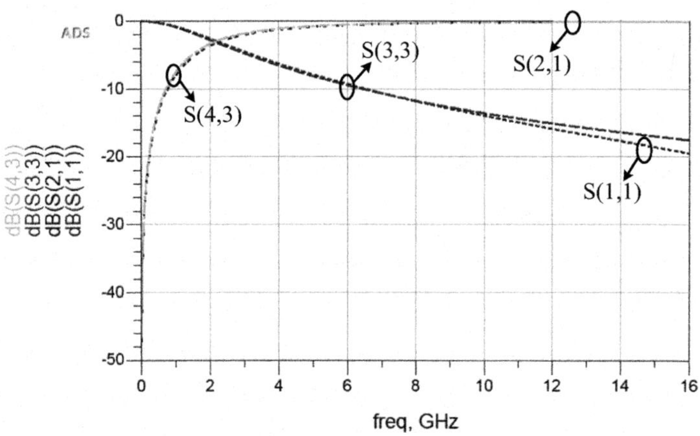

Fig. 3.73 S-parameter results of the capacitor co-simulation

3.3.4 Layout of Spiral Inductor

1. Layout Drawing of Spiral Inductor

The inductance value of a spiral inductor is related to its inner radius, number of turns, width, and the space of winding. The spiral inductor with the inductance of 1.36 nH is taken as an example to describe its drawing steps in detail below.

(1) Create a new layout. In the "Input_Absorptive_Bandstop_Filter_wrk" main interface of the workspace, execute menu commands [File] → [New] → [Layout…] and modify the "Cell" name to "spiral inductor" in the opened [New Layout] dialog box. Click the [Create Layout] button, a new layout window will pop up.
(2) Insert spiral inductor. Select the [TLines-Microstrip] from the drop-down menu in the left component palette list, as shown in Fig. 3.74. Click the [Microstrip Round Spiral Inductor] icon, then input the number of turns N = 2.5, the inner radius Ri = 70 μm, the conductor width W = 15 μm, and the conductor spacing S = 15 μm in the displayed [Edit Instance Parameters] dialog box, as shown in Fig. 3.75. Click the [OK] button to add a spiral inductor in the drawing area, then press "Esc" on the keyboard to exit.
(3) Layer modification of spiral inductor. Select the inserted spiral inductor, execute menu commands [Edit] → [Component] → [Flatten…], and click the [OK] button in the pop-up dialog box. Execute menu commands [Edit] → [Merge] → [Union] to unite the spiral inductor layer. Then select "bond: drawing" in the drop-down menu of the [All Shapes] → [Layer] under [Properties] at the right side of the window.
(4) Spiral inductor lamination. Execute menu commands [Edit] → [Copy/Paste] → [Copy To Layer…], select [text: drawing] in the pop-up [Copy to Layer] dialog box. Click the [Apply] button to copy a text layer in the original position. Similarly, copy a leads layer, a symbol layer, and a packages layer in the original position, respectively. After copying all the layers, click the [Cancel] button to close this dialog box.
(5) Air bridge construction. In order to make the circuits inside spiral inductors connect with the peripheral circuits, air bridges are constructed. Execute menu

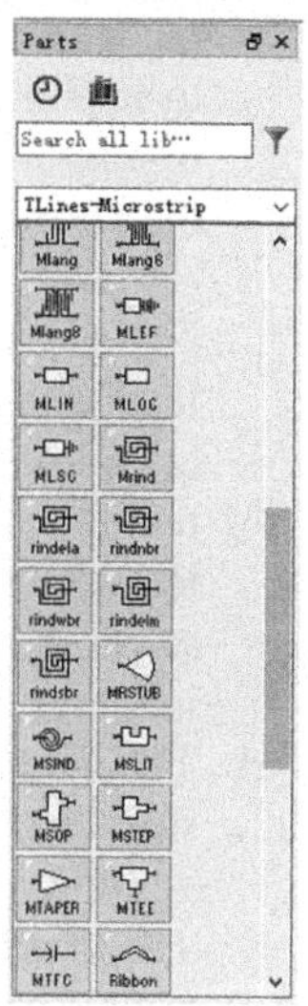

Fig. 3.74 [TLines-Microstrip] list

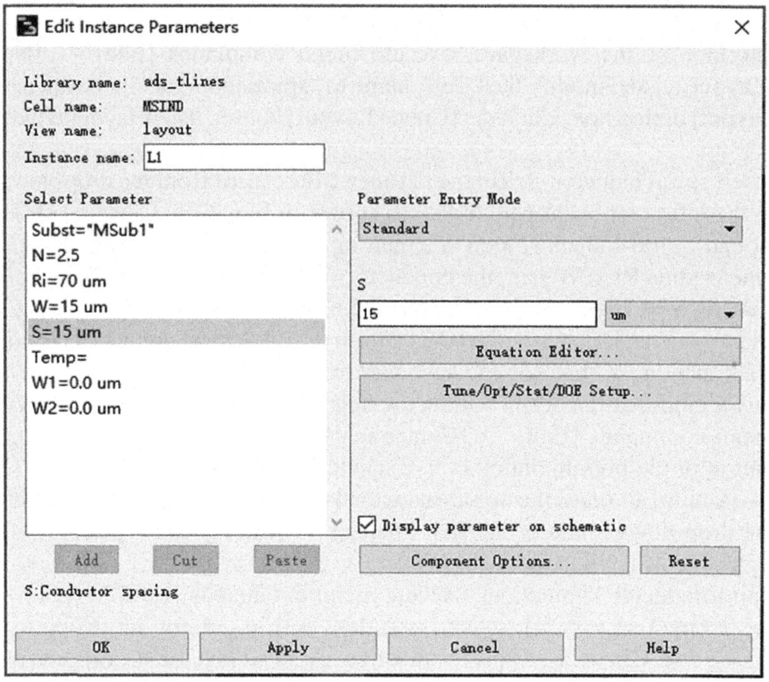

Fig. 3.75 [Edit Instance Parameters] dialog box

commands [Insert] → [Rectangle] to insert a rectangle in the layout, then press "Esc" to exit. Select the inserted rectangle, select "cond: drawing" in the drop-down menu of the [All Shapes] → [Layer] under [Properties] at the right side of the window, and set the [Rectangles] → [Width] to an arbitrary value, [Height] to 40 μm. Select and long press the modified rectangle with the left mouse button to move it to the air bridge construction position, as shown in Fig. 3.76. Press the shortcut "Ctrl + A" to select all layers. Execute menu commands [Edit] → [Boolean Logical…] in the pop-up [Boolean Logical Operation Between Layers] dialog box to modify the Boolean logical operation between the bond layer and the cond layer according to Fig. 3.77, then click the [Apply] button to complete the operation. Similarly, complete the Boolean operations among the text layer, symbol layer, packages layer, and the cond layer, only retain the top leads metal layer, and click the [Cancel] button to close the dialog box. Then select the cond layer and press "Delete" on the keyboard to delete it. The spiral inductor after the construction of an air bridge is shown in Fig. 3.78.

(6) The indentation of layers. There are different indentations between the layers of spiral inductor. Since the layer has been disconnected after the Boolean logical operation, execute menu commands [View] → [Layer View] → [By Name…]. Then, in the [Layout Layers] dialog box shown in Fig. 3.79, select [text: drawing] to only display the text layer of the spiral inductor. Press the shortcut "Ctrl + A"

Fig. 3.76 Air bridge construction position

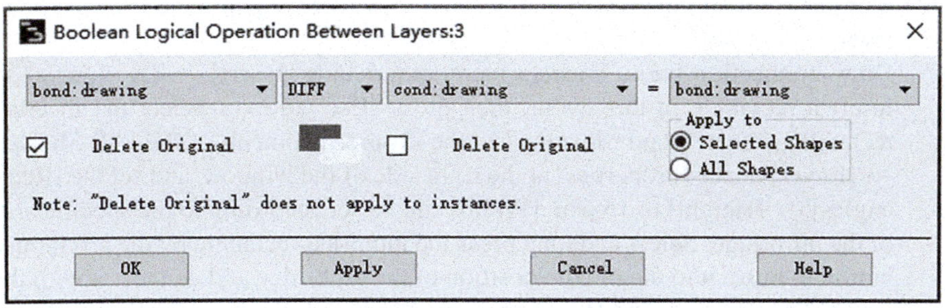

Fig. 3.77 [Boolean Logical Operation Between Layers] dialog box

Fig. 3.78 Spiral inductor after the construction of an air bridge

to select the text layer and execute menu commands [Edit] → [Scale/Oversize] → [Oversize…] to open the dialog box. Since the text layer is indented by 2 μm compared with the bond layer, enter-2 in the "Oversize(+)/Undersize(-) " column of the pop-up [Oversize] dialog box and click the [Apply] button.

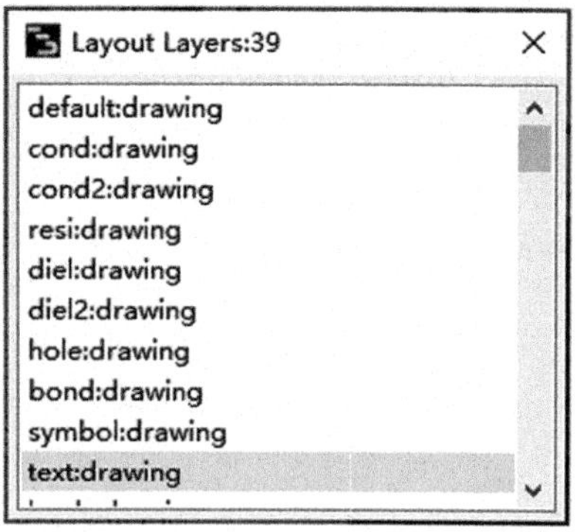

Fig. 3.79 [Layout Layers] dialog box

Similarly, indent the symbol and packages layers by 2 μm relative to the bond layer.

(7) Draw connection parts. Execute menu commands [Insert] → [Rectangle] to insert a rectangle in the layout, then press "Esc" to exit. Select the inserted rectangle, select "bond: drawing" in the drop-down menu of the [All Shapes] → [Layer] under [Properties] at the right side of the window, and set the [Rectangles] → [Height] to 15 μm. [Width] can be set according to the specific size of the air bridge. Select and long press the modified rectangle by the left mouse button to move it to the middle position of the air bridge and connect it with the original bond layer. The layout of a spiral inductor has been drawn, as shown in Fig. 3.80.

2. Layout Simulation of Spiral Inductor

(1) Insert ports. Execute menu commands [Insert] → [Pin] and click the left mouse button to insert two pins into the I/O ports of the spiral inductor, as shown in Fig. 3.81.
(2) Modify simulation settings. Execute menu commands [EM] → [Simulation Settings…], click the [Create EM Setup View] button in the pop-up the [New

Fig. 3.80 Final layout of the spiral inductor

Fig. 3.81 Insert pins on spiral inductor

EM Setup View] dialog box, and the [EM Setup for simulation] window will pop up. Select the second method "Momentum Microwave" of the EM simulator at first. Then select the [Frequency Plan] tab to set the simulation frequency. Select "Adaptive" in the "Type" column and set the "Fstart" to 0 GHz, the "Fstop" to 16 GHz, and the "Npts" to 50. Finally, select the [Options] tab. Click "Preprocessor", select the "User specified snap distance" option under the "Heal the layout", and set the distance to 2.5 μm. Click "Mesh" and select the "Edge Mesh" option. Keep other settings in default. After finishing all the steps, close the window and click the [OK] button to save the settings.

(3) Layout simulation. Execute menu commands [EM] → [Simulate] to simulate the layout of the spiral inductor. During the simulation, a status window will pop up to display the current item's progress. After the finishing of the simulation, the data displayed window will be opened automatically. Refer to the steps in Sect. 3.2.2 to view and handle traces dB(S(1,1)) and dB(S(2,1)), the final results are shown in Fig. 3.82.

3. Co-simulation of Spiral Inductor

To verify whether the inductance of the drawn spiral inductor is 1.36 nH, the co-simulation of inductor schematic and layout is required.

(1) Create EM model and symbol of the spiral inductor. In the layout window, execute menu commands [EM] → [Component] → [Create EM Model And Symbol…], then click the [OK] button in the pop-up dialog boxes. Execute menu commands [Edit] → [Component] → [Update Component Definitions…], then click the [OK] button in the pop-up dialog box to complete the creation of the EM model and the symbol in the spiral inductor.
(2) Create a new schematic and insert the spiral inductor component. Return to the "Input_Absorptive_Bandstop_Filter_wrk" main interface of the workspace. Execute menu commands [File] → [New] → [Schematic…], modify the "Cell" name to "spiral inductor-cosimulation" in the opened [New Schematic] dialog box, and click the [Create Schematic] button to create a new schematic. Click the [Open the Library Browser] icon at the left side of the schematic window,

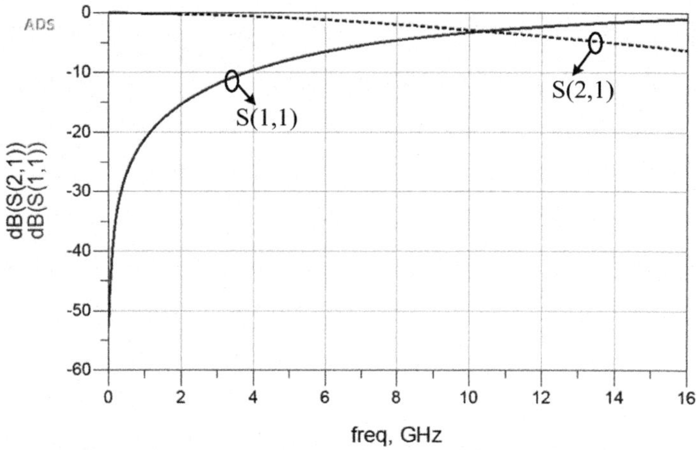

Fig. 3.82 Final simulated results of the drawn spiral inductor

then select the "spiral inductor" component under the [Workspace Libraries] in the displayed [Component Library] window. Click the right mouse button, then click the [Place Component] in the pop-up menu to insert a spiral inductor component in the schematic. Press "Esc" to exit.

(3) Insert ideal inductor. Select the [Lumped-Components] from the drop-down menu in the left component palette list and click the [Inductor] icon to insert an inductor. Press "Esc" on the keyboard to exit. Double-click the inductor and modify L = 1.36 nH in the [Edit Instance Parameter] pop-up dialog box (check whether the unit setting is consistent), then click the [OK] button to save the parameter.

(4) Insert the S-parameter simulator, term ports, and grounds. Select the [Simulation-S_Param] from the drop-down menu in the left component palette list and click the [S-parameter Simulator] icon to insert an S-parameter simulator. Click the [Port Impedance Termination for S-Parameters] icon to insert four term ports, then press "Esc" to exit. Execute menu commands [Insert] → [GROUND] to place four grounds (Or click the [TermG Port Impedance Termination for S-Parameters with Grounded Reference] icon to insert four term ports with the grounded reference). Execute menu commands [Insert] → [Wire] to connect the inserted components. After finishing all the steps, press "Esc" to exit.

(5) Set simulation frequency. Double-click the S-parameter simulator S-PARAMETERS. Set the "Start" to 0 GHz, the "Stop" to 16 GHz, and the "Step-size" to 0.01 GHz, and click the [OK] button. The final inductor co-simulation circuit is obtained, as shown in Fig. 3.83.

(6) Co-simulation. Execute menu commands [Simulate] → [Simulate]. After the simulation completes, the data display window will be opened automatically. Refer to the steps in Sect. 3.2.2 to view and handle traces of dB(S(1,1)),

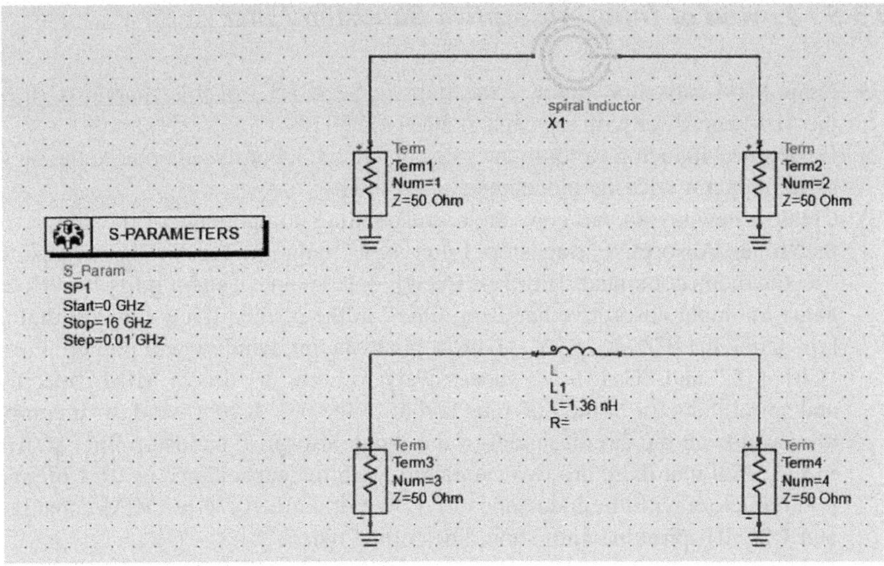

Fig. 3.83 Inductor co-simulation circuit

dB(S(2,1)), dB(S(3,3)), and dB(S(4,3)), the final results are shown in Fig. 3.84. It can be seen that traces dB(S(1,1)) and dB(S(3,3)), traces dB(S(2,1)) and dB(S(4,3)) are almost coincident, indicating that the inductance of the drawn spiral inductor is approximately equal to 1.36 nH. If the differences between these traces are large, go back to modify the spiral inductor layout and repeat the steps above until the error of two traces is in the range of acceptance (within an acceptable range).

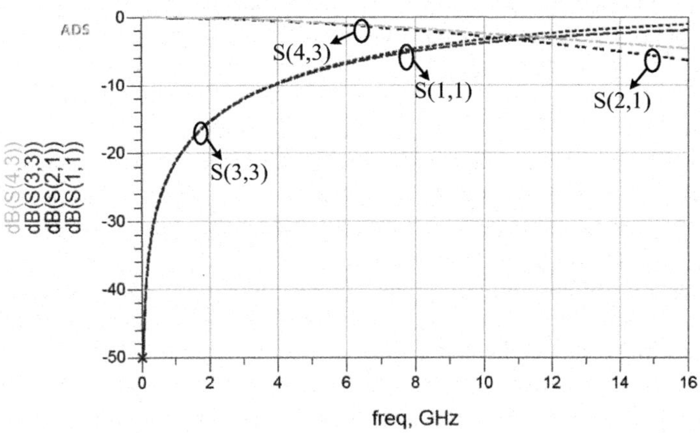

Fig. 3.84 S-parameter results of the inductor co-simulation

3.3.5 Layout of Input Absorptive Bandstop Filter

(1) Draw MIM capacitor. Refer to the steps in Sect. 3.3.3 of this chapter to draw the MIM capacitor with the capacitance of 0.90 pF.
(2) Draw spiral inductor. Refer to the steps in Sect. 3.3.4 of this chapter to the draw spiral inductor with the inductance of 1.13 nH.
(3) Create a new layout and copy the corresponding components of the layout. In the "Input_Absorptive_Bandstop_Filter_wrk" main interface of the workspace, execute menu commands [File] → [New] → [Layout...] and modify the "Cell" name to "input absorptive bandstop filter" in the opened [New Layout] dialog box. Click the [Create Layout] button, a new layout window will pop up. Press "Ctrl + C" and "Ctrl + V" successively to copy the drawn MIM capacitor and spiral inductor in the previous layout to the new layout window. It can be observed from the circuit model of the input absorptive bandstop filter shown in Fig. 3.29 that there are two capacitors with the capacitance of 0.74 pF and two inductors with the inductance of 1.36 nH, so the 0.74 pF MIM capacitor and 1.36 nH spiral inductor should be copied twice.
(4) Draw signal pads. Signal pads should be added to the I/O ports in layout because the "ground-single-ground" ("GSG") probes are used in the measurement. The overall size of the signal pad is 100 μm × 100 μm. Meanwhile, the outer ring ground is added to make the probe measurement more convenient. The specific steps for drawing pads and outer ring ground are listed as follows: Execute menu commands [Insert] → [Rectangle] to insert a rectangle in the layout, then press "Esc" to exit. Select the inserted rectangle, select "bond: drawing" in the drop-down menu of the [All Shapes] → [Layer] under [Properties] at the right side of the window. Then set the [Rectangles] → [Width] and [Height] to appropriate values. Execute menu commands [Edit] → [Copy/Paste] → [Copy To Layer...], select [text: drawing] in the pop-up [Copy to Layer] dialog box, and click the [Apply] button to copy a test layer in the original position. Similarly, copy a leads layer, a symbol layer, and a packages layer in the original position, respectively. After copying all the needed layers, click the [Cancel] button to close this dialog box. Then indent the layers. Select the text layer of the pads and outer ring ground, execute menu commands [Edit] → [Scale/Oversize] → [Oversize...]. Since the text layer is indented by 2 μm compared with the bond layer, enter-2 in the "Oversize(+)/Undersize(−)" column of the pop-up [Oversize] dialog box and click the [Apply] button. Similarly, indent the symbol layer and packages layer by 2 μm relative to the bond layer.
(5) Layout adjustment and component connection. The layout is adjusted in the consideration of the circuit size, the layout beautification, and other factors. According to the simulated circuit model in Fig. 3.29, all components are connected by microstrip lines. The drawing of microstrip lines is similar to that of pads. For more details, please refer to step (4), which is not repeated here. The final layout of the input absorptive bandstop filter is obtained after the optimizations in several times, as shown in Fig. 3.85 (unit: μm).

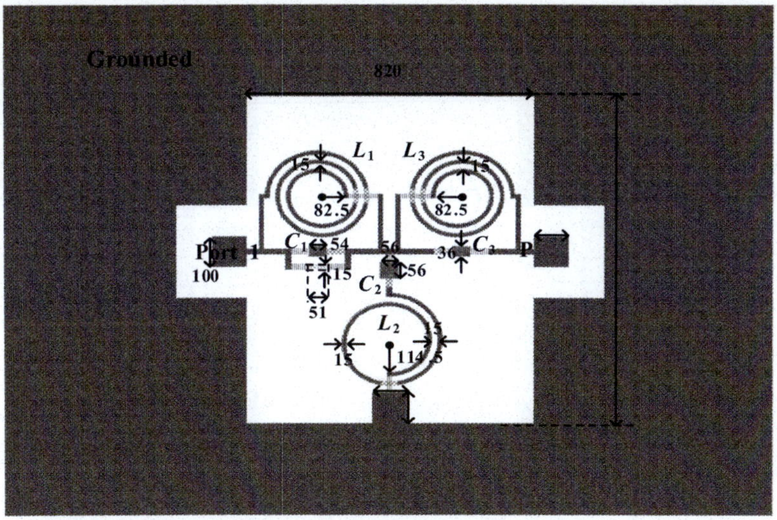

Fig. 3.85 Final layout of the input absorptive bandstop filter

3.3.6 Layout Simulation

(1) Insert pins. Execute menu commands [Insert] → [Pin], then click the left mouse button to insert pins into the I/O and grounded ports of the input absorptive bandstop filter, as shown in Fig. 3.86. Since the probe spacing used in the measurement is 300 μm, the vertical distance between the pins is 300 μm. Long press the "Ctrl" key and select all pins, then select "leads: drawing" in the drop-down menu of the [All Shapes] → [Layer] under [Properties] at the right side of the window.

(2) Edit ports. Click the [Port Editor] icon in the toolbar, and the [Port Editor] dialog box shown in Fig. 3.87 will pop up. The "GSG" structure is consist of ports P1, P3, and P4. P1 is the input port. P3 and P4 are grounding ports. Select ports P3 and P4 with the left mouse button and drag them to the "Gnd" position of port P1, respectively. Similarly, the "GSG" structure is consist of ports P2, P5, and P6. P2 is the output port. P5 and P6 are grounding ports. Select ports P5 and P6 and drag them to the "Gnd" position of port P2, respectively. The [Port Editor] dialog box after the editing is shown in Fig. 3.88.

(3) Modify simulation settings. Execute menu commands [EM] → [Simulation Settings…], click the [Create EM Setup View] button in the pop-up the [New EM Setup View] dialog box, and the [EM Setup for simulation] window will be displayed. Select the second method "Momentum Microwave" of the EM simulator at first. Then select the [Frequency Plan] tab to set the simulation frequency. Select "Adaptive" in the "Type" column and set the "Fstart" to 0 GHz, the "Fstop" to 16 GHz, and the "Npts" to 50. Finally, select the [Options] tab. Click "Preprocessor", select the "User specified snap distance" option under

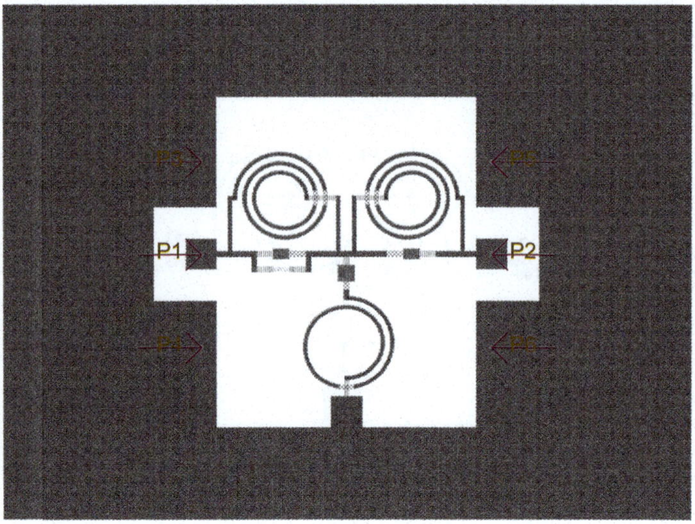

Fig. 3.86 Insert pins on the final layout

Fig. 3.87 [Port Editor] dialog box before the editing

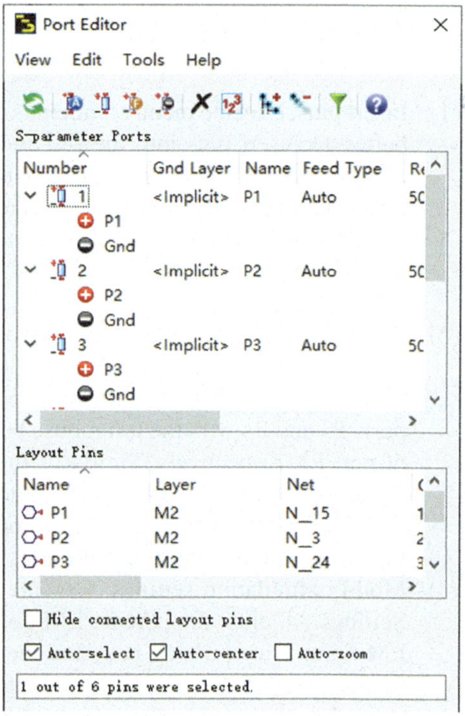

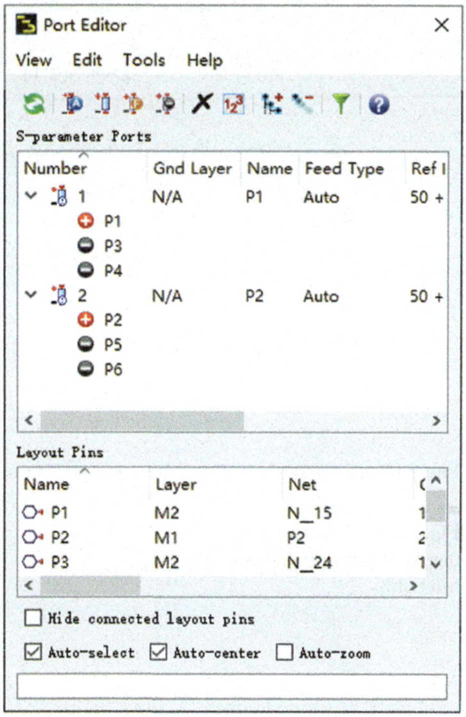

Fig. 3.88 [Port Editor] dialog box after the editing

the "Heal the layout", and set the distance to 2.5 μm. Click "Mesh" and select the "Edge Mesh" option. Keep other settings in default. After finishing all the steps, close the window and click the [OK] button to save the settings.

(4) Layout simulation. Execute menu commands [EM] → [Simulate] to simulate. During the simulation, a status window will pop up to display the current item's progress. After the finishing of the simulation, the data display window will be opened automatically. Refer to the steps in Sect. 3.2.2 to view and handle traces of dB(S(1,1)), dB(S(2,1)), and dB(S(2,2)), the results are shown in Fig. 3.89.

3.4 Chip Measurement

1. Measured Results

The probe station and vector network analyzer are used to measure the S-parameter of the input absorptive bandstop filter chip, as shown in Fig. 3.90. The measured results are shown in Fig. 3.91.

2. Result Analysis

It can be seen from Fig. 3.91 that the measured results are mostly consistent with the simulated results. The deviations are caused by the mechanical error and inaccurate

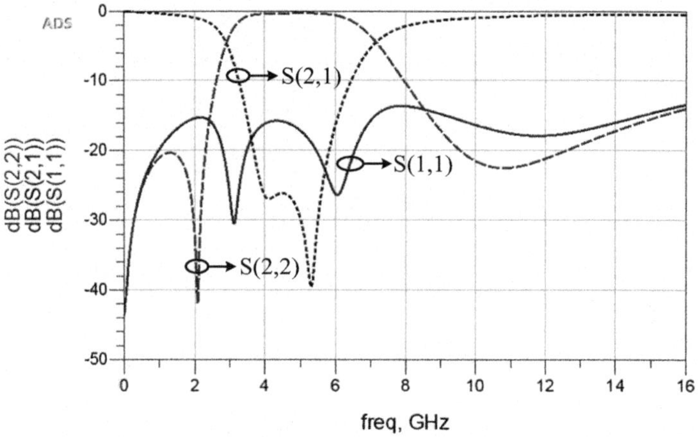

Fig. 3.89 Simulated results of the input absorptive bandstop filter

Fig. 3.90 Measurement of the input absorptive bandstop filter chip

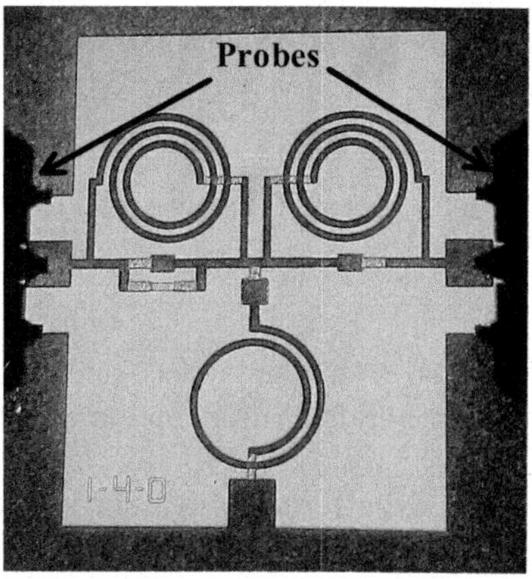

Fig. 3.91 Measured S-parameter results of the input absorptive bandstop filter chip

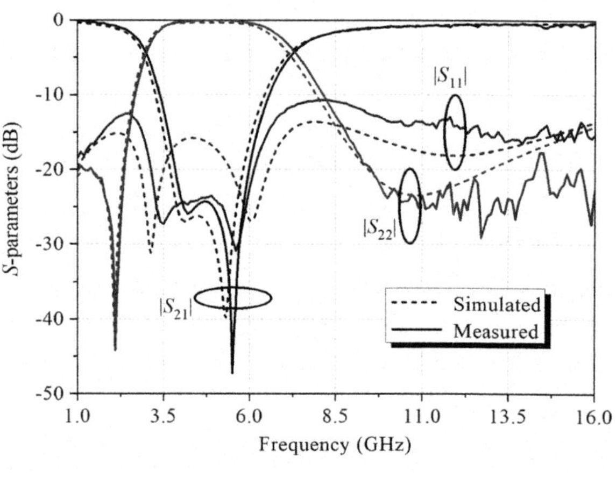

dielectric constants of industrial materials. The measured input return loss ($|S_{11}|$) is better than 10.7 dB from 1 to 16 GHz. The 10-dB stopband relative bandwidth is 63.8% from 3.38 to 6.57 GHz, while the maximum stopband rejection ($|S_{21}|$) is 47.3 dB at 5.5 GHz.

Reference

1. M. Kong, Y. Wu, Z. Zhuang, W. Wang, C. Wang, Ultra-miniaturized wideband input-absorptive bandstop filter based on TFIPD technology, in *IEEE Transactions on Circuits and Systems II: Express Briefs*, vol. 68, no. 7 (2021), pp. 2414–2418

Chapter 4
Design and Simulation of Impedance-Transforming Power Divider

Power dividers are widely applied in the antenna arrays, mixers, and the power amplifiers as a crucial passive component in modern microwave and communication systems. In order to adapt to the broadband requirement of the modern RF/Microwave communication systems, various wideband power dividers have been extensively studied over the past few years [1-3]. In this chapter, we are going to introduce an ultraminiaturized impedance-transforming power divider based on the TFIPD technology [4] and how to use the ADS software to build, simulate, and optimize its ideal simulated model and full-wave electromagnetic simulated model. Finally, the manufactured impedance-transforming power divider chip is packaged, and the measured results are also analyzed.

4.1 Overview of Power Divider

4.1.1 Theoretical Basis

1. Introduction

   Power dividers are three-port or multi-port passive components. Its main function is to divide one signal into two or more channels in equal or unequal ways. It can also be used in reverse. Two or more channels of signal power can be combined into one channel for output, which can be called the combiner. The isolation should be ensured between the output ports of the power divider.
2. Performance Index

   The evaluation indexes of power dividers include the insertion loss, return loss, and the isolation between output ports, etc. If port 1 is the input port of the power divider, port 2 and port 3 are the output ports of the power divider, its performance index parameters can be expressed as follow.

Y. Wu and W. Wang, *Microwave and Millimeter-Wave Chips Based on Thin-Film Integrated Passive Device Technology*,
https://doi.org/10.1007/978-981-99-1455-5_4

a. Insertion Loss (IL): $\text{IL(dB)} = -20\log_{10}|S_{21}|$.
b. Return Loss (RL): $\text{RL(dB)} = -20\log_{10}|S_{11}|$.
c. Isolation: $\text{Isolation(dB)} = -20\log_{10}|S_{23}|$.

4.1.2 Propaedeutics

1. Quasi-Chebyshev Impedance-Transforming Power Divider

Figure 4.1 shows the circuit structure of the quasi-Chebyshev wideband power divider with the impedance-transforming function [4]. In order to achieve the input matching of the power divider, the input impedance in Fig. 4.1 should satisfy the following condition:

$$R_{in} = 2R_S, \tag{4.1}$$

where R_S is the terminal impedance of Port 1. Therefore, the design of the power divider can be considered as one of the impedance-transforming networks from $2R_S$ to R_L.

For performing the desired specifications including the terminal impedances (R_S, R_L), cutoff frequencies (f_L, f_U), and the achievable maximum level of input return loss ($S_{11\max}$), the order 2m should meet

$$m \geq \frac{\cosh^{-1}\frac{r-1}{2\sqrt{r\left(10^{0.1\alpha_{\max}-1}\right)}}}{\cosh^{-1}\frac{f_U^2+f_L^2}{f_U^2-f_L^2}}, \tag{4.2}$$

where r denotes the impedance-transforming ratio ($r = 2R_S/R_L$), $\alpha_{\max}$ is the maximal allowable ripple. For clearly depicting the performance of power divider,

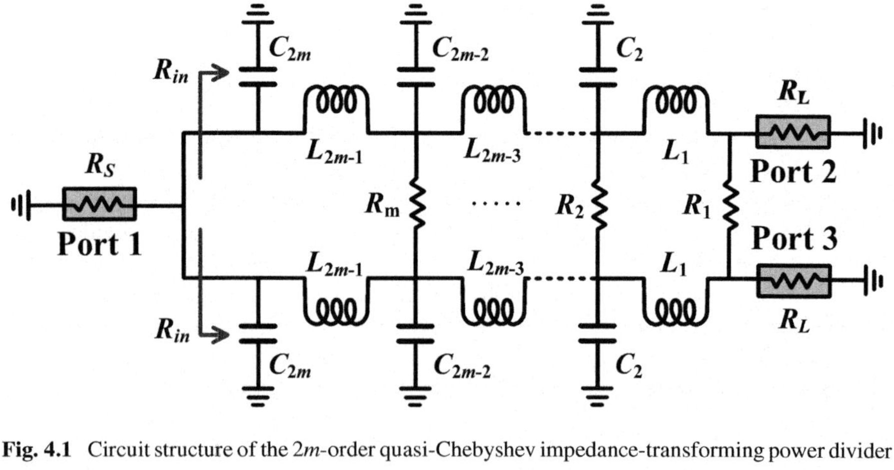

Fig. 4.1 Circuit structure of the $2m$-order quasi-Chebyshev impedance-transforming power divider

the value of α_{max} can be equivalent to the achievable maximum level of input return loss S_{11max} as

$$\alpha_{max} = 10\log_{10}\left(1 - 10^{S_{11\,max}/10}\right). \tag{4.3}$$

Thus, the element values g_i of Chebyshev impedance-transforming network can be obtained by the specified impedance ratio, cutoff frequencies, and the level of return loss using the low-pass impedance transformation network design equations. Furthermore, the values of L and C can be transformed from the obtained element values g_i as

$$\begin{cases} L_i = \frac{g_i R_L}{2\pi f_0} \\ C_i = \frac{g_i}{2\pi f_0 R_L}, \quad i = 1, 2, \ldots, 2m \end{cases}. \tag{4.4}$$

Since m is a positive integer, the value of A can be calculated by (4.5) when m is determined by (4.2).

$$A = \frac{1}{\cosh\left(\frac{1}{m}\cosh^{-1}\frac{r-1}{2\sqrt{r\left(10^{0.1\alpha_{max}-1}\right)}}\right)}. \tag{4.5}$$

To keep the value of the desired S_{11max}, the specified lower and upper cutoff frequencies need to be recalculated by

$$\begin{cases} f_{Lr} = f_0\sqrt{1 - A} \\ f_{Ur} = f_0\sqrt{1 + A} \end{cases}, \tag{4.6}$$

in which

$$f_0 = \sqrt{\frac{f_U^2 + f_L^2}{2}}. \tag{4.7}$$

It can be found that the desired bandwidth would expand a little since m is larger than the preset value by (4.2). The final bandwidth of the desired S_{11max} can be given as

$$BW = f_{Ur} - f_{Lr}. \tag{4.8}$$

To further achieve the output matching and isolation between the output ports, the isolation resistors R_i ($i = 1, 2, \ldots, m$) are loaded between path 1 and path 2.

2. Quasi-Elliptic Impedance-Transforming Power Divider

In order to further improve the stopband rejection, the shunt capacitor C_{2m} near the input port is replaced by the series-resonant branch (L_{e1}, C_{e1}), which generates

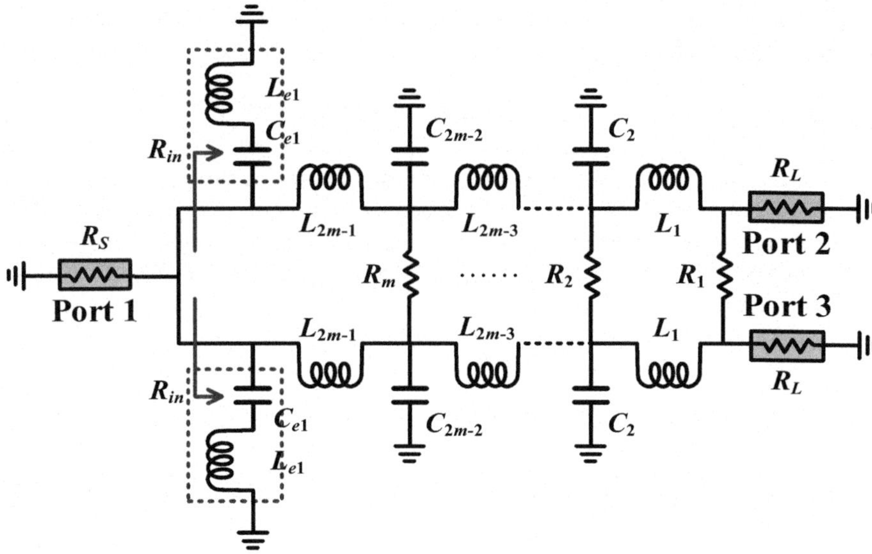

Fig. 4.2 Circuit structure of the $2m$-order quasi-elliptic impedance-transforming power divider

with an extra out-of-band transmission zero and form the quasi-elliptic low-pass response, while other LC parameters remain unchanged. The corresponding transforming equations can be given as

$$\begin{cases} L_{e1} = \dfrac{1}{C_{2m}\left[(2\pi f_Z)^2 - (2\pi f_C)^2\right]} \\ C_{e1} = \dfrac{C_{2m}\left[(2\pi f_Z)^2 - (2\pi f_C)^2\right]}{(2\pi f_Z)^2} \end{cases}. \tag{4.9}$$

where f_Z is the specified transmission zero of the quasi-elliptic response and f_C is the cutoff frequency. Figure 4.2 shows the circuit structure of $2m$-order impedance-transforming quasi-elliptic power divider.

4.2 Schematic of Impedance-Transforming Power Divider

ADS can realize the parameterized model simulation. Take a fourth-order quasi-Chebyshev power divider as an example to introduce how to build and simulate its ideal circuit model, which has the terminal impedances $R_S = R_L = 50\ \Omega$, cutoff frequencies $f_L = 3.3$ GHz, $f_U = 5.0$ GHz, and the achievable maximum level of 20 dB input return loss $S_{11\max}$. The element values L_1, C_2, L_3, C_4, and two isolation resistors R_i ($i = 1$ and 2) of the fourth-order quasi-Chebyshev impedance-transforming power divider can be parameterized in the circuit model.

4.2.1 New Workspace and Circuit Model

1. New Workspace

    (1) Double-click the ADS shortcut icon, and click the [OK] in the displayed dialog box to start the ADS software. After ADS runs, the [Get Started] dialog box will pop up automatically. Click the [Close] button in the lower right corner to enter the main interface [Advanced Design System 2020 (Main)], as shown in Fig. 4.3.
    (2) Create a new workspace to store all files of the design simulation. Execute menu commands [File] → [New] → [Workspace…] to open the new workspace dialog box shown in Fig. 4.4, where you can set the workspace name [Name] and the working path [Create in]. Here, change the workspace name to "Impedance_Transforming_Power_Divider_wrk" and keep the working path in the default setting. Click the [Create Workspace] button to complete the creation of a new workspace.
    (3) The [Folder View] in the ADS main interface can display the name and working path of the created workspace, as shown in Fig. 4.5. The workspace name is "Impedance_Transforming_Power_Divider_wrk", the corresponding path is "D:\ADS\Impedance_Transforming_Power_Divider_wrk", and a subfolder named "Impedance_Transforming_Power_Divider_wrk" in the ADS folder of disk D can be found in your computer.

2. New Circuit Model

Fig. 4.3 Main interface of ADS

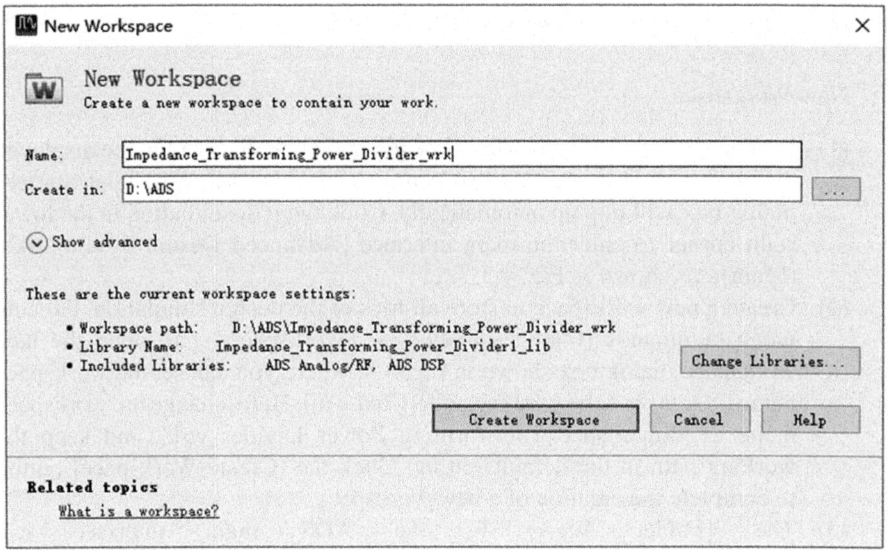

Fig. 4.4 [New workspace] dialog box

Fig. 4.5 Folder view of the new workspace

(1) Create a new schematic. Execute menu commands [File] → [New] → [Schematic…] to open the [New Schematic] dialog box shown in Fig. 4.6. Modify the name of "Cell" to "circuit structure" and click the [Create Schematic] button to complete the creation of a new schematic, as shown in Fig. 4.7.

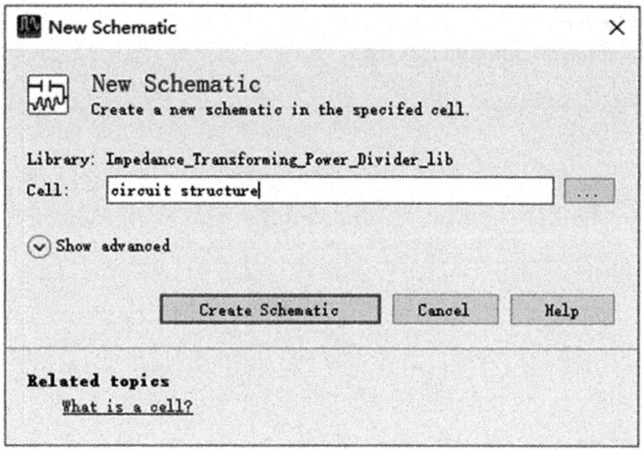

Fig. 4.6 [New schematic] dialog box

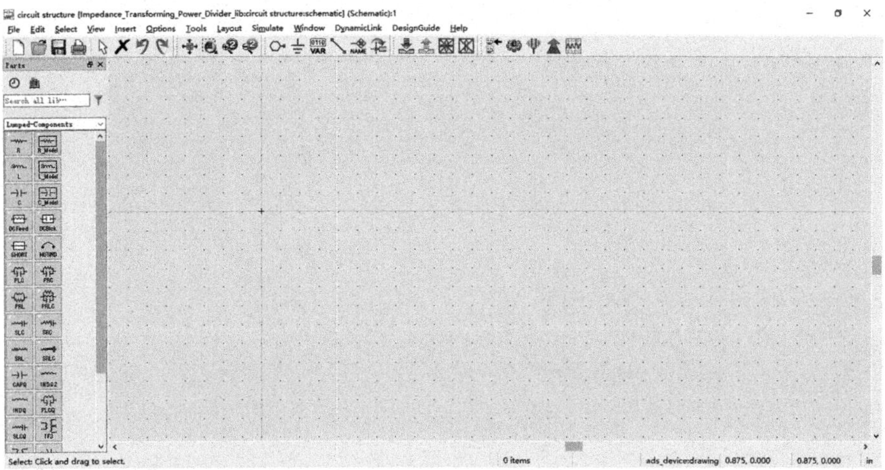

Fig. 4.7 New schematic window

(2) Insert inductors. Select the [Lumped-Components] from the drop-down menu in the left component palette list, as shown in Fig. 4.8, which includes some commonly used ideal lumped-element models, such as the capacitor, the inductor, and the resistor, etc. Click the [Inductor] icon under the [Lumped-Components] to insert four inductors, then press "Esc" on the keyboard to exit, as shown in Fig. 4.9.

(3) Insert and rotate capacitors. Click the [Capacitor] icon under the [Lumped-Components] to insert four capacitors, then press "Esc" on the keyboard to exit, as shown in Fig. 4.10. Right-click the capacitors to be rotated, select [Rotate] from the pop-up menu to rotate them in 90° clockwise. Capacitors have been inserted and rotated, as shown in Fig. 4.11.

Fig. 4.8 [Lumped-Components] list

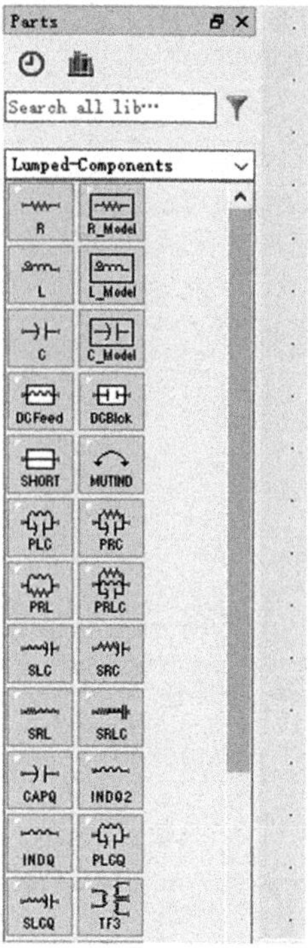

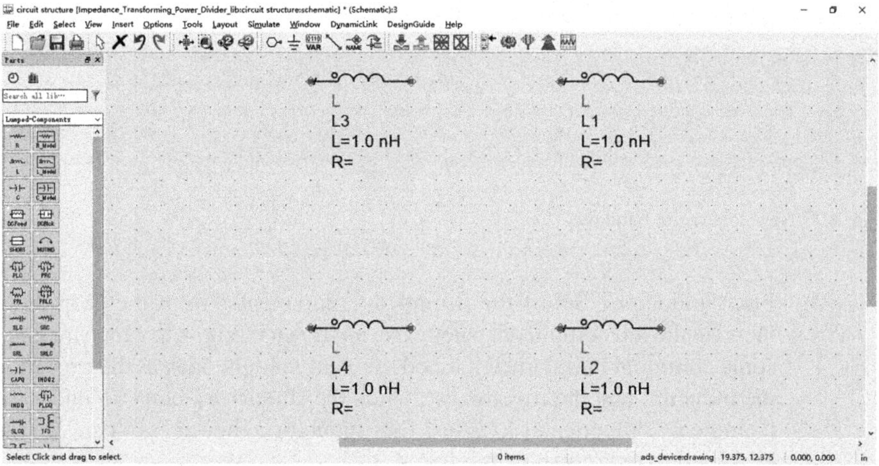

Fig. 4.9 Insertion of inductors

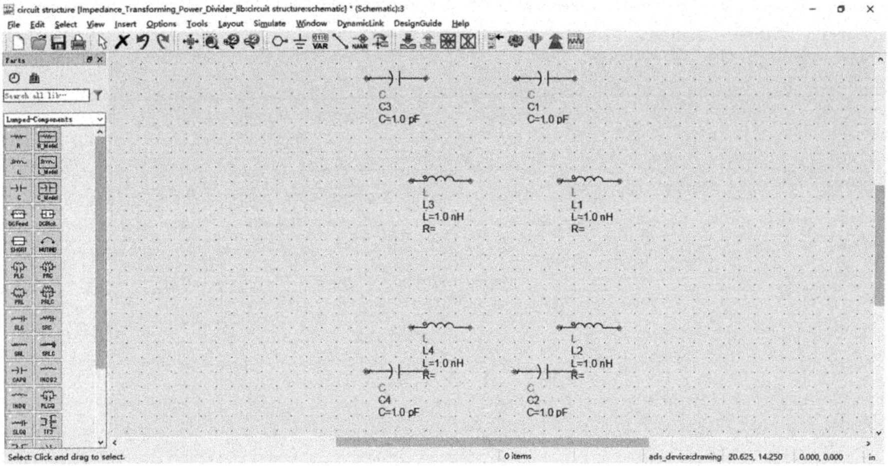

Fig. 4.10 Insertion of capacitors

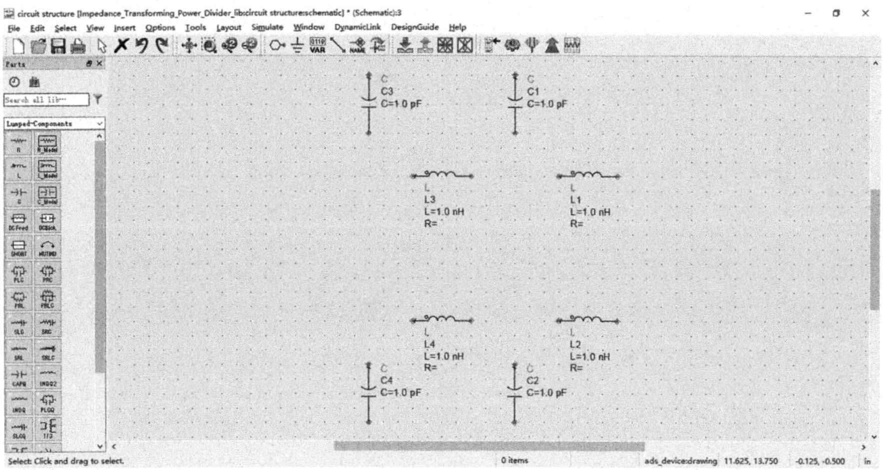

Fig. 4.11 Rotation of capacitors

(4) Insert and rotate resistors. Click the [Resistor] icon  under the [Lumped-Components] to insert two resistors, then press "Esc" on the keyboard to exit. Resistors have been inserted, as shown in Fig. 4.12. Right-click the resistors to be rotated, select [Rotate] from the pop-up menu to rotate them in 90° clockwise. Resistors have been inserted and rotated, as shown in Fig. 4.13.

(5) Place and rotate grounds. Execute menu commands [Insert] → [GROUND] to place four grounds, then press "Esc" on the keyboard to exit, as shown in Fig. 4.14. Right-click the ground symbols to be rotated,

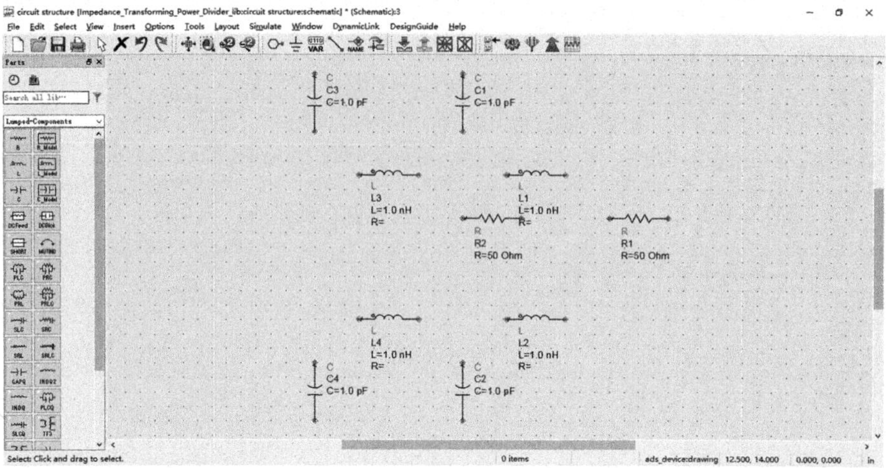

Fig. 4.12 Insertion of resistors

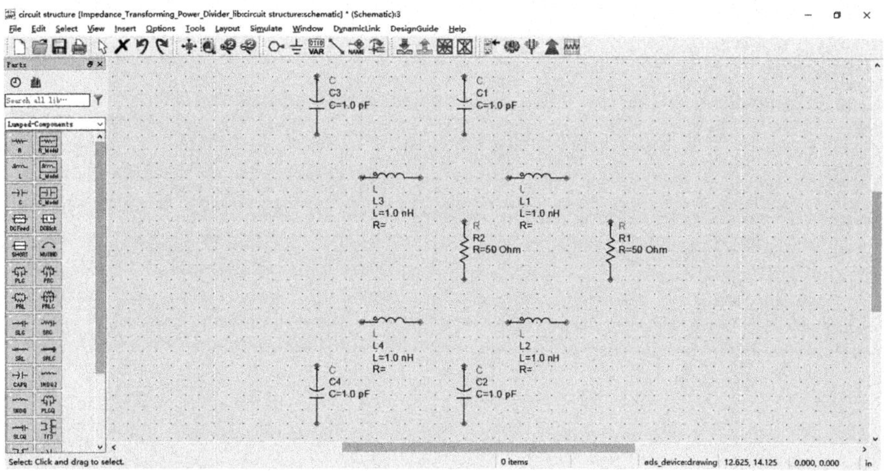

Fig. 4.13 Rotation of resistors

select [Rotate] from the pop-up menu to rotate them in 180° clockwise. Ground symbols have been inserted and rotated, as shown in Fig. 4.15.

(6) Connect components. Execute menu commands [Insert] → [Wire] to connect each component according to the circuit structure shown in Fig. 4.1. Figure 4.16 shows the circuit structure after the components are connected.

(7) Modify circuit model parameters. Double-click the inductor L1, then modify the parameter value to L1 (nH) in the pop-up [Edit Instance Parameters] dialog box, as shown in Fig. 4.17 (Check whether the unit settings

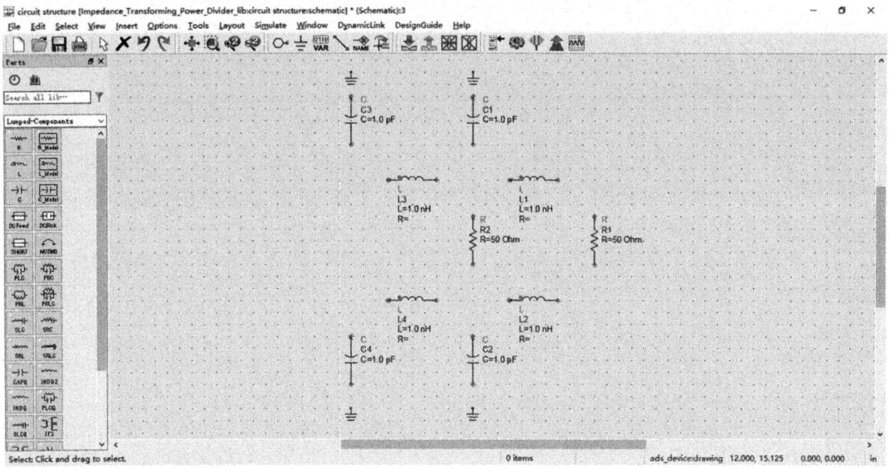

Fig. 4.14 Placement of grounds

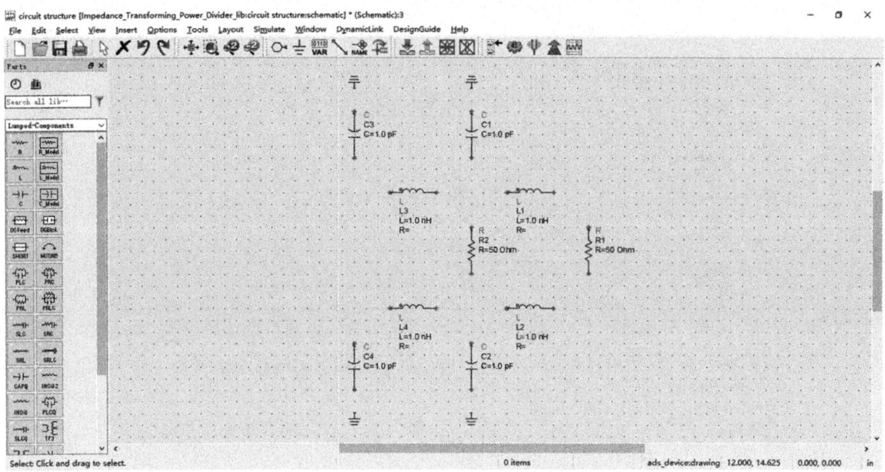

Fig. 4.15 Rotation of grounds

are consistent). Click the [OK] button to save the parameter and close the dialog box. Similarly, modify the parameter value of L2 to L1 (nH), the parameter values of L3 and L4 to L3 (nH) (taking L3 as an example, as shown in Fig. 4.18), the parameter value of C1 and C2 to C2 (pF) (taking C1 as an example, as shown in Fig. 4.19), the parameter value of C3 and C4 to C4 (pF) (taking C3 as an example, as shown in Fig. 4.20), the parameter value of R1 to R1 (Ohm), as shown in Fig. 4.21, and the parameter value of R1 to R (Ohm), as shown in Fig. 4.22.

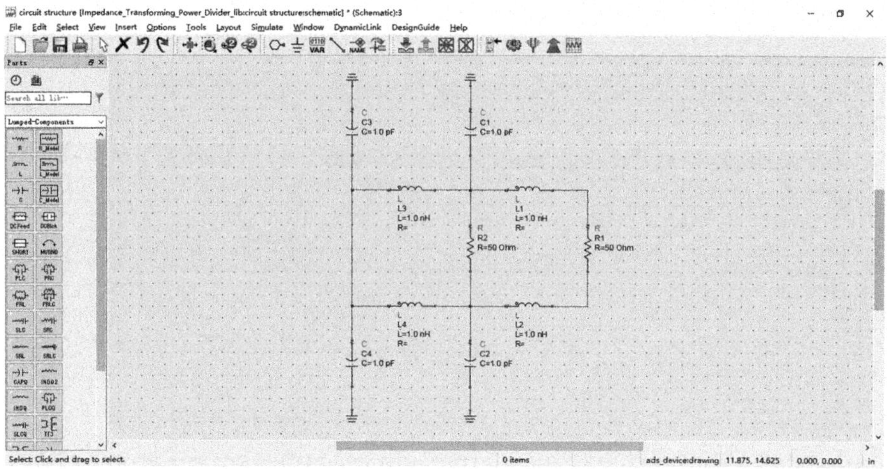

Fig. 4.16 Connection of components

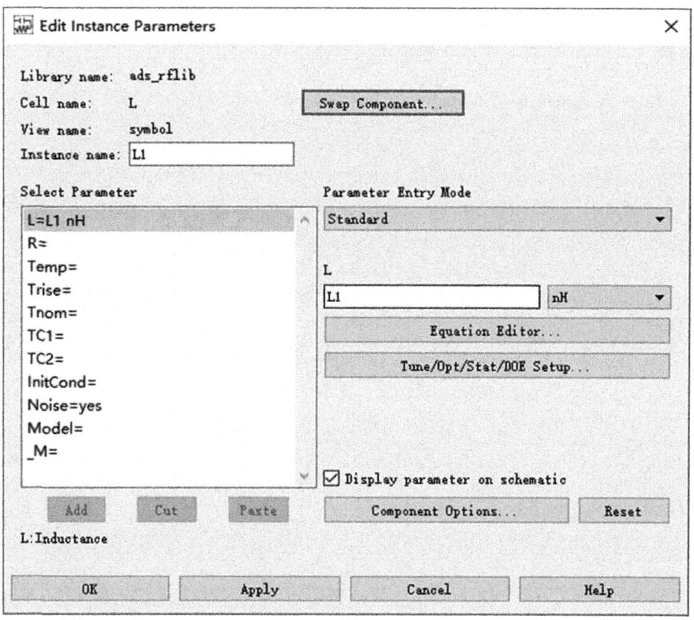

Fig. 4.17 Parameter setting of the inductor L1

(8) Define parameter values of variables. Click the [Insert VAR: Variable Equations] icon  in the toolbar to insert a VAR item in the schematic. Double-click the component  to open the [Edit Instance Parameter] dialog box. The "Variable or Equation Entry Mode" is "Standard" by default. Input the

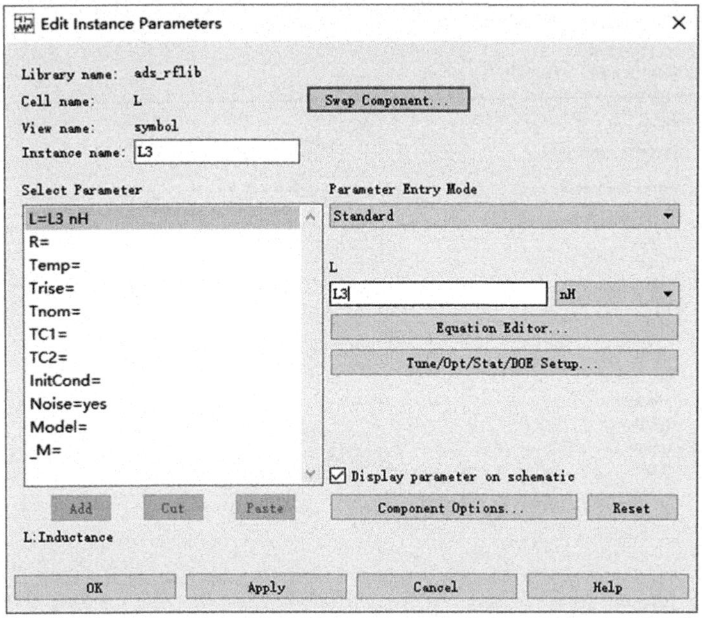

Fig. 4.18 Parameter setting of the inductor L3

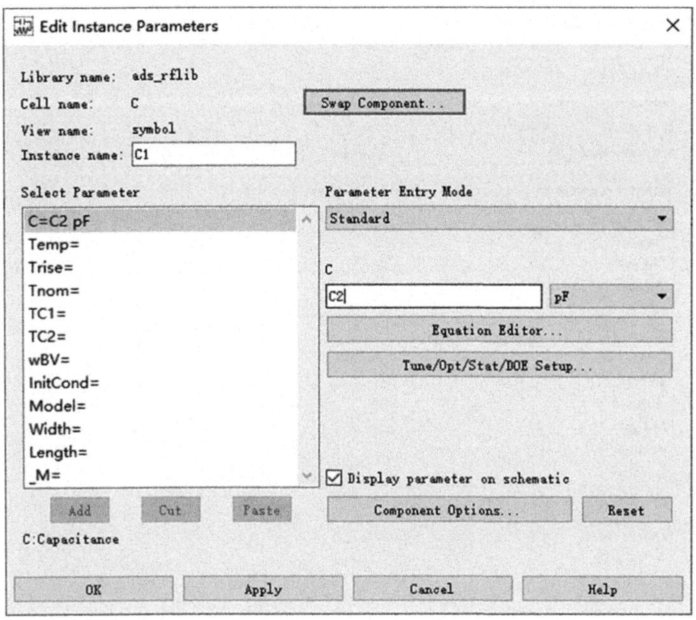

Fig. 4.19 Parameter setting of the capacitor C1

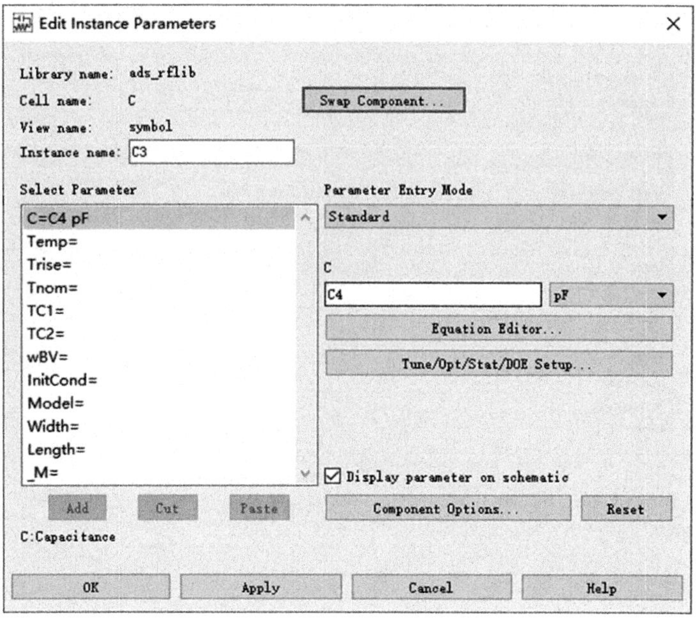

Fig. 4.20 Parameter setting of the capacitor C3

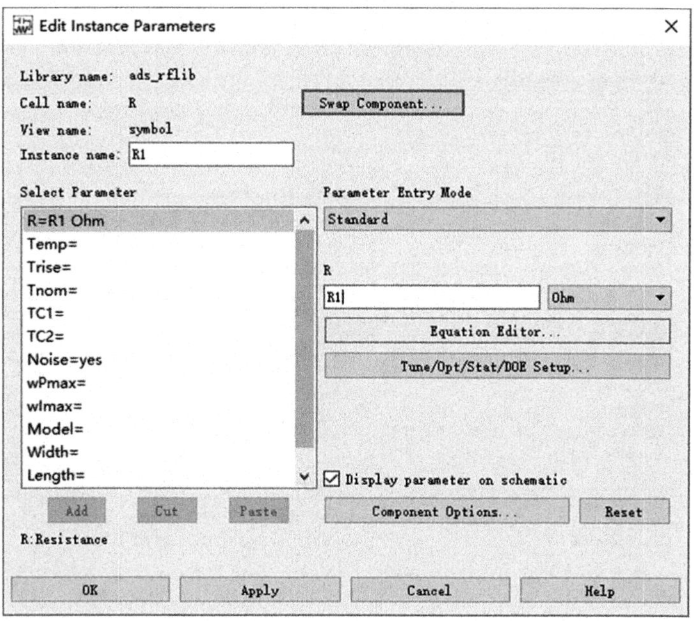

Fig. 4.21 Parameter setting of the resistor R1

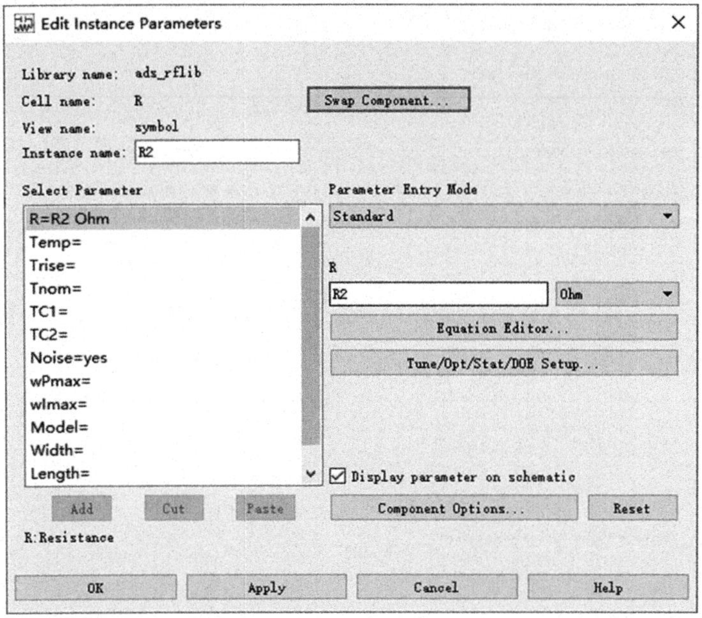

Fig. 4.22 Parameter setting of the resistor R2

variable name "L1" in the "Name" column, the variable value "1.74" in the "Variable Value" column. Click the [Apply] button, and the dialog box after the setting is shown in Fig. 4.23. If you click [OK], the dialog box can be closed directly. Then define other variables in turn, including C2 = 0.65, L3 = 3.25, C4 = 0.35, R1 = 220, and R2 = 230, as shown in Fig. 4.24. While defining other variables, click the [Add] button to add them. If you click the [Apply] button, the selected variable in the parameter list on the left can be replaced directly. In addition, there is no need to set units because the unit of each variable has already been defined in the component model.

(9) The VAR item provides three methods to add variables, including "Standard", "Name=Value", and "File Based". You can select the most suitable method in the "Variable or Equation Entry Mode" column. The second method is introduced below. Select the "Name=Value" mode, as shown in Fig. 4.25. Input "L1=1.74" under the "Variable Value" column, then click the [Apply] button. Similarly, when defining other variables, click the [Add] button to add them.

(10) Complete parameter definition. After the definition of all variables, click the [OK] button, the final circuit model of the fourth-order quasi-Chebyshev impedance-transforming power divider defined by the ideal parameters is shown in Fig. 4.26. Readers can compare the parameters in

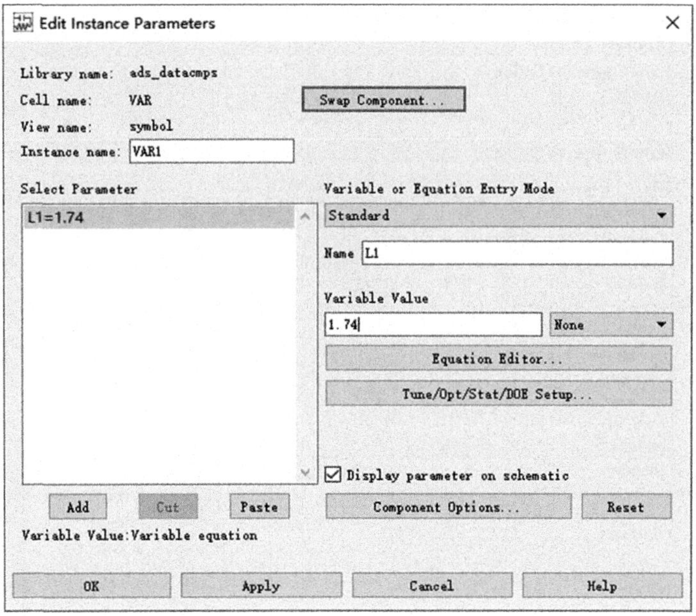

Fig. 4.23 Definition of the variable L1

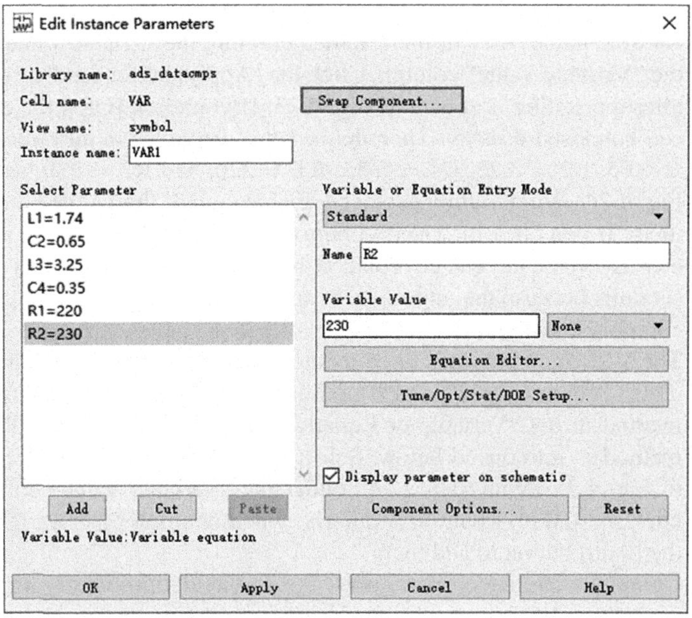

Fig. 4.24 Definitions of all variables in the circuit

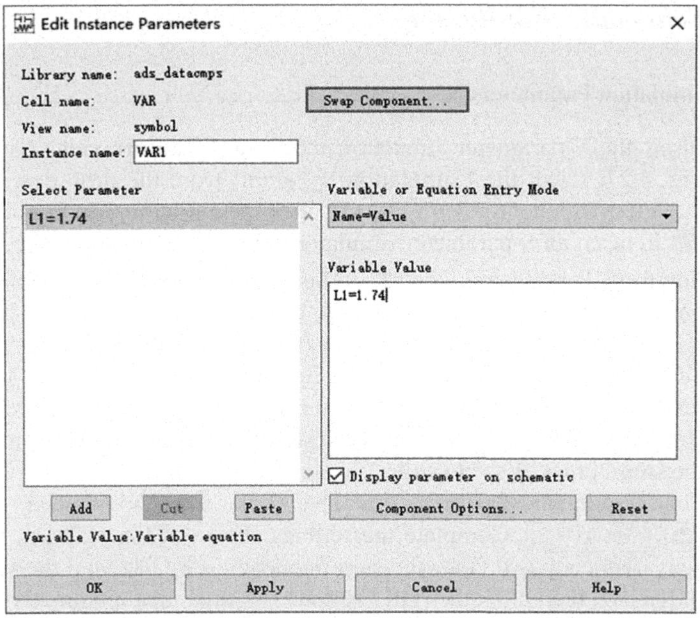

Fig. 4.25 The second method to define the variable L1

the lumped-element circuit structure to check whether all parameters are set correctly in detail.

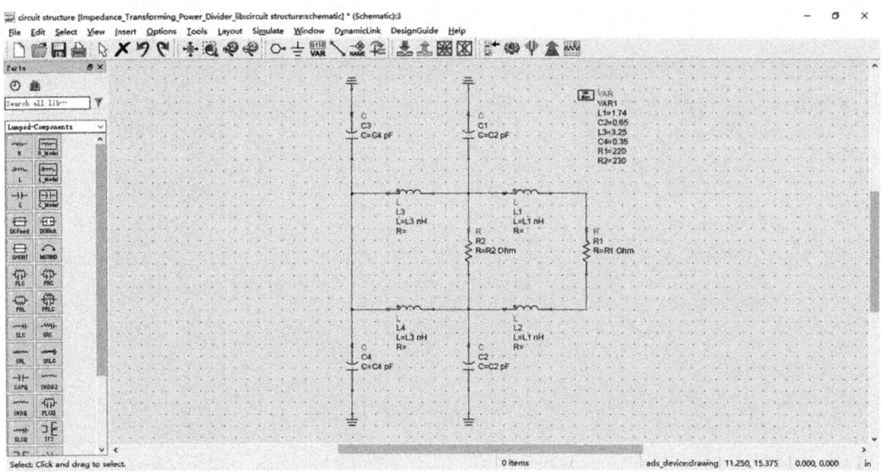

Fig. 4.26 Fourth-order quasi-Chebyshev impedance-transforming power divider circuit schematic with the defined ideal parameters

4.2.2 *Schematic Simulation*

1. Set Simulation Parameters

   (1) Insert the S-parameter simulator, term ports, and grounds. As shown in Fig. 4.27, select the [Simulation-S_Param] from the drop-down menu in the left component palette list and click the [S-parameter Simulator] icon to insert an S-parameter simulator. Click the [Port Impedance Termination for S-Parameters] icon to insert three term ports, then press "Esc" on the keyboard to exit. After that, execute menu commands [Insert] → [GROUND] to place three grounds (Or click the [TermG Port Impedance Termination for S-Parameters with Grounded Reference] icon to insert three term ports with the grounded reference). Execute menu commands [Insert] → [Wire] to connect the inserted components. After finishing all the steps, press "Esc" to exit.
   (2) Set simulation frequency. Double-click the S-parameter simulator S-PARAMETERS. Complete the settings shown in Fig. 4.28, in which the start frequency is 0 GHz, the stop frequency is 8 GHz, and the step-size is 0.01 GHz, then click the [OK] button. The final ideal parameter simulated

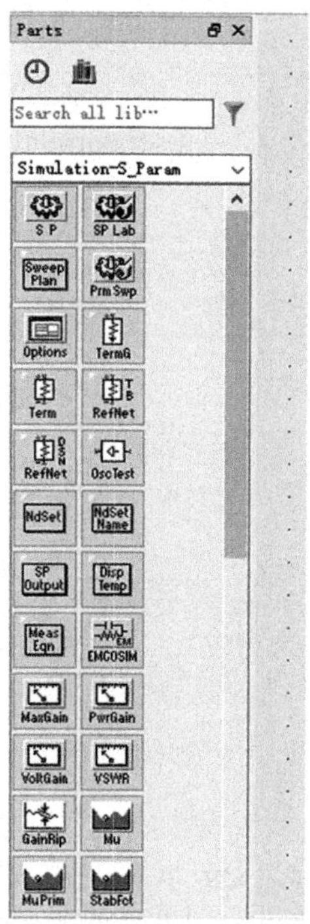

Fig. 4.27 [Simulation-S_Param] list

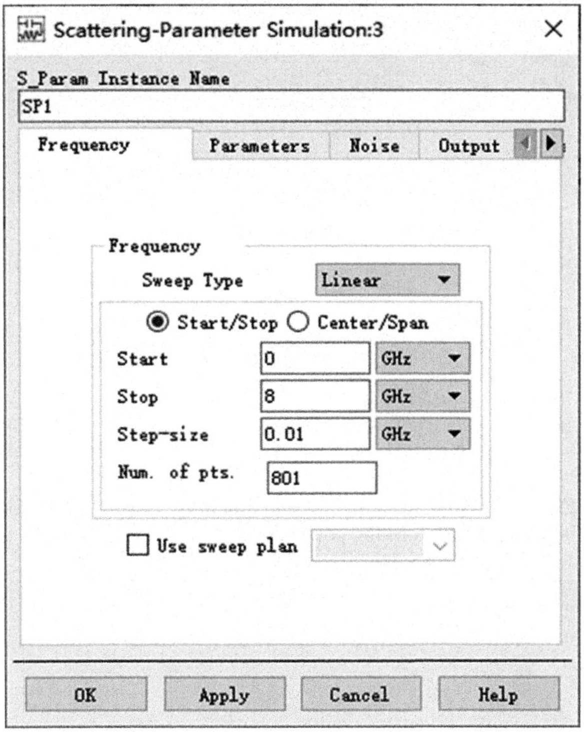

Fig. 4.28 Frequency settings of the S-parameter simulation

circuit model of the fourth-order quasi-Chebyshev impedance-transforming power divider is obtained, as shown in Fig. 4.29.

2. View Simulated Results

(1) Execute menu commands [Simulate] → [Simulate]. After the simulation completes, the data display window will be opened, as shown in Fig. 4.30.
(2) Click the [Rectangular Plot] icon 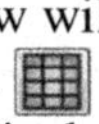 in the [Palette], then click the left mouse button at the blank data display area to open the [Plot Traces & Attributes] dialog box shown in Fig. 4.31, where the parameter traces can be set and plotted.
(3) Long press the "Ctrl" key to select S(1,1) and S(2,1), then click the [>>Add>>] button. Choose [dB] in the pop-up dialog box shown in Fig. 4.32, then click the [OK] button. It can be seen that dB(S(1,1)) and dB(S(2,1)) have been added to the [Traces] list box, as shown in Fig. 4.33.
(4) Click the [OK] button in Fig. 4.33, the traces of dB(S(1,1)) and dB(S(2,1)) can be displayed in the data display area, the ordinate is the dB value, as shown in Fig. 4.34.
(5) Similarly, the traces of dB(S(2,2)) and dB(S(3,2)) can also be displayed in the data display area, the ordinate is the dB value, as shown in Fig. 4.35.

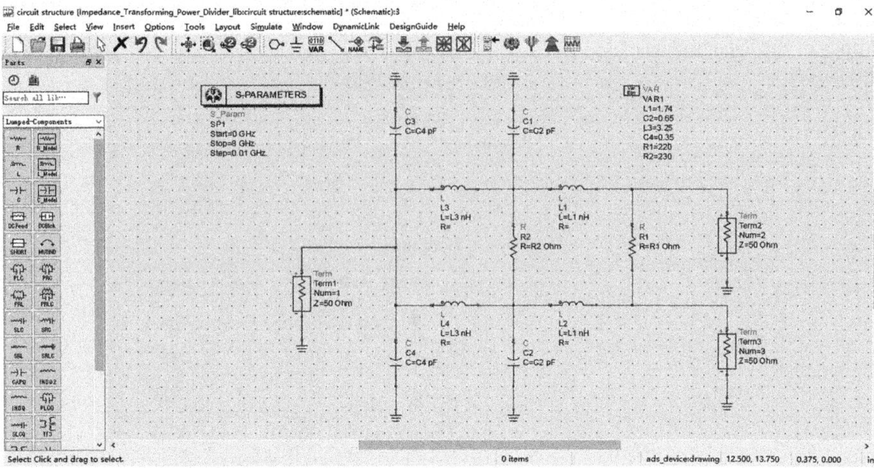

Fig. 4.29 Final ideal parameter simulated circuit model of the fourth-order quasi-Chebyshe impedance-transforming power divider

Fig. 4.30 Data display window

3. Handle Traces

Next, traces of dB(S(1,1)) and dB(S(2,1)) are taken as examples to introduce the curves handling. Similar steps can also be applied in other traces.

(1) By changing the frequency position of the marker, the values of any point on the trace can be read. Execute menu commands [Marker] → [New…] to open the dialog box shown in Fig. 4.36. Select a trace and click the left mouse button to insert a marker on it, as shown in Fig. 4.37. Similarly, insert

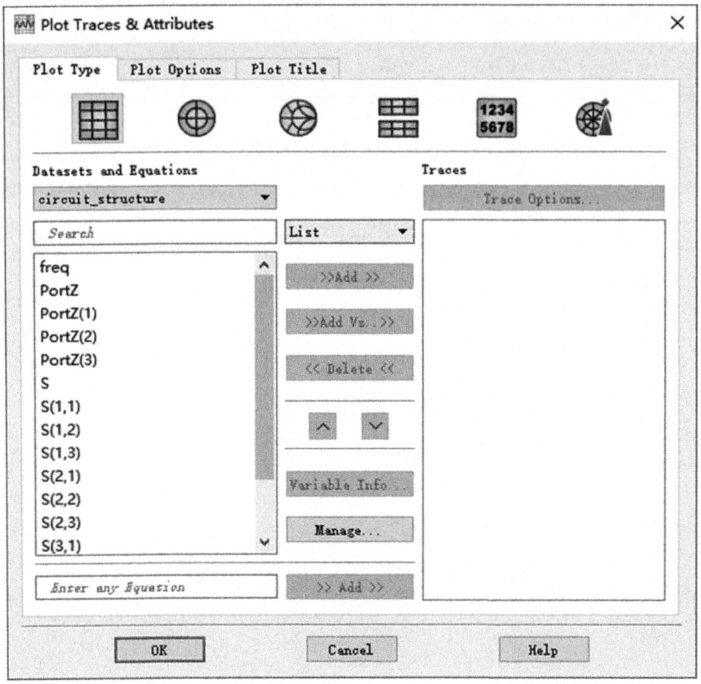

Fig. 4.31 [Plot Traces & Attributes] dialog box

Fig. 4.32 Handling of the data

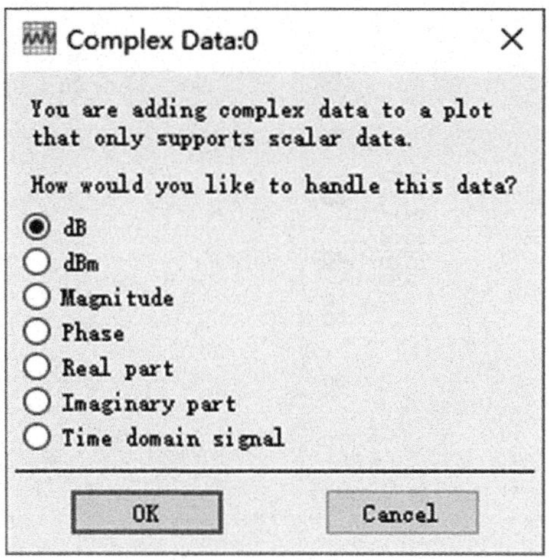

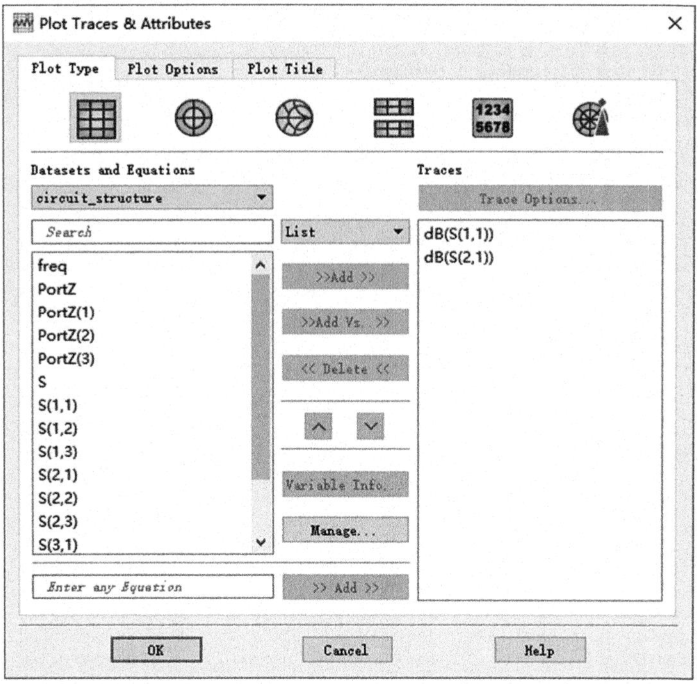

Fig. 4.33 Add of S(1,1) and S(2,1)

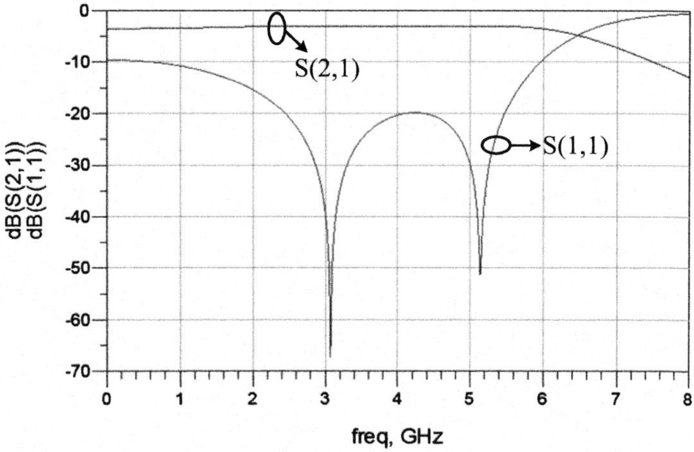

Fig. 4.34 Traces of dB(S(1,1)) and dB(S(2,1))

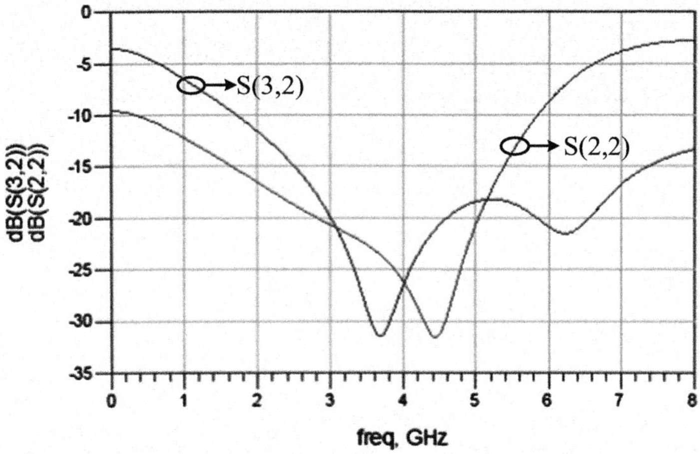

Fig. 4.35 Traces of dB(S(2,2)) and dB(S(3,2))

a marker for another trace dB(S(2,1)). In addition, you can long press the left mouse button to move the position of the data frame.

(2) After selecting the inserted marker, the left and right arrow keys on the keyboard can be used to adjust the position of the abscissa (freq). You can also click the position shown in Fig. 4.38 with the left mouse button to modify the specific frequency you want to view directly. Here, the values of dB(S(1,1)) and dB(S(2,1)) at the center frequency of 4.24 GHz are displayed, as shown in Fig. 4.39.

Fig. 4.36 [Insert Marker] dialog box

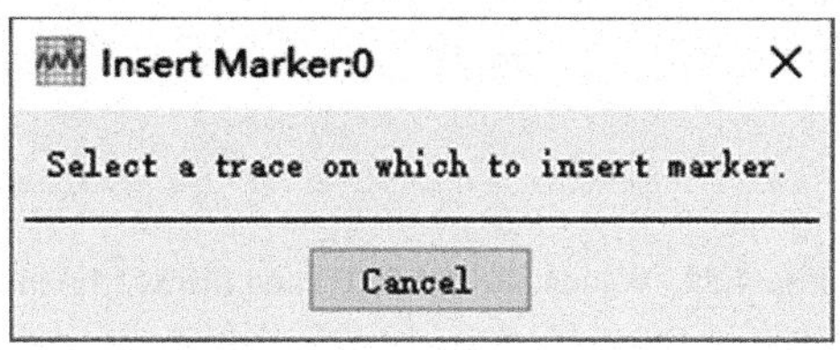

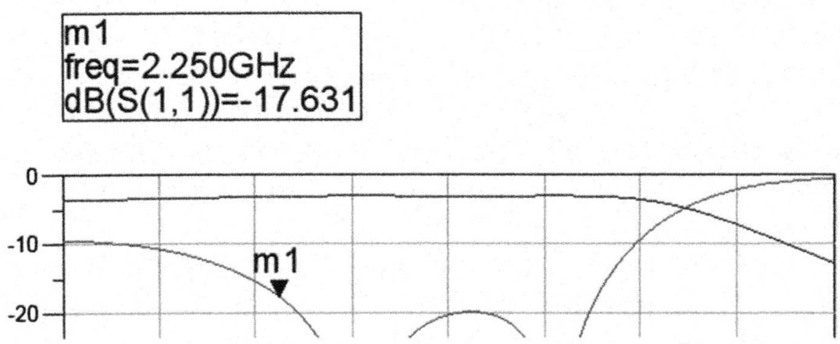

Fig. 4.37 Insertion of marker on the trace

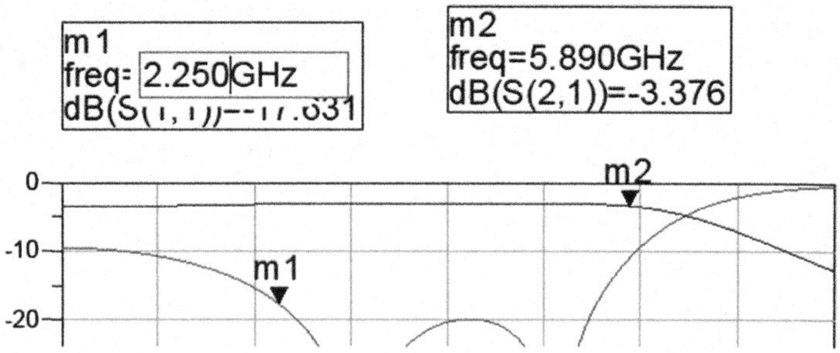

Fig. 4.38 Modification of the abscissa (freq) value

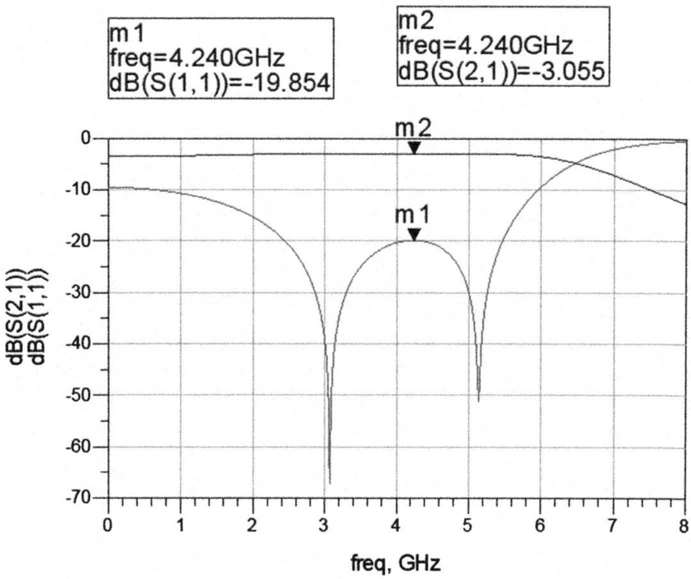

Fig. 4.39 Values of dB(S(1,1)) and dB(S(2,1)) at the center frequency

(3) The editing function of the data display is introduced below. Take the modification and beautification of the Y axis as examples. Double-click the simulated result figure, then the [Plot Traces & Attributes] dialog box pops up. Click the [Plot Options] tab to deselect the "Auto Scale" (The automatic adjustment scale of the ADS software is not used). Adjust the Y axis scale shown in Fig. 4.40, then the adjusted S-parameter results in Fig. 4.41 can be obtained.

(4) In addition, the type, color, thickness, and symbol of the trace can also be modified. Double-click the dB(S(1,1)) to open the [Trace Options] dialog box and modify the trace, as shown in Fig. 4.42 (Keep the trace color as default). Similarly, modify the traces of dB(S(2,1)) and dB(S(2,2)) shown

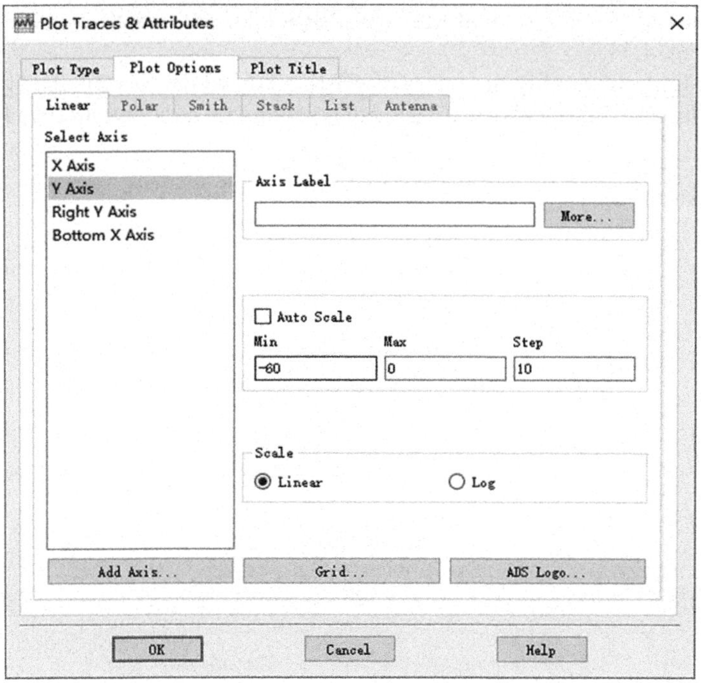

Fig. 4.40 Adjustment of the Y axis scale

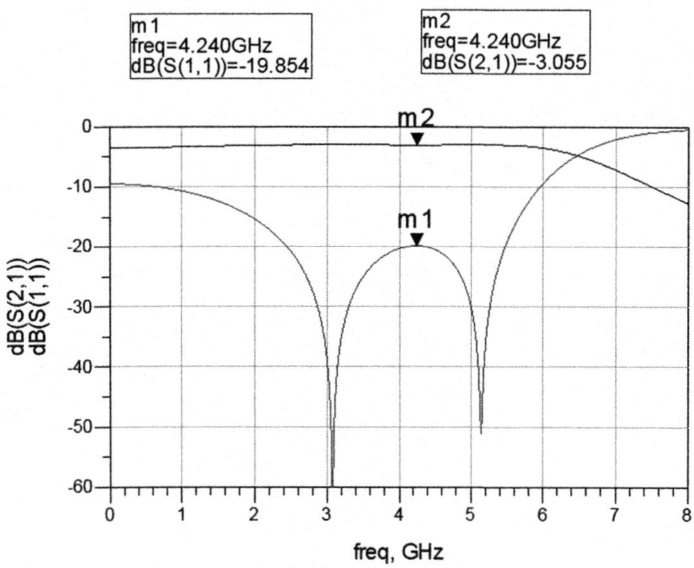

Fig. 4.41 The adjusted S-parameter results

in Fig. 4.43. The final traces of dB(S(1,1)) and dB(S(2,1)) are shown in Fig. 4.44.

(5) Similarly, modify the traces of dB(S(2,2)) and dB(S(3,2)). The final traces of dB(S(2,2)) and dB(S(3,2)) can be obtained, as shown in Fig. 4.45.

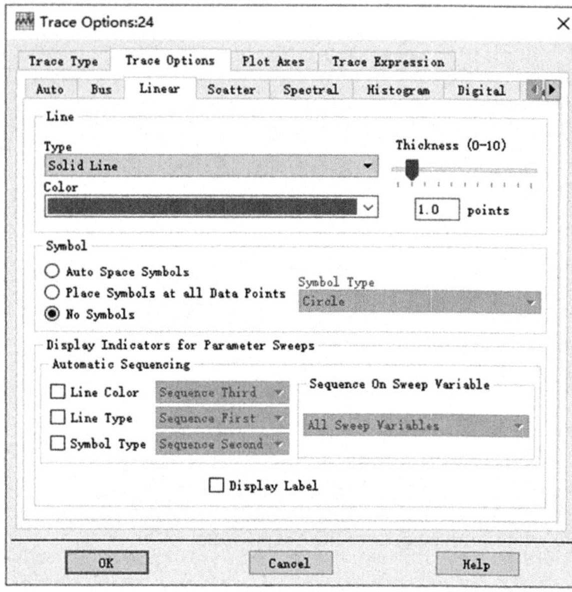

Fig. 4.42 Modification of the dB(S(1,1)) trace

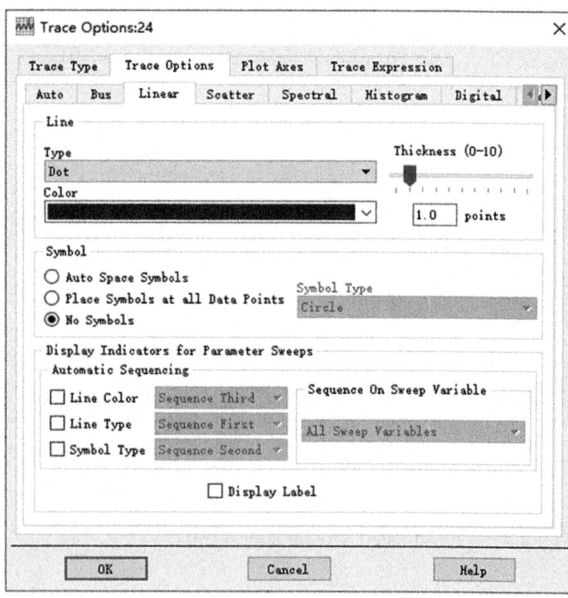

Fig. 4.43 Modification of the dB(S(2,1)) trace

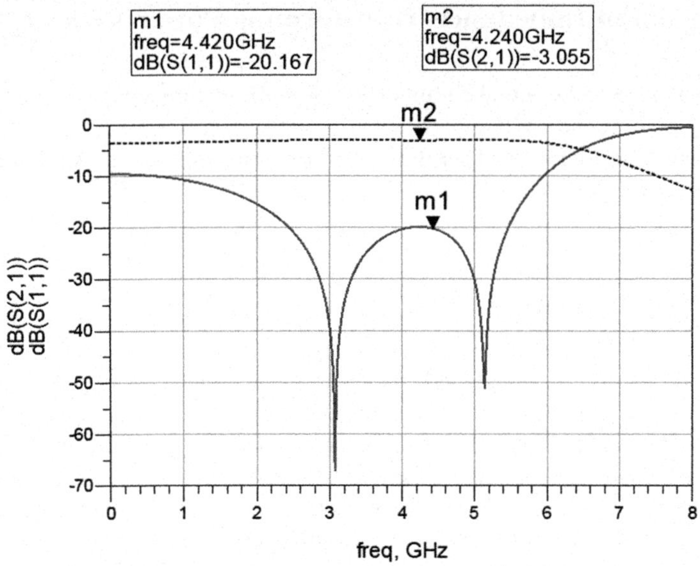

Fig. 4.44 The final traces of dB(S(1,1)) and dB(S(2,1))

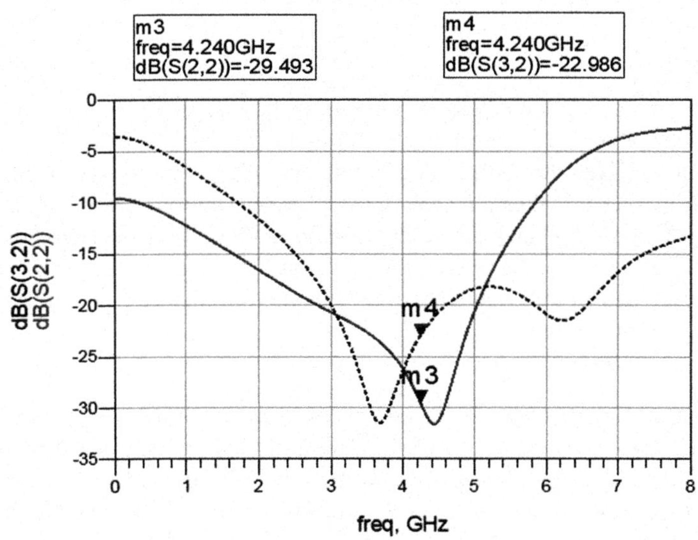

Fig. 4.45 The final dB(S(2,2)) and dB(S(3,2)) traces

4.3 Layout of Impedance-Transforming Power Divider

The performance of the actual circuit often has some differences from the theoretical results, so the influence of the interference, coupling, and some other factors should be taken into consideration. Therefore, it is necessary to use the ADS software for the layout simulation.

4.3.1 New Substrate File

All circuit elements are constructed on the gallium arsenide (GaAs) substrate with the thickness of 200 μm, the relative dielectric constant of 12.85, and the loss tangent of 0.006. The 75-nm-thick nickel chromium alloy (NiCr) layer with a block resistance of about 25 Ω/sq is used to realize the thin-film resistance, while the top and bottom 5-μm-thick metals and the 0.2-μm-thick intermediate Si_3N_4 layer construct the MIM capacitor. In addition, an air bridge can be built between the two copper layers to connect the circuits inside the spiral inductor and the peripheral circuits, which can make the layout design more flexible.

Before drawing the layout, the substrate file needs to be set according to the TFIPD technology.

1. New Layout

   Return to the "Impedance_Transforming_Power_Divider_wrk" main interface of the workspace. Execute menu commands [File] → [New] → [Layout...] to open the [New Layout] dialog box in Fig. 4.46. Modify the "Cell" name to "thin film resistor", then click the [Create Layout] button. After that, the [Choose Layout Technology] dialog box shown in Fig. 4.47 will pop up. Here, select "Standard ADS Layers, 0.001 micron layout resolution", which is 0.001 μm (note that the unit is μm in this chapter), click [Finish], and the new layout window will pop up, as shown in Fig. 4.48.
2. New Substrate

   (1) Create a new substrate. Execute menu commands [EM] → [Substrate...] in the layout window, then click [OK] in the dialog box which pops up. The [New Substrate] dialog box in Fig. 4.49 will be displayed, in which "File name" and "Template" can be modified. Keep the default file name here. Because the used TFIPD technology is based on the GaAs substrate, select "100μmGaAs" in the "Template" column and click the [Create Substrate] button. Then execute menu commands [View] → [View All] in the pop-up window to view all substrate items, as shown in Fig. 4.50.

   (2) Add conductors. Execute menu commands [Technology] → [Material Definitions...] to open the [Material Definitions] dialog box shown in Fig. 4.51. Select the [Conductors] tab and define the relevant conductors in Fig. 4.52. The specific steps are listed as follows: Click the [Add From Database...]

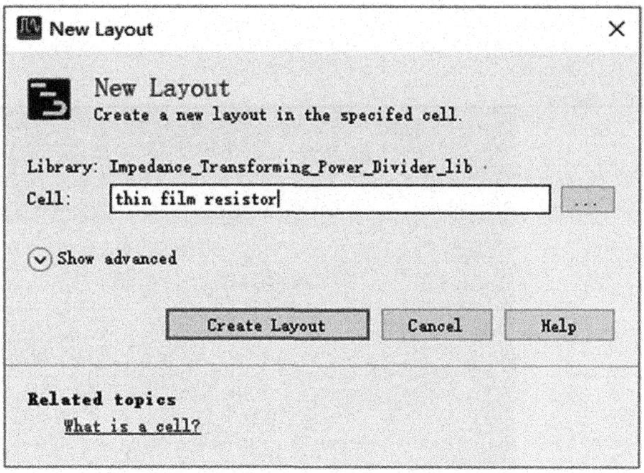

Fig. 4.46 [New Layout] dialog box

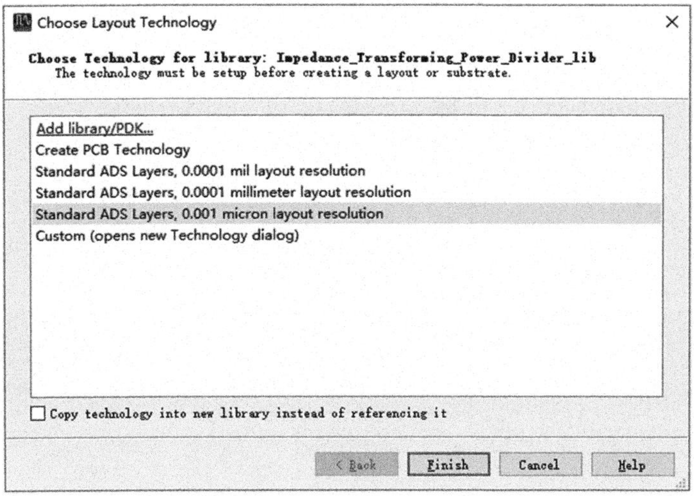

Fig. 4.47 [Choose Layout Technology] dialog box

button in the lower right corner of Fig. 4.52. If the conductor to be added exists in the pop-up [Add Materials From Database] dialog box shown in Fig. 4.53, select the conductor and click [OK] to add it. If the conductor to be added does not exist in the [Add Materials From Database] dialog box, return to the [Material Definitions] dialog box and click the [Add Conductor] button to add a conductor and modify its properties. In addition, you can remove unwanted conductors by clicking the [Remove Conductor] button in the [Material Definitions] dialog box. After finishing all the steps, click the [Apply] button.

(3) Add dielectrics. Select the [Dielectrics] tab and define the relevant dielectrics in Fig. 4.54. The specific steps are listed as follows: Click the [Add From

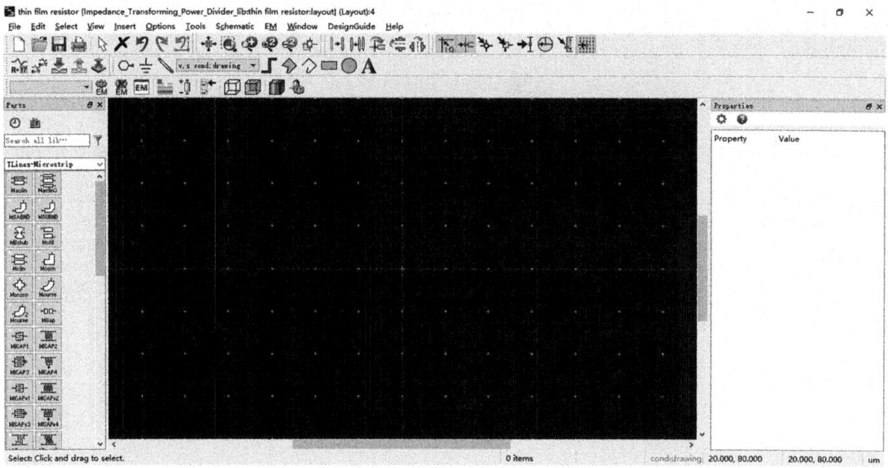

Fig. 4.48 New layout window

Fig. 4.49 [New Substrate] dialog box

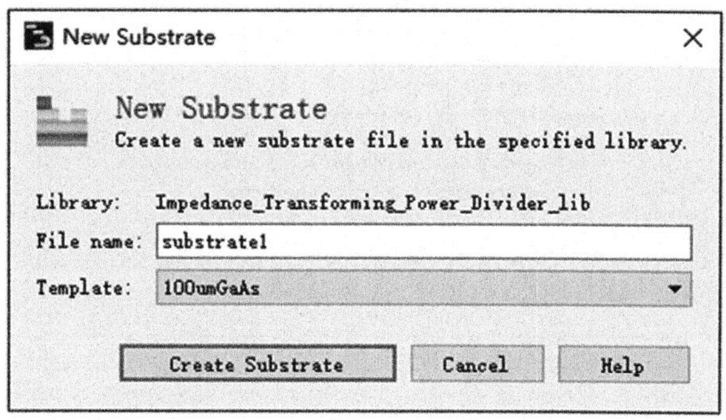

Database…] button in the lower right corner of Fig. 4.54. If the dielectric to be added exists in the pop-up [Add Materials From Database] dialog box shown in Fig. 4.55, select the dielectric and click [OK] to add it. If the dielectric to be added does not exist in the [Add Materials From Database] dialog box, return to the [Material Definitions] dialog box and click the [Add Dielectric] button to add a dielectric and modify its properties. In addition, you can remove unwanted dielectrics by clicking the [Remove Dielectric] button in the [Material Definitions] dialog box. After finishing all the steps, click the [OK] button to close the [Material Definitions] dialog box.

3. Edit Substrate Items

(1) Select the "cond" layer and click the right mouse button to pop up 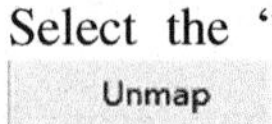

. Click the "Unmap" to delete this layer (Or select the "cond"

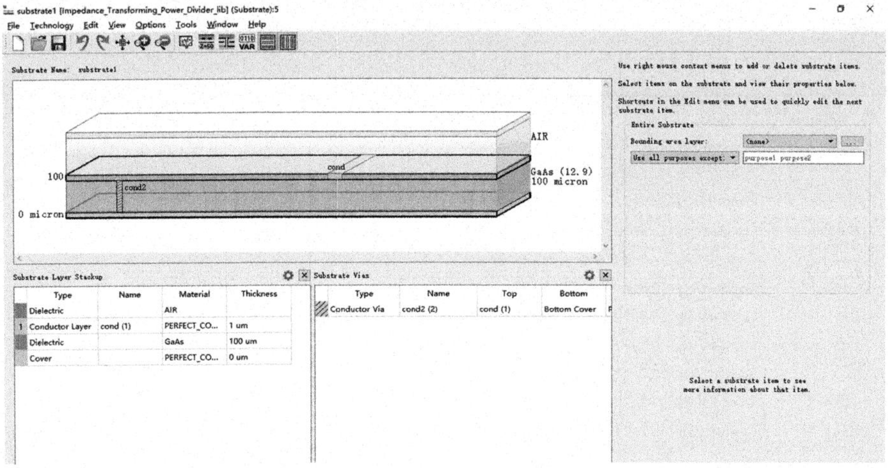

Fig. 4.50 View of all substrate items

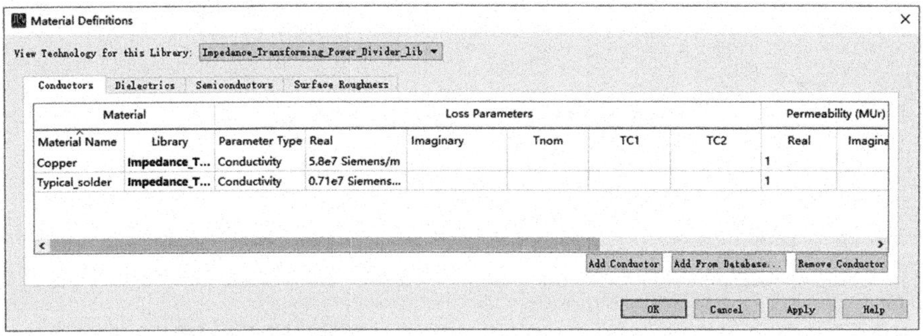

Fig. 4.51 [Material Definitions] dialog box

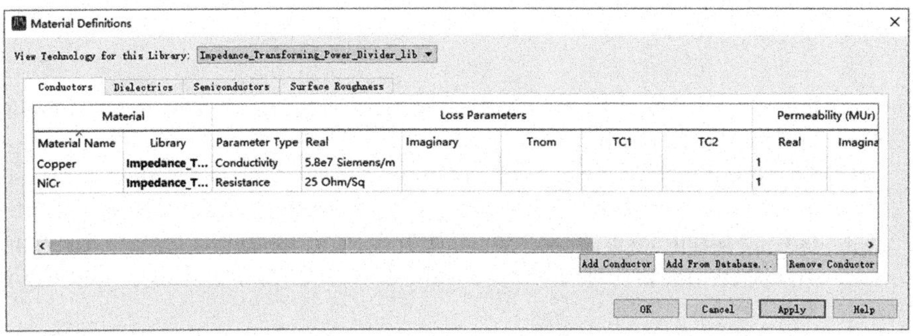

Fig. 4.52 Add of conductors and modification of their properties

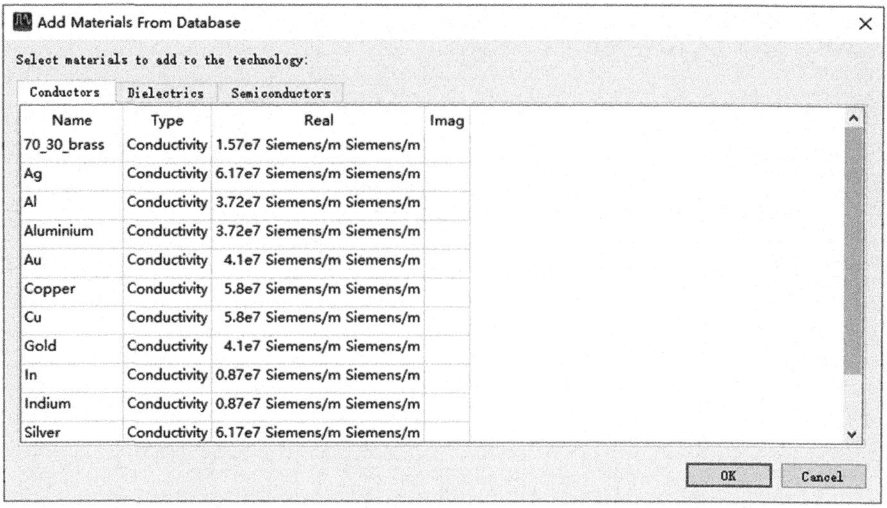

Fig. 4.53 [Conductors] tab of the [Add Materials From Database] dialog box

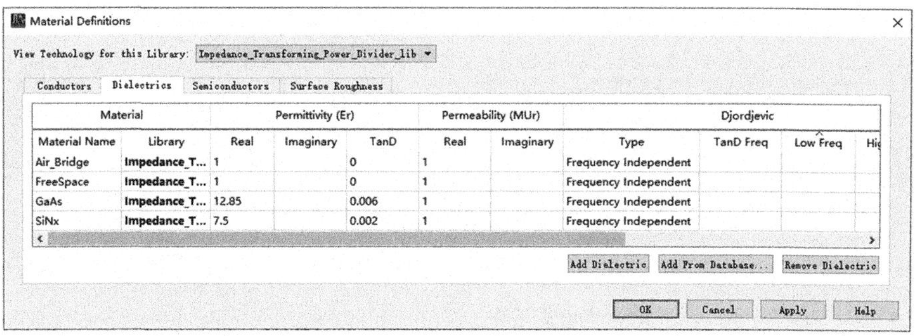

Fig. 4.54 Add dielectrics and modification of their properties

layer, then press "Delete" on the keyboard to delete it). Delete the "cond2" layer in the same way.

(2) Edit substrate layers. Select an existing substrate layer, right-click the mouse, and choose the [Insert Substrate Layer] from the pop-up menu to insert a new substrate layer. Select the substrate layer to be modified to view and modify its properties under the [Substrate Layer] at the right side of the window.

(3) Edit conductor layers. Select the surface of the substrate layer on which you want to insert a conductor layer, right-click the mouse, and choose the [Map Conductor Layer] from the pop-up menu to insert a new conductor layer. Select the conductor layer to be modified to view and modify its properties under the [Conductor Layer] at the right side of the window.

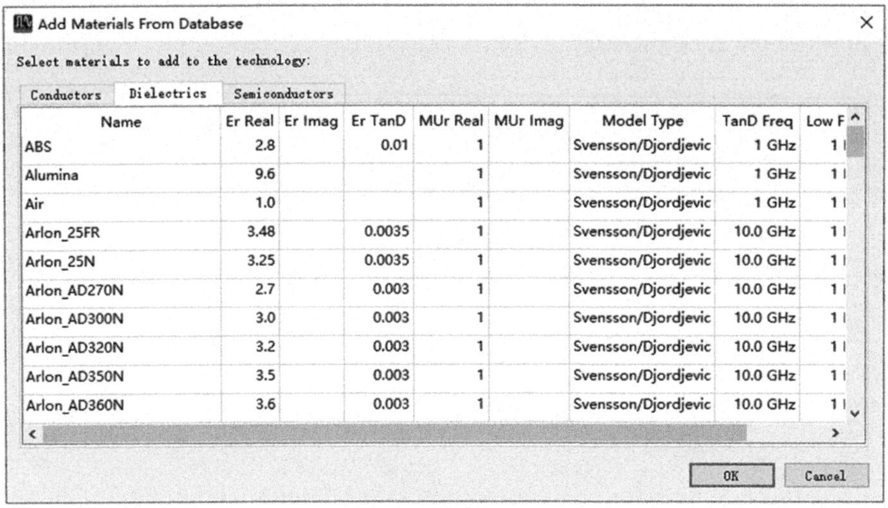

Add Materials From Database

Select materials to add to the technology:

Conductors | Dielectrics | Semiconductors

Name	Er Real	Er Imag	Er TanD	MUr Real	MUr Imag	Model Type	TanD Freq	Low F
ABS	2.8		0.01	1		Svensson/Djordjevic	1 GHz	1
Alumina	9.6			1		Svensson/Djordjevic	1 GHz	1
Air	1.0			1		Svensson/Djordjevic	1 GHz	1
Arlon_25FR	3.48		0.0035	1		Svensson/Djordjevic	10.0 GHz	1
Arlon_25N	3.25		0.0035	1		Svensson/Djordjevic	10.0 GHz	1
Arlon_AD270N	2.7		0.003	1		Svensson/Djordjevic	10.0 GHz	1
Arlon_AD300N	3.0		0.003	1		Svensson/Djordjevic	10.0 GHz	1
Arlon_AD320N	3.2		0.003	1		Svensson/Djordjevic	10.0 GHz	1
Arlon_AD350N	3.5		0.003	1		Svensson/Djordjevic	10.0 GHz	1
Arlon_AD360N	3.6		0.003	1		Svensson/Djordjevic	10.0 GHz	1

OK | Cancel

Fig. 4.55 [Dielectrics] tab of the [Add Materials From Database] dialog box

(4) Edit conductor vias. Select the substrate layer into which you want to insert a conductor via, right-click the mouse, and choose the [Map Conductor Via] from the pop-up menu to insert a new conductor via. Select the conductor via to be modified to view and modify its properties under the [Conductor Via] at the right side of the window.

(5) All substrate items are shown in Fig. 4.56. The bottom layer is "Cover", the thickness of the GaAs substrate layer is 200 μm, and the thickness of the first SiNx layer is 0.1 μm. For the diel layer, the "Process Role" is selected as "Conductor", the "Material" is selected as "NiCr", the "Operation" is selected as "Sheet", and the thickness is 75 nm. For the bond layer, the "Process Role" is selected as "Conductor", the "Material" is selected as "Copper", the "Operation" is selected as "Expand the substrate", the "Position" is selected as "Above interface", and the thickness is 5 μm. The thickness of the second SiNx layer is 0.2 μm. For the text layer, the "Process Role" is selected as "Conductor", the "Material" is selected as "Copper", the "Operation" is selected as "Expand the substrate", the "Position" is selected as "Above interface", and the thickness is 0.5 μm. The thickness of the Air_Bridge layer is 3 μm. For the leads layer, the "Process Role" is selected as "Conductor", the "Material" is selected as "Copper", the "Operation" is selected as "Intrude the substrate", the "Position" is selected as "Above interface", and the thickness is 5 μm. The top layer is the "FreeSpace". For the symbol layer, the "Process Role" is selected as "Conductor Via" and the "Material" is selected as "Copper". For the packages layer, the "Process Role" is selected as "Conductor Via" and the "Material" is selected as "Copper".

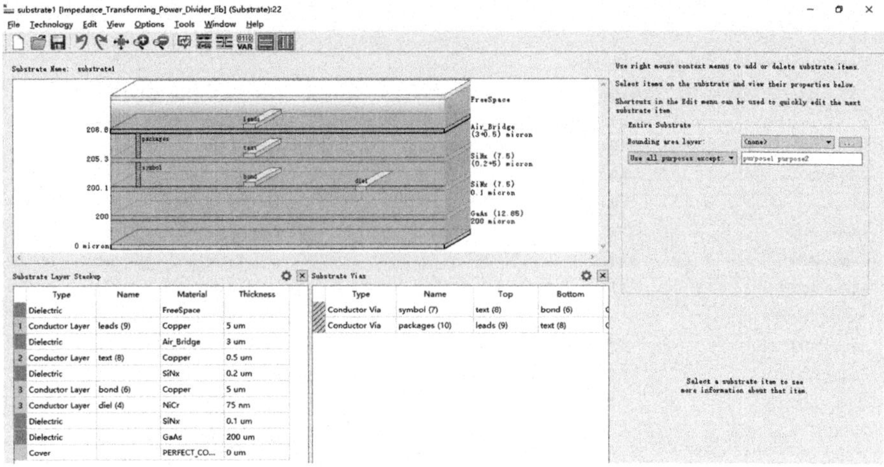

Fig. 4.56 Substrate layer stackup and substrate vias

(6) Moreover, execute menu commands [Technology] → [Layer Definitions…] to open the [Layer Definitions] dialog box shown in Fig. 4.57, where you can modify the display properties of each layer such as color and pattern, etc. Here, keep all the settings in default.

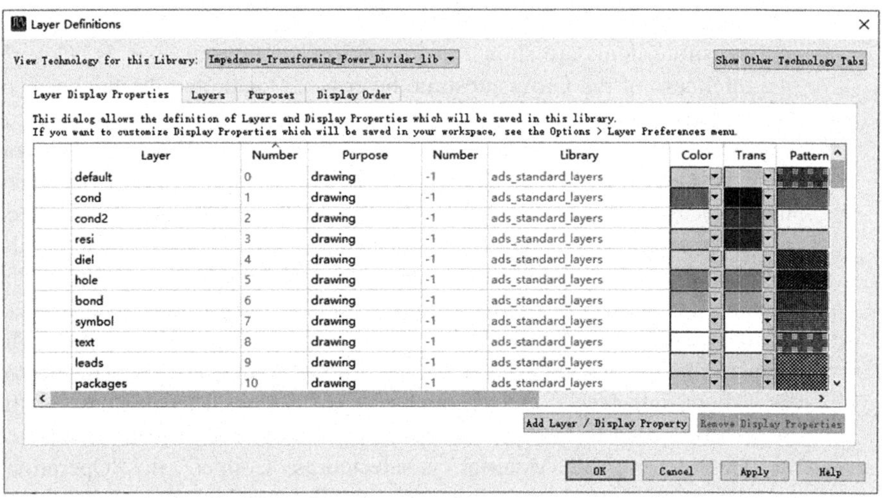

Fig. 4.57 [Layer Display Properties] tab of the [Layer Definitions] dialog box

4.3.2 *Layout of Thin-Film Resistor*

To facilitate the layout drawing, select the function key in the box of Fig. 4.58 in the “thin film resistor” cell. Execute menu commands [Options] → [Preferences…], select the [Grid/Snap] tab in the pop-up [Preferences for Layout] dialog box to modify the display grid in the layout. Set the “Snap Grid Distance (in layout units)”, “Snap Grid Per Minor Display Grid”, and “Minor Grid Per Major Display Grid” to appropriate values, as shown in Fig. 4.59. (Or right-click in the layout drawing area and select “<0.1-1-100>” under [Grid Spacing…]; Or use the shortcut key “Ctrl + Shift + 8”.)

1. Layout Drawing of Thin-Film Resistor

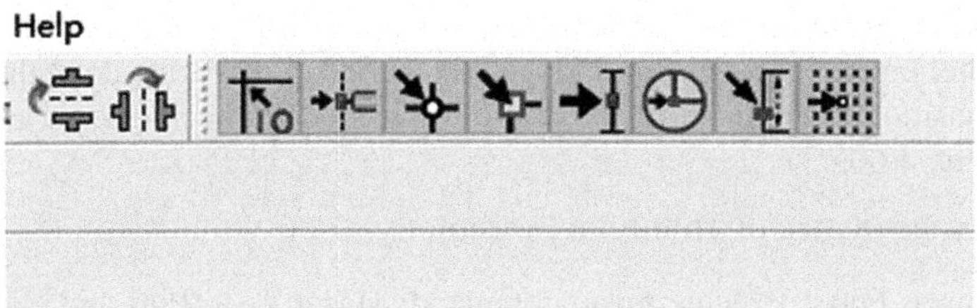

Fig. 4.58 Selection of the function key in the box

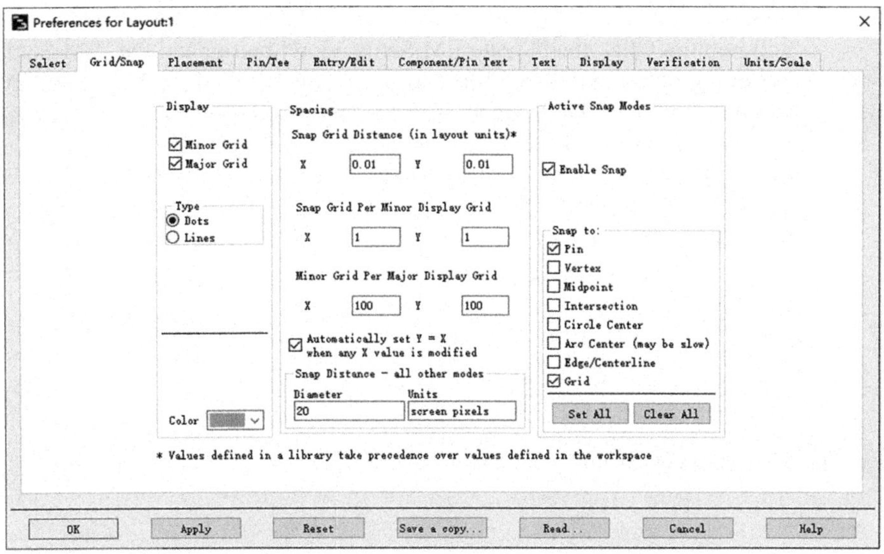

Fig. 4.59 Modification of the display grid in the layout

The thin-film resistance is realized by the 75-nm-thick NiCr layer with a block resistance of about 25 Ω/sq. The thin-film resistor with the resistance of 220 Ω is taken as an example to describe its drawing steps in detail below.

(1) Execute menu commands [Insert] → [Rectangle] to insert a rectangle into the layout, then press "Esc" to exit. Select the inserted rectangle, select "diel: drawing" in the drop-down menu of the [All Shapes] → [Layer] under [Properties] at the right side of the window, and set the [Rectangles] → [Width] to 108 μm, [Height] to 10 μm.
(2) Execute menu commands [Insert] → [Rectangle] to insert a rectangle into the layout, then press "Esc" to exit. Select the inserted rectangle, select "bond: drawing" in the drop-down menu of the [All Shapes] → [Layer] under [Properties] at the right side of the window, and set the [Rectangles] → [Width] to 40 μm, [Height] to 20 μm. Select and long press the modified rectangle to move it to the middle position of 15 μm wide overlap with the diel layer rectangle. Similarly, insert a bond layer rectangle at the opposite position. The layout of the thin-film resistor has been drawn, as shown in Fig. 4.60.

2. Layout Simulation of Thin-Film Resistor

(1) Insert ports. Execute menu commands [Insert] → [Pin] and click the left mouse button to insert two pins into the I/O ports of the thin-film resistor, as shown in Fig. 4.61.
(2) Modify simulation settings. Execute menu commands [EM] → [Simulation Settings…] and the [New EM Setup View] dialog box shown in Fig. 4.62 will pop up. Click the [Create EM Setup View] button and the [EM Setup for simulation] window will pop up, as shown in Fig. 4.63. Select the EM simulator at first. The second method "Momentum Microwave" is used generally, because this method runs fast and its high accuracy can also meet the application requirement (The first method "Momentum RF" is the fastest with the least accuracy. The third method "FEM" has the highest accuracy, but it shows the slowest simulation speed, so it is mainly used for some complex three-dimensional structure simulations). Select the [Frequency

Fig. 4.60 Final layout of a thin-film resistor

Fig. 4.61 Insertion of pins

Plan] tab to set the simulation frequency. Select "Adaptive" in the "Type" column, then set the "Fstart" to 0 GHz, the "Fstop" to 8 GHz, and the "Npts" to 50. Finally, select the [Options] tab. Click "Preprocessor", select the "User specified snap distance" option under the "Heal the layout", and set the distance to 2.5 μm. Click "Mesh" and select the "Edge Mesh" option. Keep other settings in default. After finishing all the steps, close the window and click the [OK] button to save the settings.

(3) Layout simulation. Execute menu commands [EM] → [Simulate] to simulate the layout of the thin-film resistor. During the simulation, a status window will pop up to display the current item's progress. After the finishing of the simulation, the data display window will be opened automatically. Refer to the steps in Sect. 4.2.2 to view and handle traces dB(S(1,1)) and dB(S(2,1)), the final results are shown in Fig. 4.64.

3. Co-Simulation of Thin-Film Resistor

To verify whether the resistance of the drawn thin-film resistor is 220 Ω, the co-simulation of the resistor schematic and layout is required.

(1) Create EM model and symbol of the thin-film resistor. In the layout window, execute menu commands [EM] → [Component] → [Create EM Model And Symbol…], then click the [OK] button in the pop-up dialog boxes. Execute menu commands [Edit] → [Component] → [Update Component Definitions…], then click the [OK] button in the pop-up dialog box to complete the creation of the EM model and the symbol in the thin-film resistor.

(2) Create a new schematic and insert the thin-film resistor component. Return to the "Impedance_Transforming_Power_Divider_wrk" main interface of the

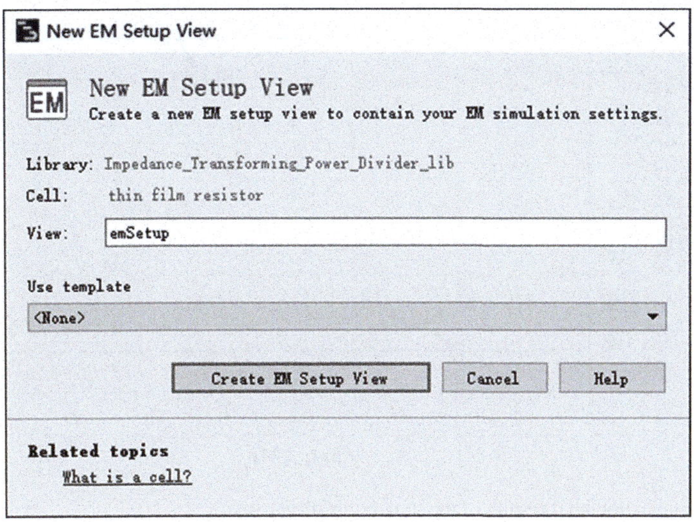

Fig. 4.62 [New EM Setup View] dialog box

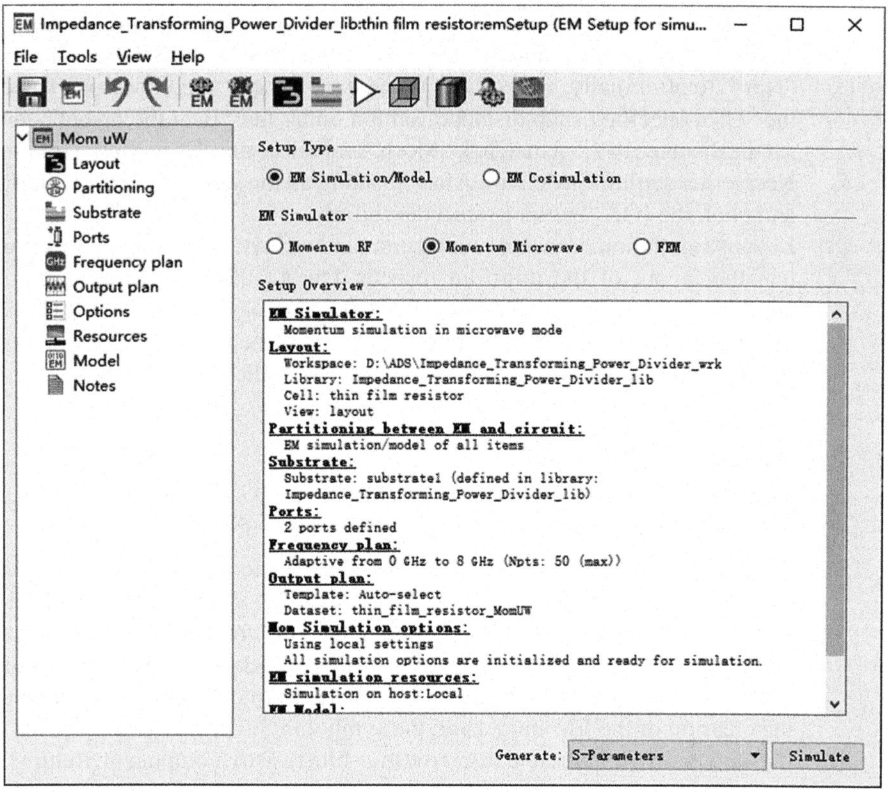

Fig. 4.63 [EM Setup for simulation] window

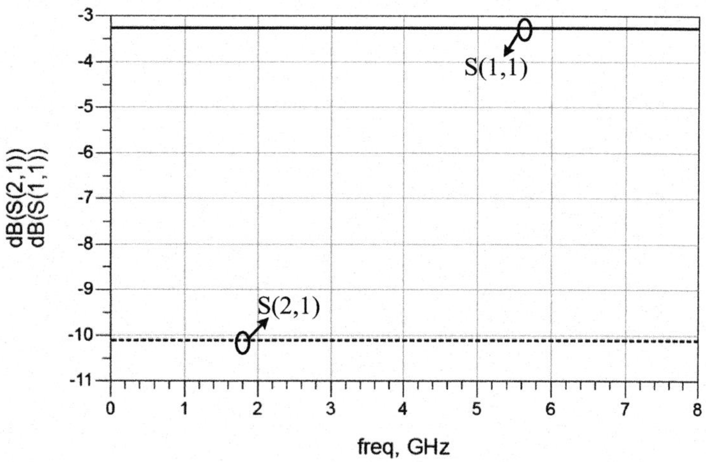

Fig. 4.64 Final simulated results of the drawn thin-film resistor

workspace. Execute menu commands [File] → [New] → [Schematic…], modify the "Cell" name to "thin film resistor-cosimulation" in the opened [New Schematic] dialog box, and click the [Create Schematic] button to create a new schematic. Click the [Open the Library Browser] icon at the left side of the schematic window, then select the "thin film resistor" component under the [Workspace Libraries] in the displayed [Component Library] window, as shown in Fig. 4.65. Click the right mouse button, then click the [Place Component] in the pop-up menu to insert a thin-film resistor component into the schematic. Press "Esc" to exit.

(3) Insert ideal resistor. Select the [Lumped-Components] from the drop-down menu in the left component palette list and click the [Resistor] icon to insert a resistor. Press "Esc" on the keyboard to exit. Double-click the resistor and modify R = 220 Ω in the [Edit Instance Parameter] pop-up dialog box (check whether the unit setting is consistent), then click the [OK] button to save the parameter.

(4) Insert the S-parameter simulator, term ports, and grounds. Select the [Simulation-S_Param] from the drop-down menu in the left component palette list and click the [S-parameter Simulator] icon to insert an S-parameter simulator. Click the [Port Impedance Termination for S-Parameters] icon to insert four term ports, then press "Esc" to exit. Execute menu commands [Insert] → [GROUND] to place four grounds (Or click the [TermG Port Impedance Termination for S-Parameters with Grounded Reference] icon to insert four term ports with the grounded reference). Execute menu commands [Insert] → [Wire] to connect the inserted components. After finishing all the steps, press "Esc" to exit.

(5) Set simulation frequency. Double-click the S-parameter simulator S-PARAMETERS. Set the "Start" to 0 GHz, the "Stop" to 8 GHz, and the "Step-size" to 0.01 GHz, and click the [OK] button. The final resistor co-simulation circuit is obtained, as shown in Fig. 4.66.

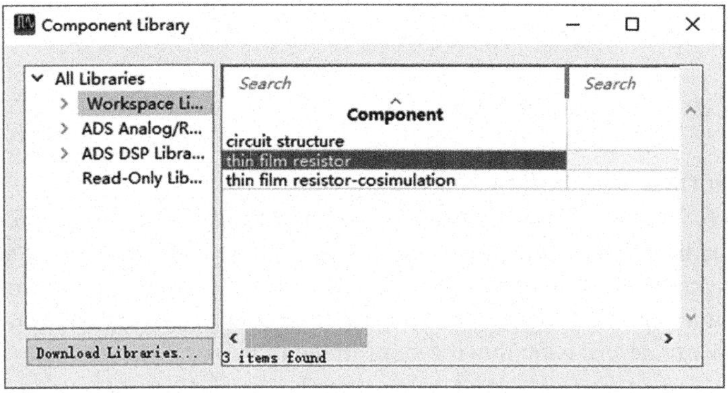

Fig. 4.65 Thin-film resistor component in the [Component Library] window

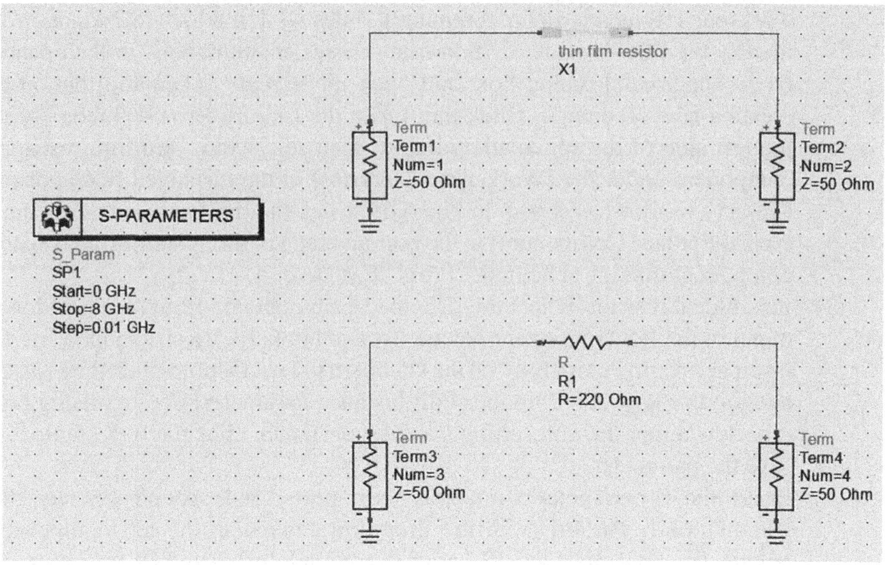

Fig. 4.66 Resistor co-simulation circuit

(6) Co-simulation. Execute menu commands [Simulate] → [Simulate]. After the simulation completes, the data display window will be opened automatically. Refer to the steps in Sect. 4.2.2 to view and handle traces of dB(S(1,1)), dB(S(2,1)), dB(S(3,3)), and dB(S(4,3)), the final results are shown in Fig. 4.67. It can be seen that traces dB(S(1,1)) and dB(S(3,3)), traces dB(S(2,1)) and dB(S(4,3)) are almost coincident, indicating that the capacitance of the drawn thin-film resistor is approximately equal to 220 Ω. If the differences between these traces are large, go back to modify the thin-film resistor layout and repeat the steps above until the error of two traces is in the range of acceptance (within an acceptable range).

4.3.3 Layout of MIM Capacitor

1. Layout Drawing of MIM Capacitor

Two 5-μm-thick copper layers (bond and leads layers) and a 0.2-μm-thick intermediate Si_3N_4 layer are used to construct the MIM capacitors. The area and the thickness of the intermediate dielectric layer determine the value of the MIM capacitor. The MIM capacitor with the capacitance of 0.35 pF is taken as an example to describe its drawing steps in detail below.

(1) Create a new layout. In the "Impedance_Transforming_Power_Divider_wrk" main interface of the workspace, execute menu commands [File] → [New] → [Layout...], and modify the "Cell" name to "MIM capacitor" in the

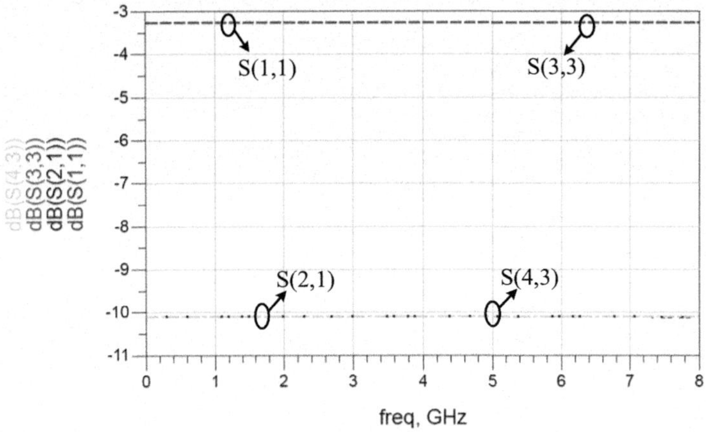

Fig. 4.67 S-parameter results of the resistor co-simulation

opened [New Layout] dialog box. Click the [Create Layout] button, then a new layout window will pop up.

(2) MIM capacitor lamination. Execute menu commands [Insert] → [Rectangle] to insert a rectangle into the layout, then press "Esc" to exit. Select the inserted rectangle, select "bond: drawing" in the drop-down menu of the [All Shapes] → [Layer] under [Properties] at the right side of the window, and set the [Rectangles] → [Width] to 43 μm, [Height] to 33 μm. Execute menu commands [Edit] → [Copy/Paste] → [Copy To Layer...], select [text: drawing] in the pop-up [Copy to Layer] dialog box, as shown in Fig. 4.68. Click the [Apply] button to copy a text layer in the original position. Similarly, copy a leads layer and a packages layer in the original position, respectively. After copying all the layers, click the [Cancel] button to close this dialog box.

Fig. 4.68 [Copy to Layer] dialog box

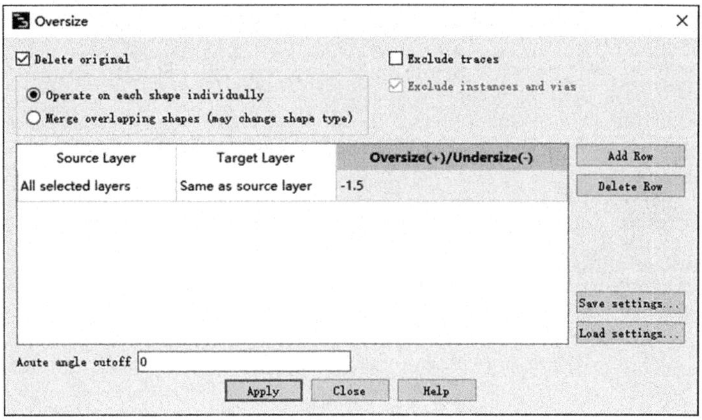

Fig. 4.69 [Oversize] dialog box

(3) The indentation of layers. There are different indentations between the layers of the MIM capacitor. Select the leads layer and execute menu commands [Edit] → [Scale/Oversize] → [Oversize…]. Since the leads layer is indented by 1.5 μm compared with the bond layer, enter − 1.5 in the "Oversize(+)/Undersize(−)" column of the pop-up [Oversize] dialog box and click the [Apply] button, as shown in Fig. 4.69. Similarly, indent the text and packages layers by 3 μm relative to the bond layer.

(4) Draw connection parts. Execute menu commands [Insert] → [Rectangle] to insert a rectangle in the layout, then press "Esc" to exit. Select the inserted rectangle, select "bond: drawing" in the drop-down menu of the [All Shapes] → [Layer] under [Properties] at the right side of the window, and set the [Rectangles] → [Width] to 40 μm, [Height] to 20 μm. Select and long press the modified rectangle by the left mouse button to move it to the middle position of the MIM capacitor and connect it with the original bond layer. Similarly, a leads layer rectangle is inserted in the opposite position and is connected to the original leads layer. The layout of a MIM capacitor has been drawn, as shown in Fig. 4.70.

2. Layout Simulation of MIM Capacitor

Fig. 4.70 Final layout of a MIM capacitor

Fig. 4.71 Insertion of pins

(1) Insert ports. Execute menu commands [Insert] → [Pin] and click the left mouse button to insert two pins into the I/O ports of the MIM capacitor, as shown in Fig. 4.71.
(2) Modify simulation settings. Execute menu commands [EM] → [Simulation Settings…], click the [Create EM Setup View] button in the pop-up the [New EM Setup View] dialog box, and the [EM Setup for simulation] window will pop up. Select the second method "Momentum Microwave" of the EM simulator at first. Then select the [Frequency Plan] tab to set the simulation frequency. Select "Adaptive" in the "Type" column and set the "Fstart" to 0 GHz, the "Fstop" to 8 GHz, and the "Npts" to 50. Finally, select the [Options] tab. Click "Preprocessor", select the "User specified snap distance" option under the "Heal the layout", and set the distance to 2.5 μm. Click "Mesh" and select the "Edge Mesh" option. Keep other settings in default. After finishing all the steps, close the window and click the [OK] button to save the settings.
(3) Layout simulation. Execute menu commands [EM] → [Simulate] to simulate the layout of the MIM capacitor. During the simulation, a status window will pop up to display the current item's progress. After the finishing of the simulation, the data display window will be opened automatically. Refer to the steps in Sect. 4.2.2 to view and handle traces dB(S(1,1)) and dB(S(2,1)), the final results are shown in Fig. 4.72.

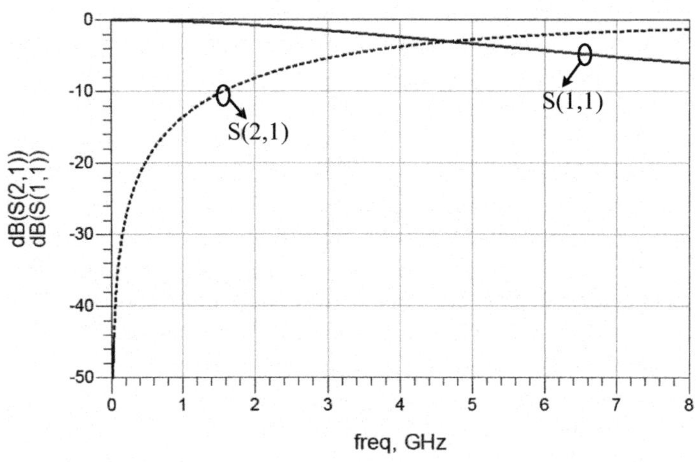

Fig. 4.72 Final simulated results of the drawn MIM capacitor

3. Co-Simulation of MIM Capacitor

To verify whether the capacitance of the drawn MIM capacitor is 0.35 pF, the co-simulation of capacitor schematic and layout is required.

(1) Create EM model and symbol of the MIM capacitor. In the layout window, execute menu commands [EM] → [Component] → [Create EM Model And Symbol…], then click the [OK] button in the pop-up dialog boxes. Execute menu commands [Edit] → [Component] → [Update Component Definitions…], then click the [OK] button in the pop-up dialog box to complete the creation of the EM model and the symbol in the MIM capacitor.

(2) Create a new schematic and insert the MIM capacitor component. Return to the "Impedance_Transforming_Power_Divider_wrk" main interface of the workspace. Execute menu commands [File] → [New] → [Schematic…], modify the "Cell" name to "MIM capacitor-cosimulation" in the opened [New Schematic] dialog box, and click the [Create Schematic] button to create a new schematic. Click the [Open the Library Browser] icon at the left side of the schematic window, then select the "MIM capacitor" component under the [Workspace Libraries] in the displayed [Component Library] window. Click the right mouse button, then click the [Place Component] in the pop-up menu to insert a MIM capacitor component in the schematic. Press "Esc" to exit.

(3) Insert ideal capacitor. Select the [Lumped-Components] from the drop-down menu in the left component palette list and click the [Capacitor] icon to insert a capacitor. Press "Esc" on the keyboard to exit. Double-click the capacitor and modify C = 0.35 pF in the [Edit Instance Parameter] pop-up dialog box (check whether the unit setting is consistent), then click the [OK] button to save the parameter.

(4) Insert the S-parameter simulator, term ports, and grounds. Select the [Simulation-S_Param] from the drop-down menu in the left component palette list and click the [S-parameter Simulator] icon to insert an S-parameter simulator. Click the [Port Impedance Termination for S-Parameters] icon to insert four term ports, then press "Esc" to exit. Execute menu commands [Insert] → [GROUND] to place four grounds (Or click the [TermG Port Impedance Termination for S-Parameters with Grounded Reference] icon to insert four term ports with the grounded reference). Execute menu commands [Insert] → [Wire] to connect the inserted components. After finishing all the steps, press "Esc" to exit.

(5) Set simulation frequency. Double-click the S-parameter simulator S-PARAMETERS. Set the "Start" to 0 GHz, the "Stop" to 8 GHz, and the "Step-size" to 0.01 GHz, and click the [OK] button. The final capacitor co-simulation circuit is obtained, as shown in Fig. 4.73.

(6) Co-simulation. Execute menu commands [Simulate] → [Simulate]. After the simulation completes, the data display window will be opened automatically. Refer to the steps in Sect. 4.2.2 to view and handle traces of dB(S(1,1)), dB(S(2,1)), dB(S(3,3)), and dB(S(4,3)), the final results are

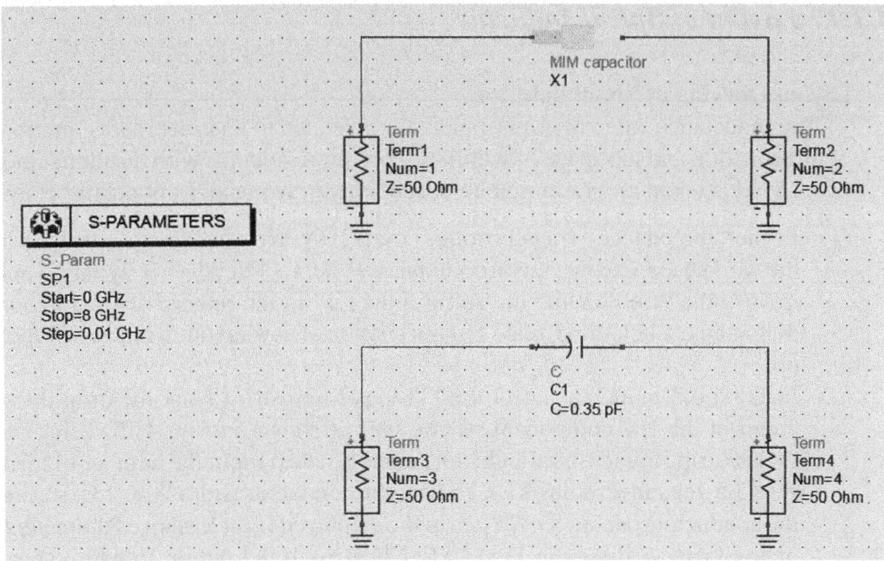

Fig. 4.73 Capacitor co-simulation circuit

shown in Fig. 4.74. It can be seen that traces dB(S(1,1)) and dB(S(3,3)), traces dB(S(2,1)) and dB(S(4,3)) are almost coincident, indicating that the capacitance of the drawn MIM capacitor is approximately equal to 0.35 pF. If the differences between these traces are large, go back to modify the MIM capacitor layout and repeat the steps above until the error of two traces is in the range of acceptance (within an acceptable range).

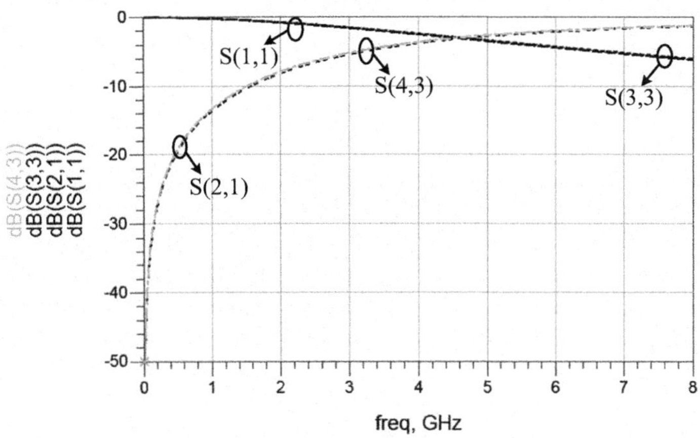

Fig. 4.74 S-parameter results of the capacitor co-simulation

4.3.4 Layout of Spiral Inductor

1. Layout Drawing of Spiral Inductor

   The inductance value of a spiral inductor is related to its inner radius, number of turns, width, and the space of winding. The spiral inductor with the inductance of 3.25 nH is taken as an example to describe its drawing steps in detail below.

   (1) In the "Impedance_Transforming_Power_Divider_wrk" main interface of the workspace, execute menu commands [File] → [New] → [Layout...] and modify the "Cell" name to "spiral inductor" in the opened [New Layout] dialog box. Click the [Create Layout] button, a new layout window will pop up.

   (2) Insert spiral inductor. Select the [TLines-Microstrip] from the drop-down menu in the left component palette list, as shown in Fig. 4.75. Click the [Microstrip Round Spiral Inductor] icon , then input the number of turns N = 3.5, the inner radius Ri = 75 μm, the conductor width W = 15 μm, and the conductor spacing S = 15 μm in the displayed [Edit Instance Parameters] dialog box, as shown in Fig. 4.76. Click the [OK] button to add a spiral inductor in the drawing area and press "Esc" on the keyboard to exit.

   (3) Spiral inductor layer modification. Select the inserted spiral inductor, execute menu commands [Edit] → [Component] → [Flatten...], and click the [OK] button in the pop-up dialog box. Execute menu commands [Edit] → [Merge] → [Union] to unite the spiral inductor layer. Then select "bond: drawing" in the drop-down menu of the [All Shapes] → [Layer] under [Properties] at the right side of the window.

   (4) Spiral inductor lamination. Execute menu commands [Edit] → [Copy/Paste] → [Copy To Layer...], select [text: drawing] in the pop-up [Copy to Layer] dialog box. Click the [Apply] button to copy a text layer in the original position. Similarly, copy a leads layer, a symbol layer, and a packages layer in the original position, respectively. After copying all the layers, click the [Cancel] button to close this dialog box.

   (5) Air bridge construction. In order to make the circuits inside spiral inductors connect with the peripheral circuits, air bridges are constructed. Execute menu commands [Insert] → [Rectangle] to insert a rectangle in the layout, then press "Esc" to exit. Select the inserted rectangle, select "cond: drawing" in the drop-down menu of the [All Shapes] → [Layer] under [Properties] at the right side of the window, and set the [Rectangles] → [Width] to an arbitrary value, [Height] to 40 μm. Select and long press the modified rectangle with the left mouse button to move it to the air bridge construction position, as shown in Fig. 4.77. Press the shortcut "Ctrl + A" to select all layers. Execute menu commands [Edit] → [Boolean Logical...] in the pop-up [Boolean Logical Operation Between Layers] dialog box to modify the Boolean logical operation between the bond layer and the cond layer according to Fig. 4.78, then click the [Apply] button to complete the operation. Similarly, complete the Boolean operations between the text layer,

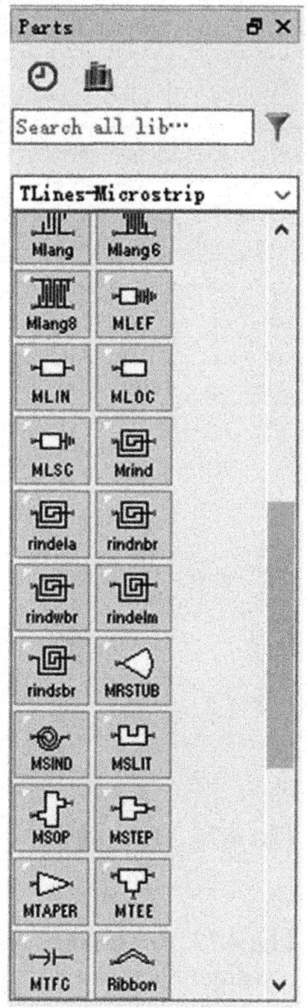

Fig. 4.75 [TLines-Microstrip] list

symbol layer, packages layer, and the cond layer, only retain the top leads metal layer, and click the [Cancel] button to close the dialog box. Then select the cond layer and press "Delete" on the keyboard to delete it. The spiral inductor after the construction of an air bridge is shown in Fig. 4.79.

(6) The indentation of layers. There are different indentations between the layers of spiral inductor. Because the layer has been disconnected after the Boolean logical operation, execute menu commands [View] → [Layer View] → [By Name…]. Then, in the [Layout Layers] dialog box shown in Fig. 4.80, select [text: drawing] to only display the text layer of the spiral inductor. Press the shortcut "Ctrl + A" to select the text layer and execute menu commands [Edit] → [Scale/Oversize] → [Oversize…] to open the dialog box. Since the text layer is indented by 2 μm compared with the bond layer, enter -2 in the "Oversize(+)/Undersize(−)" column of the pop-up [Oversize] dialog box and click the [Apply] button. Similarly, indent the symbol and packages layers by 2 μm relative to the bond layer.

(7) Draw connection parts. Execute menu commands [Insert] → [Rectangle] to insert a rectangle in the layout, then press "Esc" to exit. Select the inserted rectangle, select "bond: drawing" in the drop-down menu of the [All Shapes]

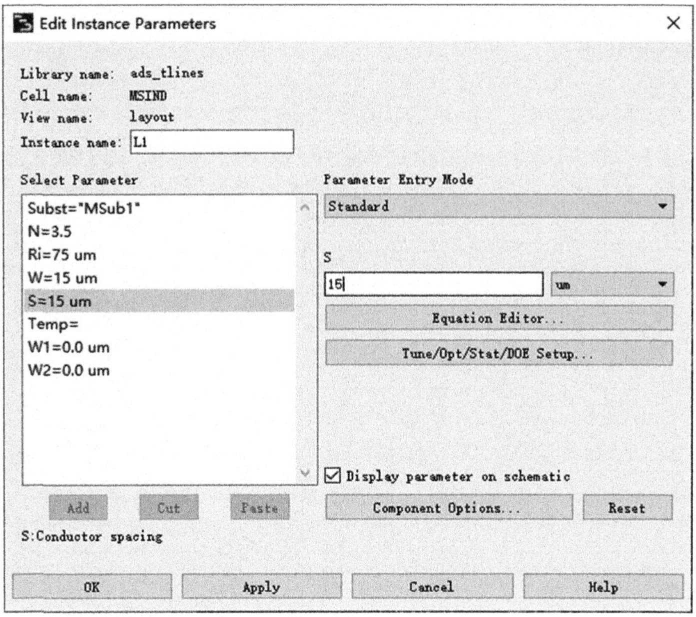

Fig. 4.76 [Edit Instance Parameters] dialog box

Fig. 4.77 Air bridge construction position

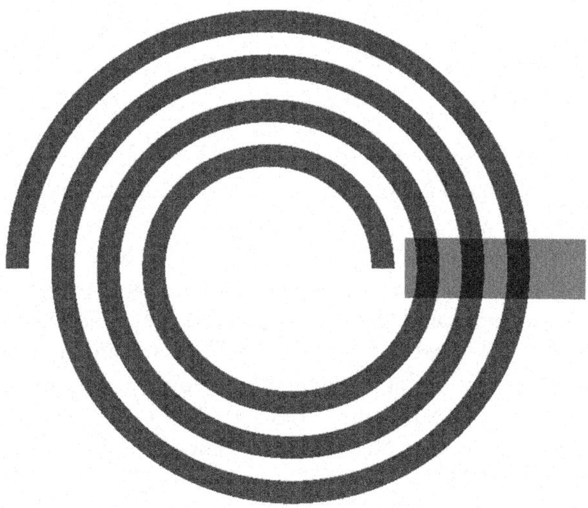

→ [Layer] under [Properties] at the right side of the window, and set the [Rectangles] → [Height] to 15 μm. [Width] can be set according to the specific size of the air bridge. Select and long press the modified rectangle by the left mouse button to move it to the middle position of the air bridge

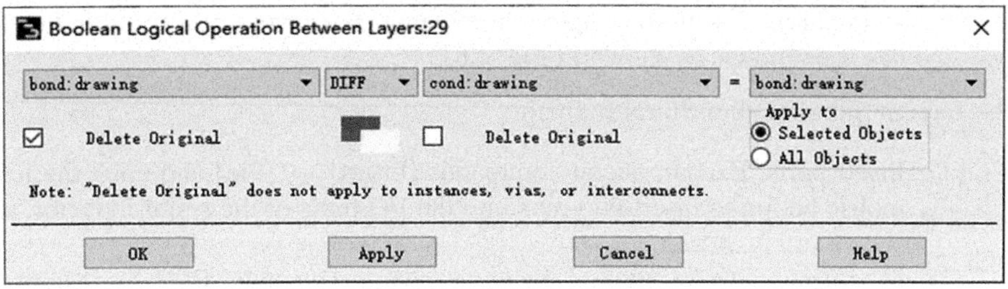

Fig. 4.78 [Boolean Logical Operation Between Layers] dialog box

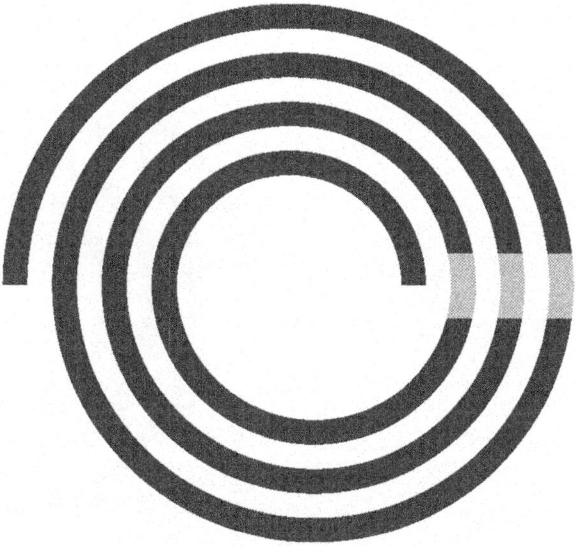

Fig. 4.79 Spiral inductor after the construction of an air bridge

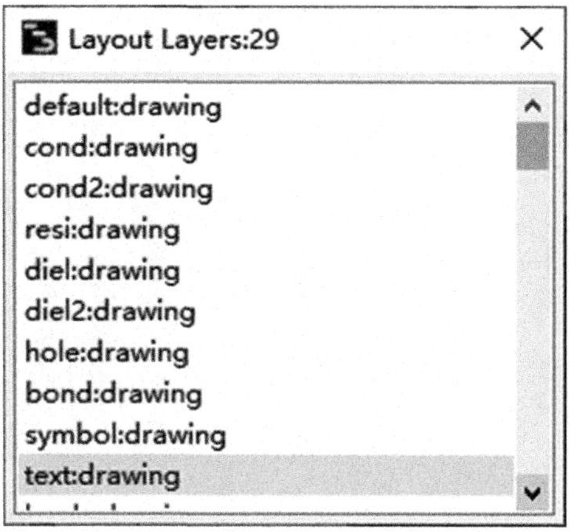

Fig. 4.80 [Layout Layers] dialog box

and connect it with the original bond layer. The layout of a spiral inductor has been drawn, as shown in Fig. 4.81.

2. Layout Simulation of Spiral Inductor

   (1) Insert ports. Execute menu commands [Insert] → [Pin] and click the left mouse button to insert two pins into the I/O ports of the spiral inductor, as shown in Fig. 4.82.
   (2) Modify simulation settings. Execute menu commands [EM] → [Simulation Settings…], click the [Create EM Setup View] button in the pop-up the [New EM Setup View] dialog box, and the [EM Setup for simulation]

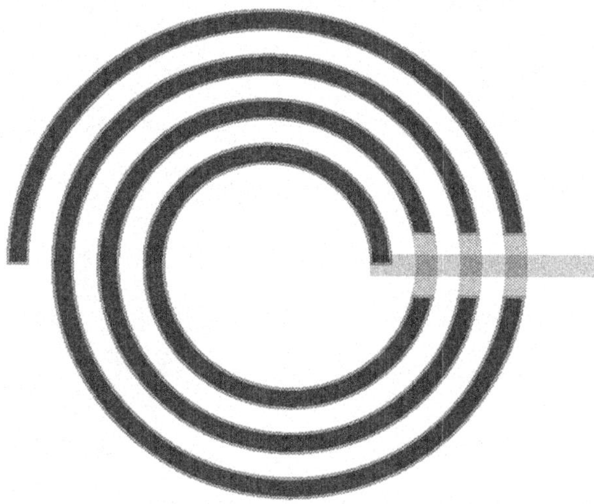

Fig. 4.81 Final layout of a spiral inductor

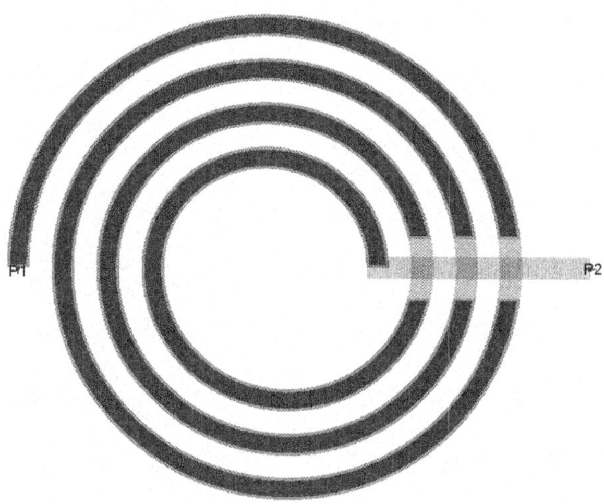

Fig. 4.82 Insertion of pins

window will pop up. Select the second method "Momentum Microwave" of the EM simulator at first. Then select the [Frequency Plan] tab to set the simulation frequency. Select "Adaptive" in the "Type" column and set the "Fstart" to 0 GHz, the "Fstop" to 8 GHz, and the "Npts" to 50. Finally, select the [Options] tab. Click "Preprocessor", select the "User specified snap distance" option under the "Heal the layout", and set the distance to 2.5 μm. Click "Mesh" and select the "Edge Mesh" option. Keep other settings in default. After finishing all the steps, close the window and click the [OK] button to save the settings.

(3) Layout simulation. Execute menu commands [EM] → [Simulate] to simulate the layout of the spiral inductor. During the simulation, a status window will pop up to display the current item's progress. After the finishing of the simulation, the data display window will be opened automatically. Refer to the steps in Sect. 4.2.2 to view and handle traces dB(S(1,1)) and dB(S(2,1)), the final results are shown in Fig. 4.83.

3. Co-simulation of Spiral Inductor

To verify whether the inductance of the drawn spiral inductor is 3.25 nH, the co-simulation of inductor schematic and layout is required.

(1) Create EM model and symbol of the spiral inductor. In the layout window, execute menu commands [EM] → [Component] → [Create EM Model And Symbol…], then click the [OK] button in the pop-up dialog boxes. Execute menu commands [Edit] → [Component] → [Update Component Definitions…], then click the [OK] button in the pop-up dialog box to complete the creation of the EM model and the symbol in the spiral inductor.

(2) Create a new schematic and insert the spiral inductor component. Return to the "Impedance_Transforming_Power_Divider_wrk" main interface of the

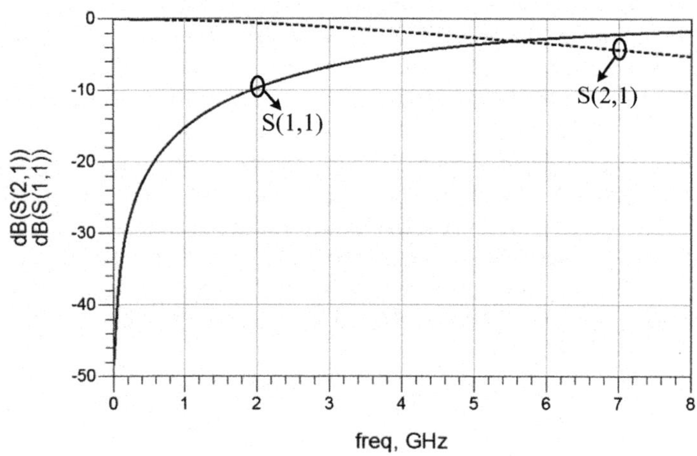

Fig. 4.83 Final simulated results of the drawn spiral inductor

workspace. Execute menu commands [File] → [New] → [Schematic…], modify the "Cell" name to "spiral inductor-cosimulation" in the opened [New Schematic] dialog box, and click the [Create Schematic] button to create a new schematic. Click the [Open the Library Browser] icon at the left side of the schematic window, then select the "spiral inductor" component under the [Workspace Libraries] in the displayed [Component Library] window. Click the right mouse button, then click the [Place Component] in the pop-up menu to insert a spiral inductor component into the schematic. Press "Esc" to exit.

(3) Insert ideal inductor. Select the [Lumped-Components] from the drop-down menu in the left component palette list and click the [Inductor] icon to insert an inductor. Press "Esc" on the keyboard to exit. Double-click the inductor and modify L = 3.25 nH in the [Edit Instance Parameter] pop-up dialog box (check whether the unit setting is consistent), then click the [OK] button to save the parameter.

(4) Insert the S-parameter simulator, term ports, and grounds. Select the [Simulation-S_Param] from the drop-down menu in the left component palette list and click the [S-parameter Simulator] icon to insert an S-parameter simulator. Click the [Port Impedance Termination for S-Parameters] icon to insert four term ports, then press "Esc" to exit. Execute menu commands [Insert] → [GROUND] to place four grounds (Or click the [TermG Port Impedance Termination for S-Parameters with Grounded Reference] icon to insert four term ports with the grounded reference). Execute menu commands [Insert] → [Wire] to connect the inserted components. After finishing all the steps, press "Esc" to exit.

(5) Set simulation frequency. Double-click the S-parameter simulator S-PARAMETERS. Set the "Start" to 0 GHz, the "Stop" to 8 GHz, and the "Step-size" to 0.01 GHz, and click the [OK] button. The final inductor co-simulation circuit is obtained, as shown in Fig. 4.84.

(6) Co-simulation. Execute menu commands [Simulate] → [Simulate]. After the simulation completes, the data display window will be opened automatically. Refer to the steps in Sect. 4.2.2 to view and handle traces of dB(S(1,1)), dB(S(2,1)), dB(S(3,3)), and dB(S(4,3)), the final results are shown in Fig. 4.85. It can be seen that traces dB(S(1,1)) and dB(S(3,3)), traces dB(S(2,1)) and dB(S(4,3)) are almost coincident, indicating that the inductance of the drawn spiral inductor is approximately equal to 3.25 nH. If the differences between these traces are large, go back to modify the spiral inductor layout and repeat the steps above until the error of two traces is in the range of acceptance (within an acceptable range).

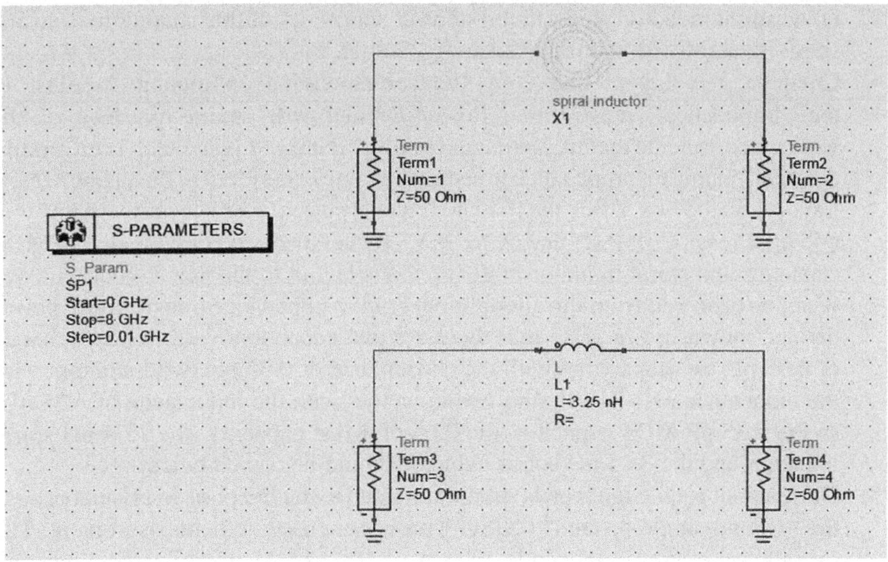

Fig. 4.84 Inductor co-simulation circuit

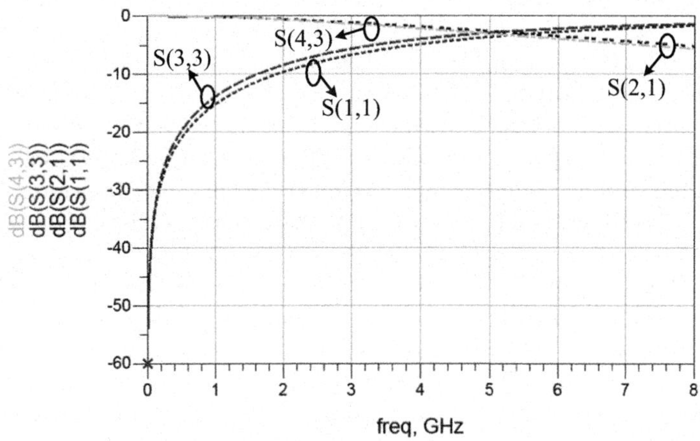

Fig. 4.85 S-parameter results of the inductor co-simulation

4.3.5 Layout of Impedance-Transforming Power Divider

(1) Draw thin-film resistor. Refer to the steps in Sect. 4.3.2 of this chapter to draw the thin-film resistor with the resistance of 230 Ω.

(2) Draw MIM capacitor. Refer to the steps in Sect. 4.3.3 of this chapter to draw the MIM capacitor with the capacitance of 0.65 pF.

(3) Draw spiral inductor. Refer to the steps in Sect. 4.3.4 of this chapter to draw the spiral inductor with the inductance of 1.74 nH.

(4) Create a new layout and copy the corresponding component layouts. In the "Impedance_Transforming_Power_Divider_wrk" main interface of the workspace, execute menu commands [File] → [New] → [Layout…] and modify the "Cell" name to "impedance transforming power divider" in the opened [New Layout] dialog box. Click the [Create Layout] button, a new layout window will pop up. Press "Ctrl + C" and "Ctrl + V" successively to copy the drawn MIM capacitor and spiral inductor in the previous layout to the new layout window. It can be observed from the circuit model of the impedance transforming power divider shown in Fig. 4.29 that there are two capacitors with the capacitance of 0.35 pF, two capacitors with the capacitance of 0.65 pF, two inductors with the inductance of 1.74 nH, and two inductors with the inductance of 3.25 nH, so the 0.35 pF MIM capacitor, the 0.65 pF MIM capacitor, the 1.74 nH spiral inductor, and the 3.25 nH spiral inductor should be copied twice.

(5) Draw signal pads. Signal pads should be added to the I/O ports in layout because the "ground-single-ground" ("GSG") probes are used in the measurement. The overall size of the signal pad is 100 μm × 100 μm. Meanwhile, the outer ring ground is added to make the probe measurement more convenient. The specific steps for drawing pads and outer ring ground are listed as follows: Execute menu commands [Insert] → [Rectangle] to insert a rectangle in the layout, then press "Esc" to exit. Select the inserted rectangle, select "bond: drawing" in the drop-down menu of the [All Shapes] → [Layer] under [Properties] at the right side of the window. Then set the [Rectangles] → [Width] and [Height] to appropriate values. Execute menu commands [Edit] → [Copy/Paste] → [Copy To Layer…], select [text: drawing] in the pop-up [Copy to Layer] dialog box, and click the [Apply] button to copy a test layer in the original position. Similarly, copy a leads layer, a symbol layer, and a packages layer in the original position, respectively. After copying all the needed layers, click the [Cancel] button to close this dialog box. Then indent the layers. Select the text layer of the pads and outer ring ground, execute menu commands [Edit] → [Scale/Oversize] → [Oversize…]. Since the text layer is indented by 2 μm compared with the bond layer, enter − 2 in the "Oversize(+)/Undersize(−)" column of the pop-up [Oversize] dialog box and click the [Apply] button. Similarly, indent the symbol layer and packages layer by 2 μm relative to the bond layer.

(6) Layout adjustment and component connection. The layout is adjusted in the consideration of the circuit size, the layout beautification, and other factors. According to the simulated circuit model in Fig. 4.29, all components are connected by microstrip lines. The drawing of microstrip lines is similar to that of pads. For more details, please refer to step (5), which is not repeated here. Considering the influences of wire bonds on the performance of the impedance-transforming power divider in chip package, the final layout is obtained after the optimizations in several times, as shown in Fig. 4.86 (unit: μm).

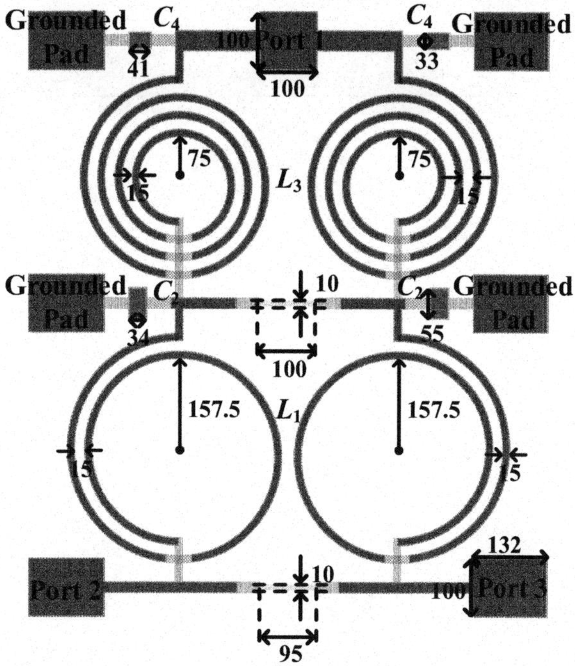

Fig. 4.86 Final layout of the impedance-transforming power divider

4.3.6 *Layout Simulation*

1. Layout Simulation of Impedance-Transforming Power Divider

    (1) Insert pins. Execute menu commands [Insert] → [Pin], then click the left mouse button to insert pins into the I/O and grounded ports of the impedance-transforming power divider, as shown in Fig. 4.87. Long press the "Ctrl" key and select all pins, then select "leads: drawing" in the drop-down menu of the [All Shapes] → [Layer] under [Properties] at the right side of the window.

    (2) Modify simulation settings. Execute menu commands [EM] → [Simulation Settings…], click the [Create EM Setup View] button in the pop-up the [New EM Setup View] dialog box, and the [EM Setup for simulation] window will be displayed. Select the second method "Momentum Microwave" of the EM simulator at first. Then select the [Frequency Plan] tab to set the simulation frequency. Select "Adaptive" in the "Type" column and set the "Fstart" to 0 GHz, the "Fstop" to 8 GHz, and the "Npts" to 50. Finally, select the [Options] tab. Click "Preprocessor", select the "User specified snap distance" option under the "Heal the layout", and set the distance to 2.5 μm. Click "Mesh" and select the "Edge Mesh" option. Keep other settings in default. After finishing all the steps, close the window and click thc [OK] button to save the settings.

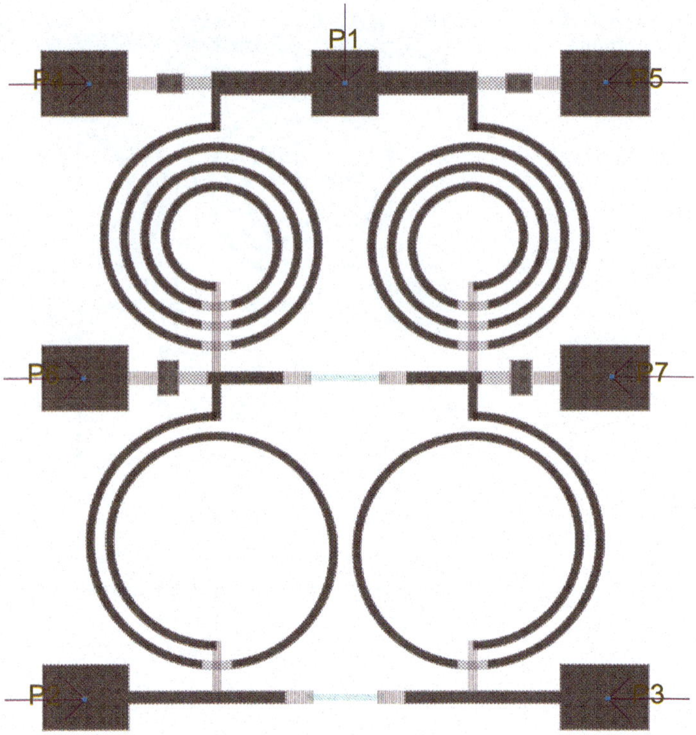

Fig. 4.87 Insertion of pins

(3) Layout simulation. Execute menu commands [EM] → [Simulate] to simulate. During the simulation, a status window will pop up to display the current item progress. The whole simulation is usually long. Since the layout is not grounded, the simulated results are not used to evaluate the performance of the impedance-transforming power divider. After the finishing of the simulation, close the data display window.

2. Co-simulation of Impedance-Transforming Power Divider

   In order to evaluate the performance of the impedance-transforming power divider, the co-simulation of the impedance-transforming power divider is carried out.

   (1) Create EM model and symbol of the impedance-transforming power divider. In the layout window, execute menu commands [EM] → [Component] → [Create EM Model And Symbol…], then click [OK] button in the pop-up dialog boxes. Execute menu commands [Edit] → [Component] → [Update Component Definitions…], then click the [OK] button in the pop-up dialog box to complete the creation of the EM model and symbol in the impedance-transforming power divider.

(2) Create a new schematic and insert the impedance-transforming power divider component. Return to the "Impedance_Transforming_Power_Divider_wrk" main interface of the workspace. Execute menu commands [File] → [New] → [Schematic...], modify the "Cell" name to "impedance-transforming power divider-cosimulation" in the opened [New Schematic] dialog box, and click the [Create Schematic] button to create a new schematic. Click the [Open the Library Browser] icon at the left side of the schematic window, then select the "balanced bandpass filter" component under the [Workspace Libraries] in the displayed [Component Library] window. Click the right mouse button, then click the [Place Component] in the pop-up menu to insert an impedance-transforming power divider component into the schematic. Press "Esc" to exit.

(3) Insert equivalent inductors of wire bonds. The equivalent inductors $L = 0.33$ nH of the wire bonds used in the chip package are inserted in the co-simulation circuit to make the co-simulation results be closer to the final chip measured results. Select the [Lumped-Components] from the drop-down menu in the left component palette list and click the [Inductor] icon to insert an inductor. Press "Esc" on the keyboard to exit. Double-click the capacitor and modify L = 0.33 nH in the [Edit Instance Parameter] dialog box that pops up, then click the [OK] button to save the parameter. Use the same steps to insert six inductors of 0.33 nH (Or select the inserted inductor, and press "Ctrl + C" and "Ctrl + V" to copy and paste).

(4) Insert the S-parameter simulator, term ports, and grounds. Select the [Simulation-S_Param] from the drop-down menu in the left component palette list and click the [S-parameter Simulator] icon to insert an S-parameter simulator. Click the [Port Impedance Termination for S-Parameters] icon to insert three term ports, then press "Esc" to exit. Execute menu commands [Insert] → [GROUND] to place seven grounds (Or click the [TermG Port Impedance Termination for S-Parameters with Grounded Reference] icon to insert three term ports with the grounded reference, then execute menu commands [Insert] → [GROUND] to place seven grounds). Execute menu commands [Insert] → [Wire] to connect all the components according to the simulated circuit model shown in Fig. 4.29. After finishing all the steps, press "Esc" to exit.

(5) Set simulation frequency. Double-click the S-parameter simulator S-PARAMETERS. Set the "Start" to 0 GHz, the "Stop" to 8 GHz, and the "Step-size" to 0.01 GHz, and click the [OK] button. The final impedance-transforming power divider co-simulation circuit is obtained, as shown in Fig. 4.88.

(6) Co-simulation. Execute menu commands [Simulate] → [Simulate]. After the simulation completes, the data display window will be opened automatically. Refer to the steps in Sect. 4.2.2 to view and handle traces of dB(S(1,1)), dB(S(2,1)), dB(S(2,2)), and dB(S(3,2)), the final results are shown in Figs. 4.89 and 4.90.

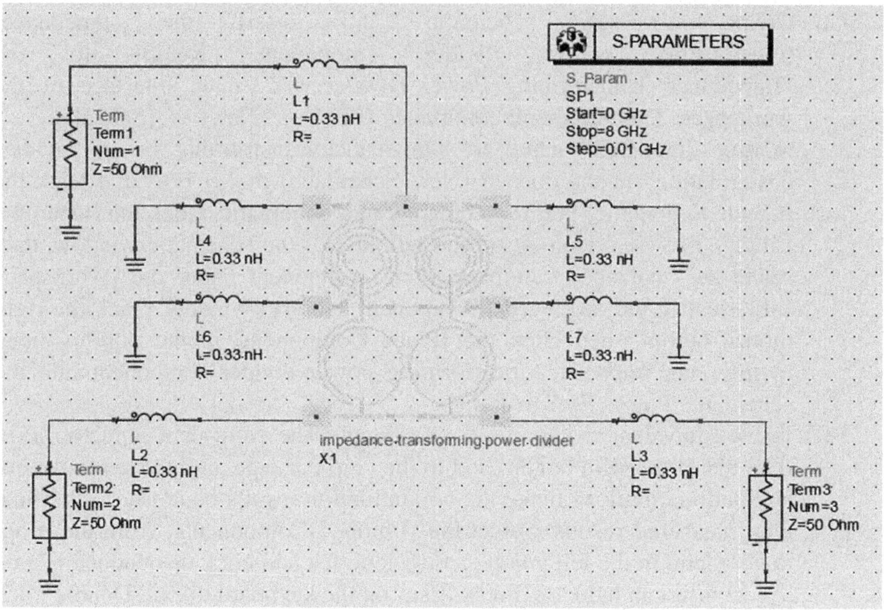

Fig. 4.88 Impedance-transforming power divider co-simulation circuit

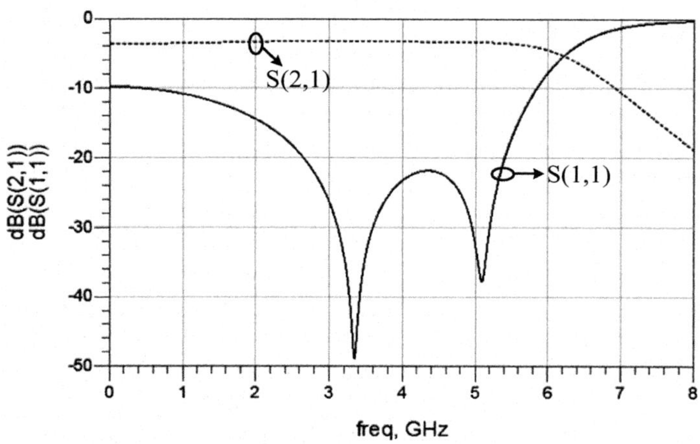

Fig. 4.89 Co-simulation of dB(S(1,1))and dB(S(2,1) traces

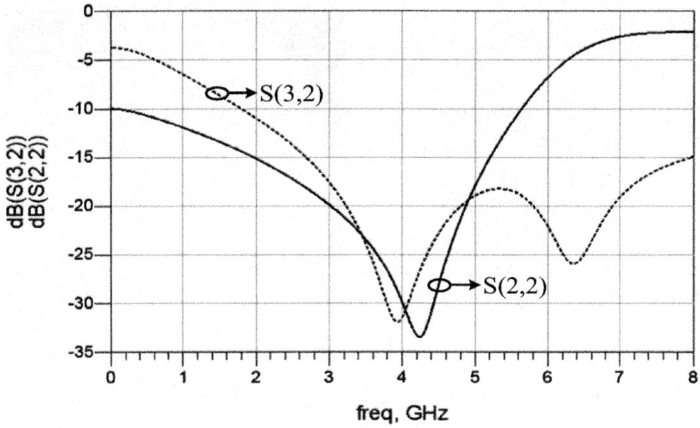

Fig. 4.90 Co-simulation of dB(S(2,2)) and dB(S(3,2)) traces

4.4 Package and Measurement

4.4.1 Chip Package

In practice, RF chips used in mobile terminals must work in coordination with other devices to form an RF front-end system and build a complete signal link. In other words, chips need to be interconnected and packaged before they become mature applications. At present, the flip-chip technology and the wire bonding are two practical methods for the integrated-circuit (IC) packages. In this chapter, the lead wire bonding technology is used to package the impedance conversion power divider chip. The wire bonding needs a variety of connecting structures such as the Au wire, the coplanar waveguide with ground (CPWG), and the microstrip transmission line. To be specific, I/O ports are linked to CPWGs and grounded pads are attached to the ground. The function of the CPWGs is extending the ports of the chip outward, so that the SMA connectors can be fixed easier at the edges of the PCB. As shown in Fig. 4.91, an impedance-transforming power divider chip is fixed on a PCB by Au wires.

4.4.2 Chip Measurement

1. Measured Results

   The measurement of the impedance-transforming power divider chip fixed on the PCB by the wire bonding is completed by using the vector network analyzer of ROHDE & SCHWARZ ZVA8. The measured results are shown in Figs. 4.92 and 4.93.

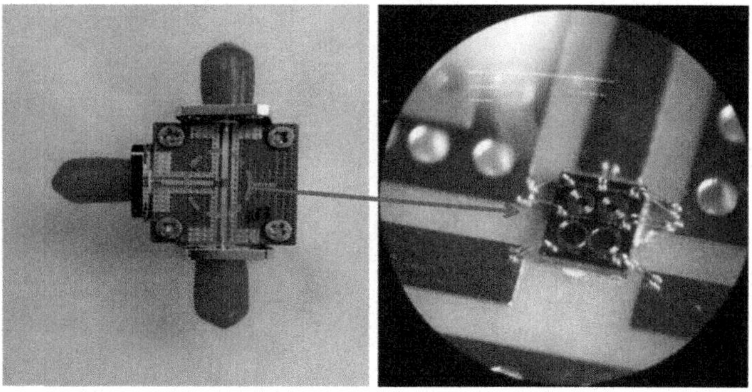

Fig. 4.91 Package of the impedance-transforming power divider chip

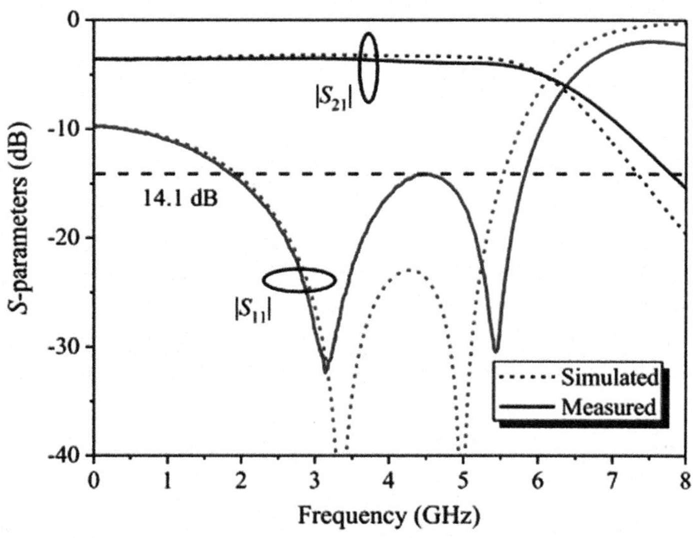

Fig. 4.92 EM simulated and measured results of the impedance-transforming power divider chip of $|S_{11}|$ and $|S_{21}|$

2. Result Analysis

   It can be seen from Figs. 4.92 and 4.93 that the measured results are mostly consistent with the simulated results. The deviations can be caused by the mechanical error and inaccurate dielectric constants of industrial materials. Within the bandwidth from 2.88 to 5.0 GHz, the measured input and output return losses ($|S_{11}|$ and $|S_{22}|$) are better than 14.10 and 15.23 dB, respectively. The measured in-band isolation between output ports $|S_{23}|$ is higher than 10.02 dB, while the measured in-band insertion loss $|S_{21}|$ is smaller than 1 dB.

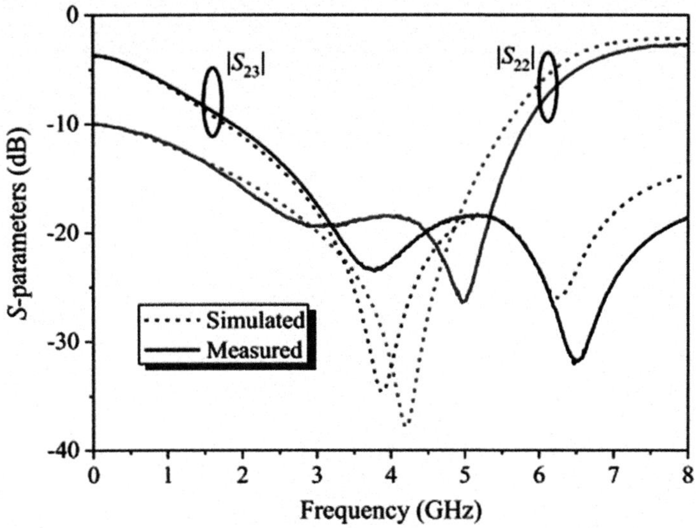

Fig. 4.93 EM simulated and measured results of the impedance-transforming power divider chip of $|S_{22}|$ and $|S_{23}|$

References

1. X. Xie, Y. Xu, L. Xia, *Microwave Integrated Circuits* (Publishing House of Electronics Industry, Beijing, 2018)
2. Q. Gu, J. Xiang, X. Yuan, *Microwave Integrated Circuit Design* (People's Posts and Telecommunications Press, Beijing, 1978)
3. R. Li, *Key Issues in RF/RFIC Circuit Design* (Higher Education Press, Beijing, 2007)
4. M. Kong, Y. Wu, Z. Zhuang, M. Kong, W. Wang, C. Wang, Ultraminiaturized wideband quasi-Chebyshev/-elliptic impedance-transforming power divider based on integrated passive device technology. IEEE Trans. Plasma Sci. **48**(4), 858–866 (2020)

Chapter 5
Design and Simulation of Bandpass Filtering Marchand Balun

In modern wireless communication systems, baluns can be used in any circuit design that requires two balanced outputs such as balanced mixers, balanced multipliers, and broadband antennas, etc. The Marchand balun is one of the most traditional baluns and has been continuously optimized [1–3]. Besides, the research on balun combined with the filtering performance is also attractive. In this chapter, we are going to introduce a new bandpass filtering Marchand balun which consists of a traditional balun and a filter as the front stage [4]. The basic theories and the methods of establishment, simulation, and optimization of the ideal simulated model and full-wave electromagnetic simulated model of the bandpass filtering Marchand balun utilizing the ADS software are demonstrated in detail. Finally, the manufactured bandpass filtering Marchand balun chip is packaged, measured, and the measured results are also analyzed.

5.1 Overview of Balun

5.1.1 Theoretical Basis

1. Propaedeutics

   Balun is a three-port device, which is a transformer essentially. Its main function is to complete the impedance conversion and matching between the balanced transmission-line circuit and the unbalanced transmission-line circuit. The balun plays a very important role in the wireless communication, especially in the broadband antenna system, which is indispensable.

   As shown in Fig. 5.1, for the three-port balun, the symmetric four-port network is used for the theoretical analysis. A three-port network can be obtained by shorting or opening any port in the symmetric four-port network.

Y. Wu and W. Wang, *Microwave and Millimeter-Wave Chips Based on Thin-Film Integrated Passive Device Technology*,
https://doi.org/10.1007/978-981-99-1455-5_5

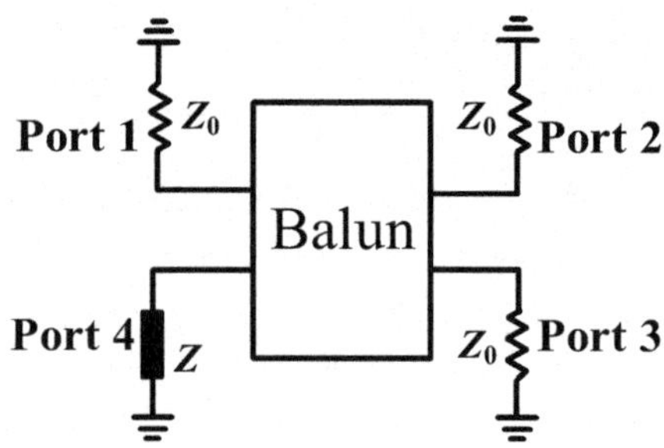

Fig. 5.1 Three-port balun network

Assume that port 1 is the input port of the balun, port 2 and port 3 are the output ports of the balun, port 4 is short or open, its ideal S-parameter is expressed as follows.

$$\begin{cases} S_{11} = 0 \\ S_{21} = -S_{31} \end{cases}, \tag{5.1}$$

where the minus sign indicates that the amplitude of two output signals is equal while the phase is opposite.

2. Performance Index

The evaluation indexes of balun include the insertion loss, the return loss, and the magnitude and phase imbalances between the output ports, etc. According to Fig. 5.1, its performance index parameters are expressed as follows.

a. Insertion Loss (IL): IL (dB) $= -20\log_{10}|S_{21}|$.
b. Return Loss (RL): RL (dB) $= -20\log_{10}|S_{11}|$.

5.1.2 *Traditional Marchand Balun*

Figure 5.2 is the circuit schematic of the traditional Marchand balun. It is composed of two coupled lines whose electrical lengths are 90° at the center frequency. According to the theory of the four-port network, the traditional balun can be obtained by the four-port network with one opening port.

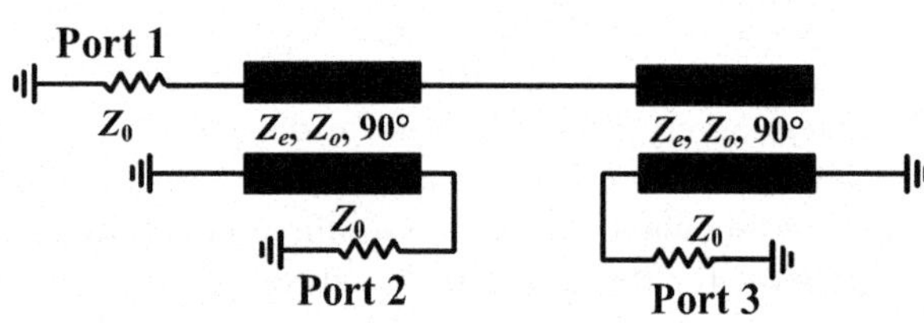

Fig. 5.2 Circuit schematic of the traditional Marchand balun

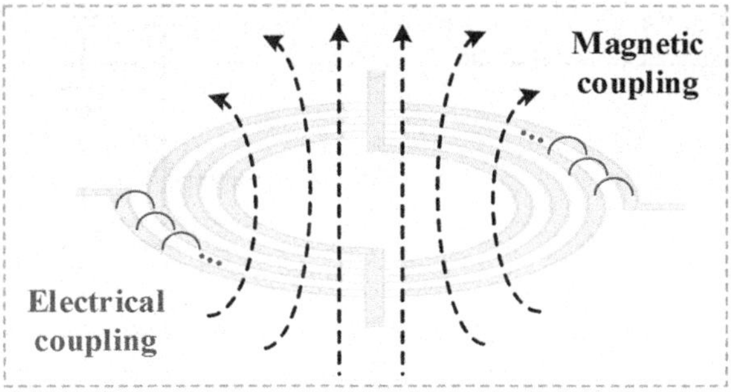

Fig. 5.3 Structure of the spiral coupled line

5.2 Bandpass Filtering Marchand Balun with Spiral Coupled Lines

5.2.1 Spiral Coupled Line

When the operation frequency is low, the length and the width of an RF chip constructed by the IPD technology are far less than the wavelength. For this reason, it is preferable to embed coupled lines in chips in the form of spiral coupled lines with smaller lengths and widths, as shown in Fig. 5.3. The lumped-circuit equivalent model of the spiral coupled line is presented in Fig. 5.4. L_s is the self-inductance of the winding, L_m is the mutual inductance, and k_L is the inductive coupling coefficient of the two windings. At the same time, C_s is the capacitance to ground used to describe the signal leakage on the substrate, C_m is the mutual capacitance used to describe the electrical coupling between the parallel windings, and k_C is the capacitive coupling coefficient. What should be emphasized is that the spiral coupled line is the coupled line with a special shape, and it still belongs to the CL, which has three key parameters including the odd-mode impedance Z_e, the even-mode impedance Z_o, and the electrical length θ. Besides, the coupling coefficient of the spiral coupled line is $k = (Z_e - Z_o)/(Z_e + Z_o)$. The coupling of the spiral coupled line can be seen as the "codirectional coupling" approximately, in which case has $k_L = k_C = k$, and the characteristic impedance of the spiral coupled line is $Z_{OCL} = \sqrt{Z_e Z_o} = \sqrt{L_s/(C_s + C_m)}$. In general, the spiral coupled line has a coupling coefficient k of about 0.7–0.8, while Z_{OCL} is normally larger than Z_0 ($Z_0 = 50\ \Omega$).

5.2.2 Modified Marchand Balun

The application of the spiral coupled line proposed in Sect. 5.2.1 can greatly reduce the circuit size of baluns. Howcvcr, coupling coefficients of the spiral coupled line are large, and it's odd- and even-mode impedances are difficult to control, which causes

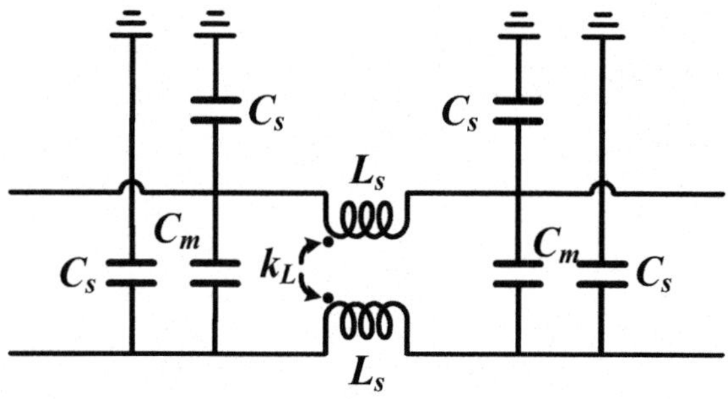

Fig. 5.4 Lumped-circuit equivalent model of the spiral coupled line

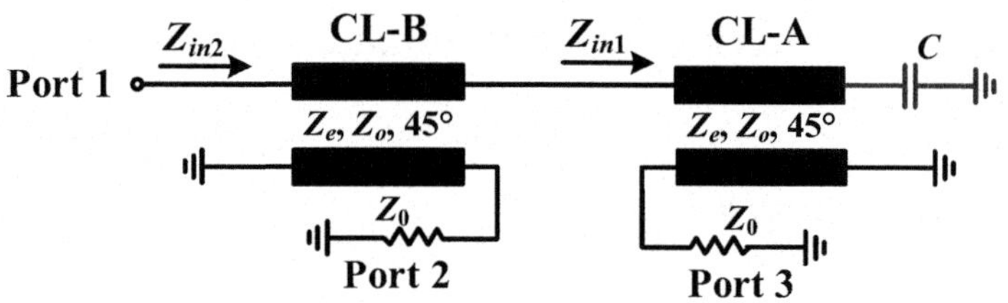

Fig. 5.5 Circuit structure of the improved Marchand balun

that input impedances of Marchand baluns are not equal to 50 Ω in some cases, so it is difficult to obtain a good termination matching at input ports. Designers have difficulty in adjusting the parameters of the spiral coupled line to meet different design requirements for the traditional Marchand balun, which results in bad performances.

In this section, an improved method is introduced. By using 1/8-wavelength coupled lines and replacing the open end in the conventional Marchand balun with the capacitor termination, the value of the input impedance can be closer to Z_0, which improves the performance and increases the flexibility of the balun design. The circuit structure of the improved Marchand balun is shown in Fig. 5.5.

5.2.3 Bandpass Filtering Marchand Balun

In this section, a Chebyshev low-pass filter is added to the front end of the modified Marchand balun. Owing to the inherent high-pass characteristic of the balun section, the bandpass response exists. The proposed bandpass filtering Marchand balun includes a balun section and a low-pass filter section. For a good impedance matching, the filter section must transform the input impedance of the balun section to the terminal impedance, which equals to Z_0. As we all know, bigger impedance transformating ratio makes the Chebyshev low-pass filter have a larger order, and the circuit becomes more complex correspondingly. However, the design of the improved Marchand balun can avoid the situation above, because the input impedance of the

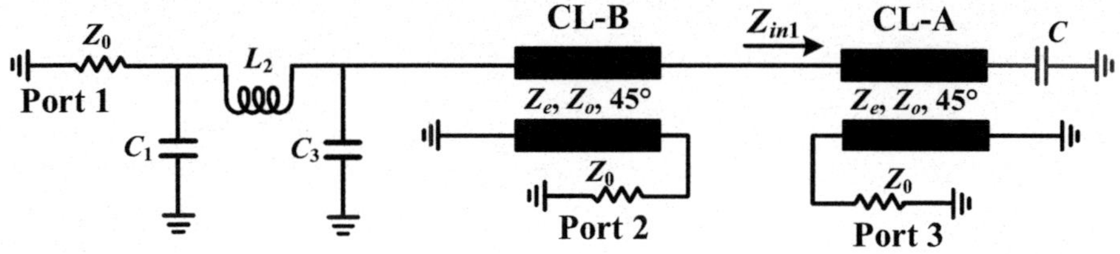

Fig. 5.6 Circuit structure of the bandpass filtering Marchand balun

balun section is already close to Z_0. The impedance transformation is not needed, and the order n can be smaller.

In order to make the circuit size as small as possible, the structure of 3-order Chebyshev low-pass filter is chosen. The final circuit structure of the proposed bandpass filtering Marchand balun is demonstrated in Fig. 5.6. The equal-ripple band edge ω_1 decides the passband of the bandpass filtering Marchand balun, as shown in (5.2).

$$\begin{cases} L_A(\omega) = 10\log_{10}\left\{1 + \varepsilon\cos^2\left[n\cos^{-1}\left(\frac{\omega}{\omega_1}\right)\right]\right\}_{\omega\leq\omega_1} \\ L_A(\omega) = 10\log_{10}\left\{1 + \varepsilon\cosh^2\left[n\cosh^{-1}\left(\frac{\omega}{\omega_1}\right)\right]\right\}_{\omega\geq\omega_1} \end{cases}. \tag{5.2}$$

5.3 Schematic of Bandpass Filtering Marchand Balun

ADS can realize the parameterized model simulation. Take a bandpass filtering Marchand balun with the center frequency $f_0 = 4$ GHz as an example to introduce how to build and simulate its ideal circuit model. The even-mode impedance Z_e, the odd-mode impedance Z_o, the electrical length θ of the balun section, and the lumped elements (C_1, L_2, C_3) of the 3-order Chebyshev low-pass filter can be parameterized in the circuit model.

5.3.1 New Workspace and Circuit Model

1. New Workspace

(1) Double-click the ADS shortcut icon , and click the [OK] in the displayed dialog box to start the ADS software. After ADS runs, the [Get Started] dialog box will pop up automatically. Click the [Close] button in the lower right corner to enter the main interface [Advanced Design System 2020 (Main)], as shown in Fig. 5.7.

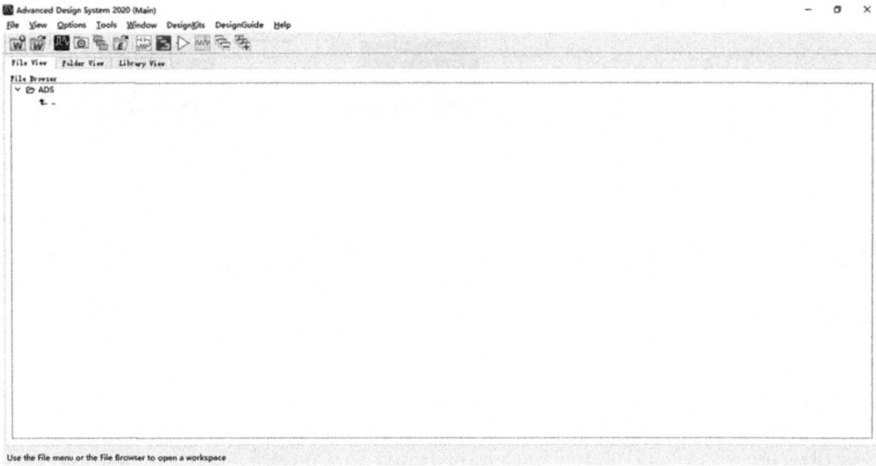

Fig. 5.7 Main interface of ADS

(2) Create a new workspace to store all files of the design simulation. Execute menu commands [File] → [New] → [Workspace…] to open the new workspace dialog box shown in Fig. 5.8, where you can set the workspace name [Name] and the working path [Create in]. Here, change the workspace name to "Bandpass_Filtering_Marchand_Balun_wrk" and keep the working path in the default setting. Click the [Create Workspace] button to complete the creation of a new workspace.
(3) The [Folder View] in the ADS main interface can display the name and working path of the created workspace, as shown in Fig. 5.9. The workspace name is "Bandpass_Filtering_Marchand_Balun_wrk", the corresponding

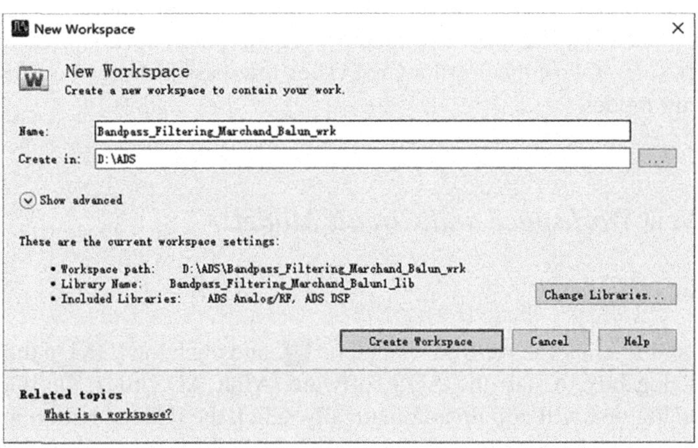

Fig. 5.8 [New workspace] dialog box

path is "D:\ADS\Bandpass_Filtering_Marchand_Balun_wrk", and a subfolder named "Bandpass_Filtering_Marchand_Balun_wrk" in the ADS folder of disk D can be found in your computer.

2. New Circuit Model

(1) Create a new schematic. Execute menu commands [File] → [New] → [Schematic…] to open the [New Schematic] dialog box shown in Fig. 5.10. Modify the name of "Cell" to "circuit structure" and click the [Create Schematic] button to complete the creation of a new schematic, as shown in Fig. 5.11.
(2) Insert coupled transmission lines. Select the [TLines-Ideal] from the drop-down menu in the left component palette list, as shown in Fig. 5.12, which

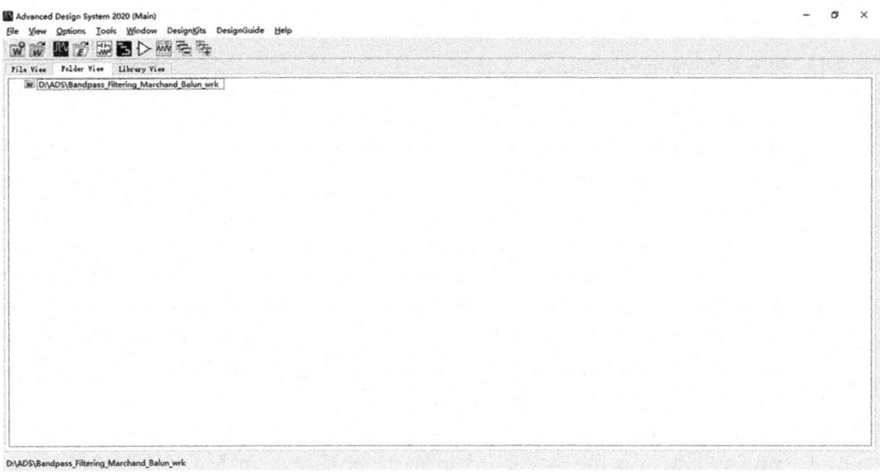

Fig. 5.9 Folder view of the new workspace

Fig. 5.10 [New schematic] dialog box

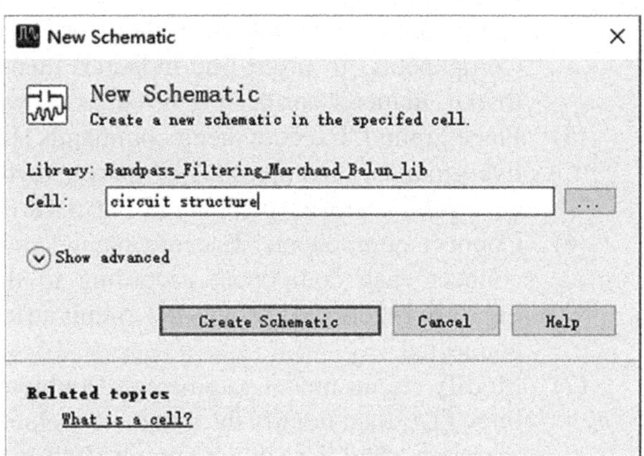

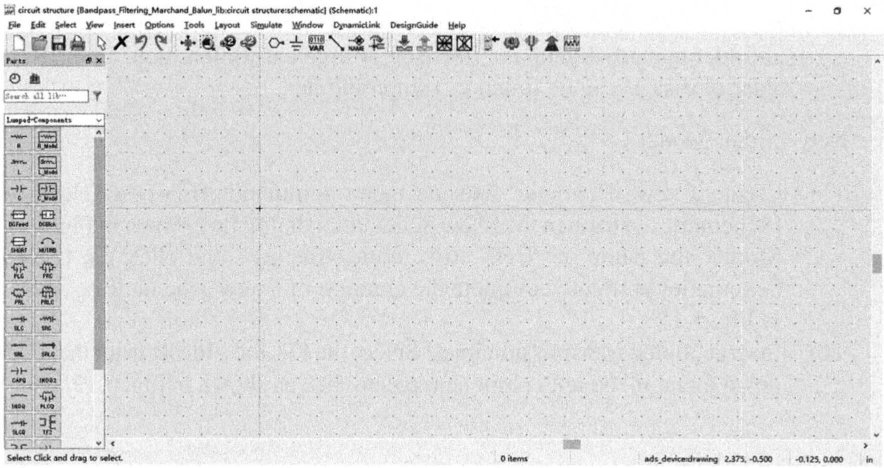

Fig. 5.11 New schematic window

includes some commonly used ideal distributed-element models, such as the transmission line and the coupled transmission line, etc. Click the [Ideal Coupled Transmission Lines] icon under the [TLines-Ideal] to insert two ideal coupled transmission lines, then press "Esc" on the keyboard to exit. Coupled transmission lines have been inserted, as shown in Fig. 5.13.

(3) Insert and rotate capacitors. Select the [Lumped-Components] from the drop-down menu in the left component palette list, as shown in Fig. 5.14, which includes some commonly used ideal lumped-element models, such as the capacitor, the inductor, and the resistor, etc. Click the [Capacitor] icon under the [Lumped-Components] to insert three capacitors, then press "Esc" on the keyboard to exit, as shown in Fig. 5.15. Right-click the capacitors to be rotated, select [Rotate] from the pop-up menu to rotate them in 90° clockwise. Capacitors have been inserted and rotated, as shown in Fig. 5.16.

(4) Insert inductor. Click the [Inductor] icon under the [Lumped-Components] to insert one inductor, then press "Esc" on the keyboard to exit. Inductor has been inserted, as shown in Fig. 5.17.

(5) Place ground. Execute menu commands [Insert] → [GROUND] to place five grounds, then press "Esc" on the keyboard to exit. Ground symbols have been inserted, as shown in Fig. 5.18.

(6) Connect components. Execute menu commands [Insert] → [Wire] to connect each component according to the circuit structure shown in Fig. 5.6. Figure 5.19 shows the circuit structure after the components are connected.

(7) Modify circuit model parameters. Double-click the coupled transmission lines TL1, then modify the parameter value to Ze (Ohm), Zo (Ohm), SitaT (deg), and f0 (GHz) in the pop-up [Edit Instance parameters] dialog box,

Fig. 5.12 [Tlines-Ideal] list

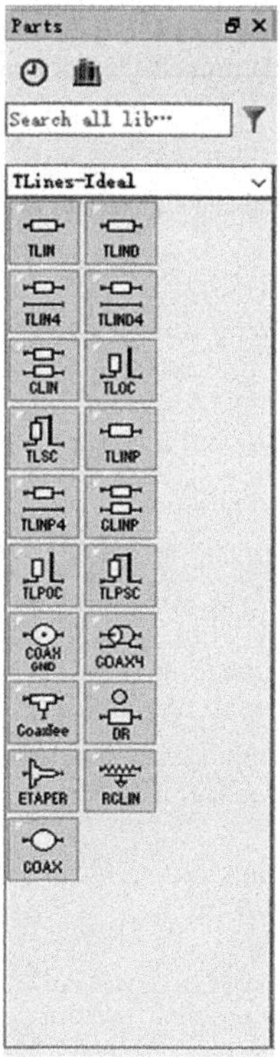

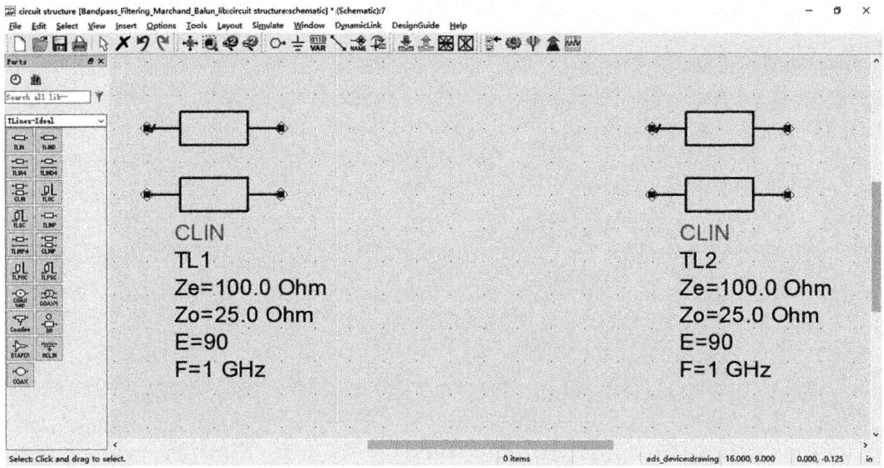

Fig. 5.13 Insertion of coupled transmission lines

Fig. 5.14
[Lumped-Components] list

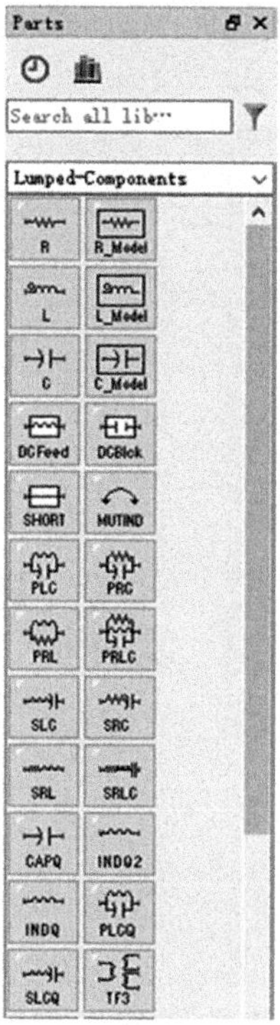

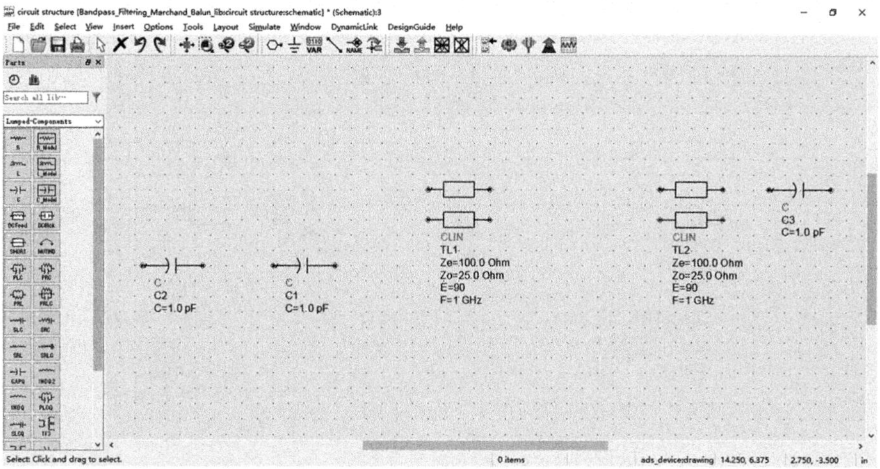

Fig. 5.15 Insertion of capacitors

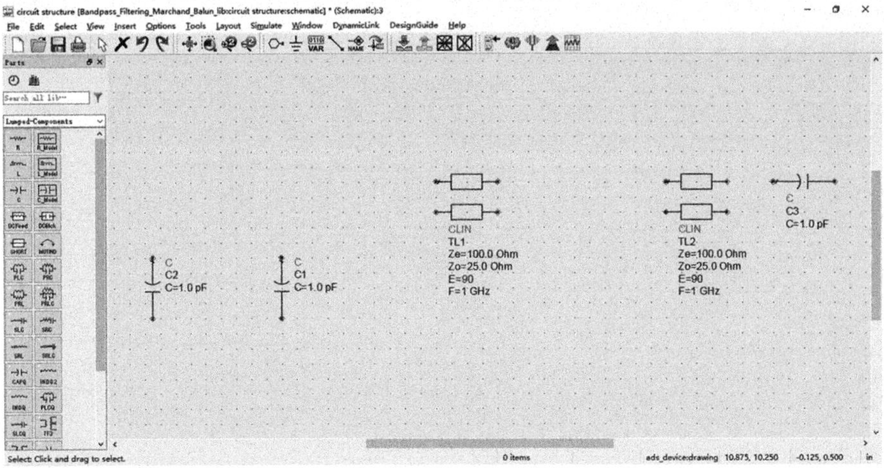

Fig. 5.16 Rotation of capacitors

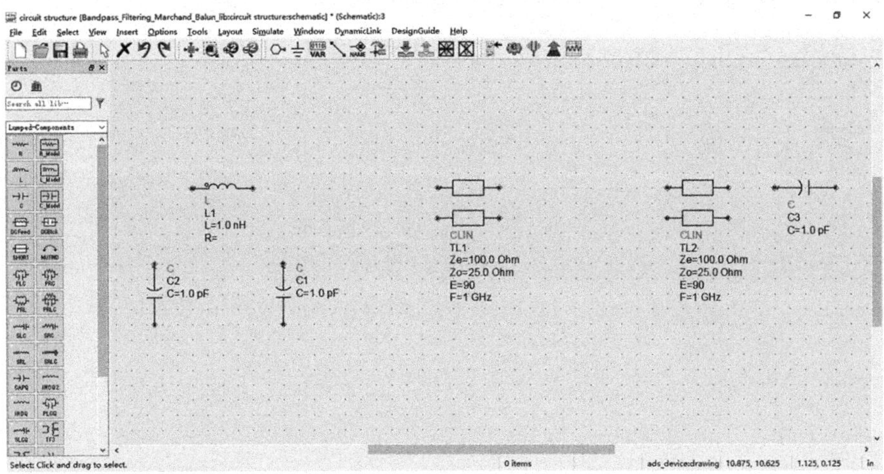

Fig. 5.17 Insertion of the inductor

as shown in Fig. 5.20 (Check whether the unit settings are consistent). Click the [OK] button to save the parameter and close the dialog box. Similarly, modify the parameter value of TL2 to Ze (Ohm), Zo (Ohm), SitaT (deg), and f0 (GHz). Modify the parameter values of C1 to C1 (pF), the parameter values of L1 to L2 (nH), the parameter value of C2 to C3 (pF), the parameter value of C3 to C (pF), as shown in Figs. 5.21, 5.22, 5.23, 5.24 and 5.25.

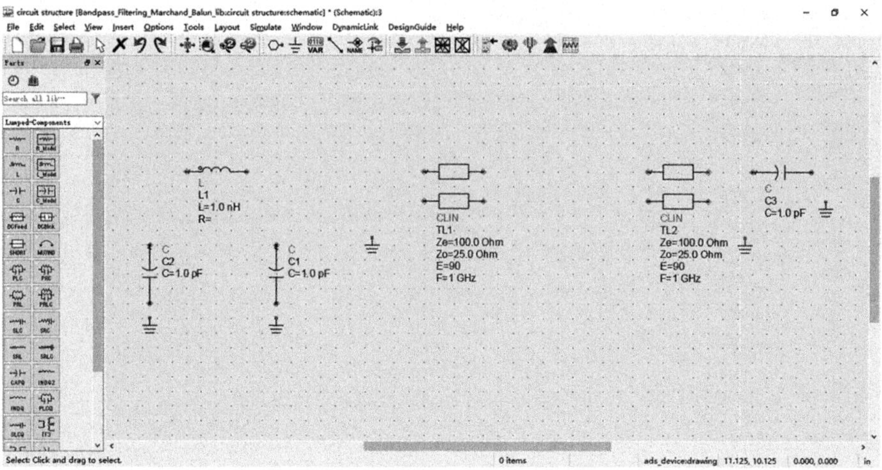

Fig. 5.18 Placement of grounds

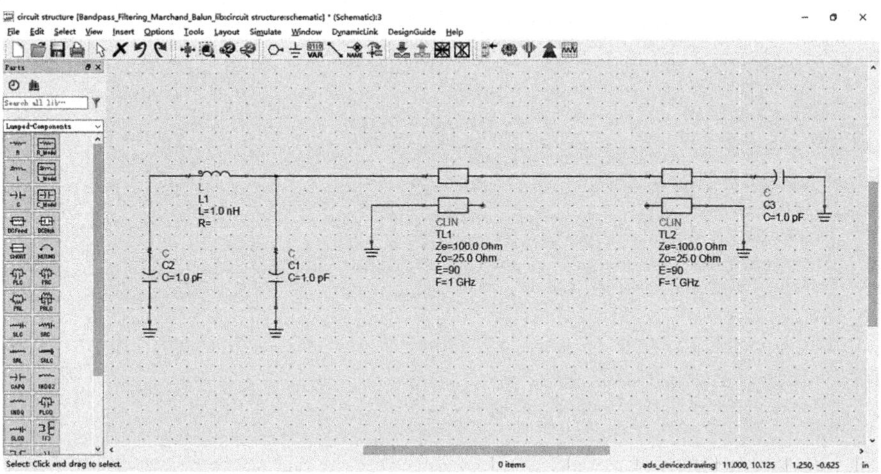

Fig. 5.19 Connection of components

(8) Define parameter values of variables. Click the [Insert VAR: Variable Equations] icon in the toolbar to insert a VAR item in the schematic. Double-click the component to open the [Edit Instance Parameter] dialog box. The "Variable or Equation Entry Mode" is "Standard" by default. Input the variable name "Ze" in the "Name" column, the variable value "232.5" in the "Variable Value" column. Click the [Apply] button, and the dialog box after the setting is shown in Fig. 5.26. If you click [OK], the dialog box can be closed directly. Then define other variables in turn, including Zo = 30.67, C1 = 0.77, L2 = 2.00, C3 = 0.72, C = 0.37, SitaT = 45, and f0 =

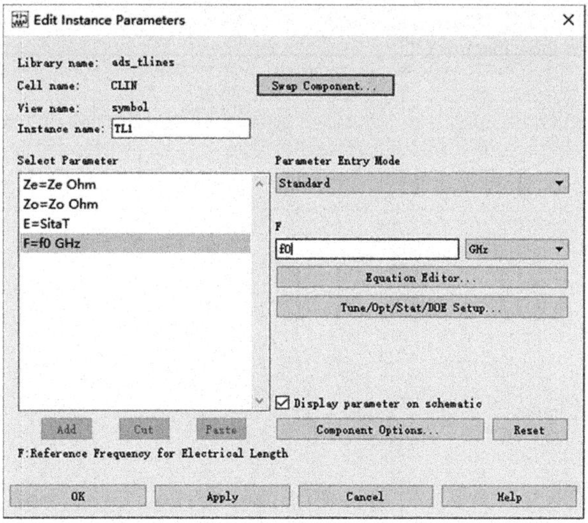

Fig. 5.20 Parameter settings of the TL1

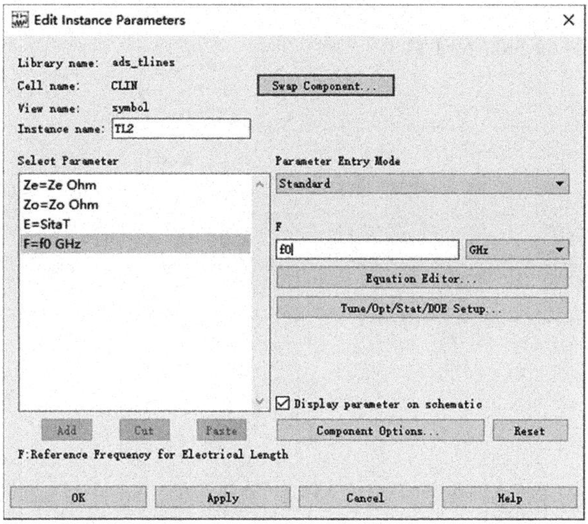

Fig. 5.21 Parameter settings of the TL2

4, as shown in Fig. 5.27. While defining other variables, click the [Add] button to add them. If you click the [Apply] button, the selected variable in the parameter list on the left can be replaced directly. In addition, there is no need to set units because the unit of each variable has already been defined in the component model.

(9) The VAR item provides three methods to add variables, including "Standard", "Name = Value", and "File Based". You can select the most suitable method in the "Variable or Equation Entry Mode" column. The second method is introduced below. Select the "Name = Value" mode, as shown

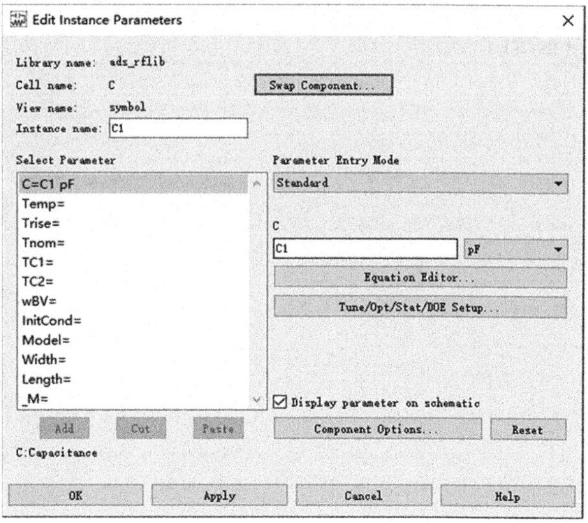

Fig. 5.22 Parameter setting of the capacitor C1

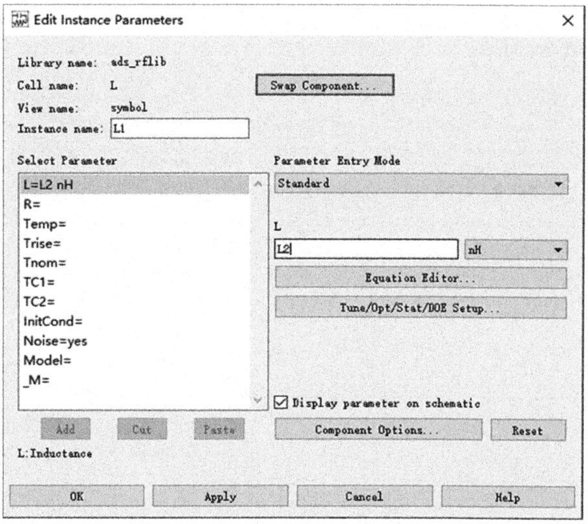

Fig. 5.23 Parameter setting of the inductor L1

in Fig. 5.28. Input "Ze = 232.5" under the "Variable Value" column, then click the [Apply] button. Similarly, when defining other variables, click the [Add] button to add them.

(10) Complete parameter definition. After the definition of all variables, click the [OK] button, and the final circuit model of the bandpass filtering Marchand balun defined by the ideal parameters is shown in Fig. 5.29. Readers can compare the parameters in the lumped-element circuit structure to check whether all parameters are set correctly in detail.

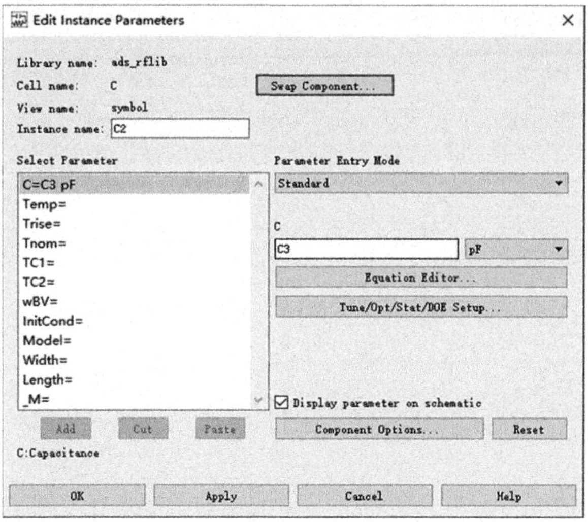

Fig. 5.24 Parameter setting of the capacitor C2

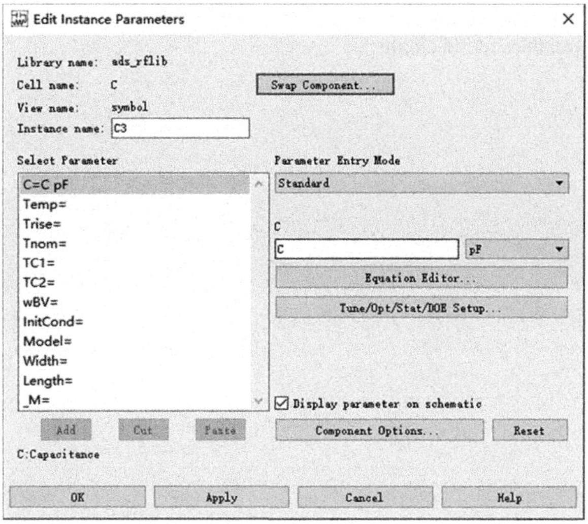

Fig. 5.25 Parameter setting of the capacitor C3

5.3.2 Schematic Simulation

1. Set Simulation Parameters

   (1) Insert the S-parameter simulator, term ports, and grounds. As shown in Fig. 5.30, select the [Simulation-S_Param] from the drop-down menu in the left component palette list and click the [S-parameter Simulator] icon  to insert an S-parameter simulator. Click the [Port Impedance Termination for S-Parameters] icon  to insert three term ports, then press "Esc"

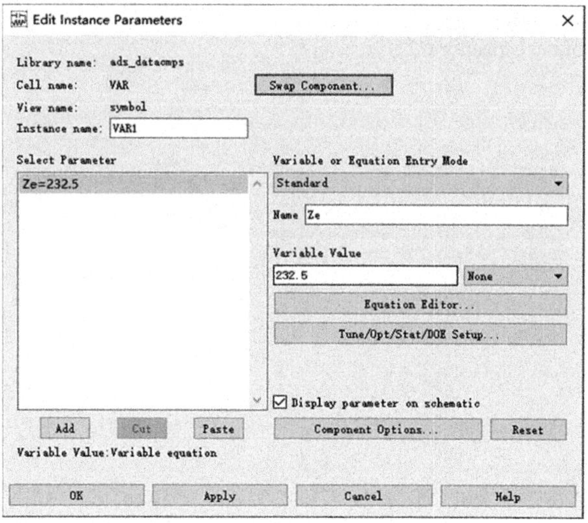

Fig. 5.26 Definition of the variable Ze

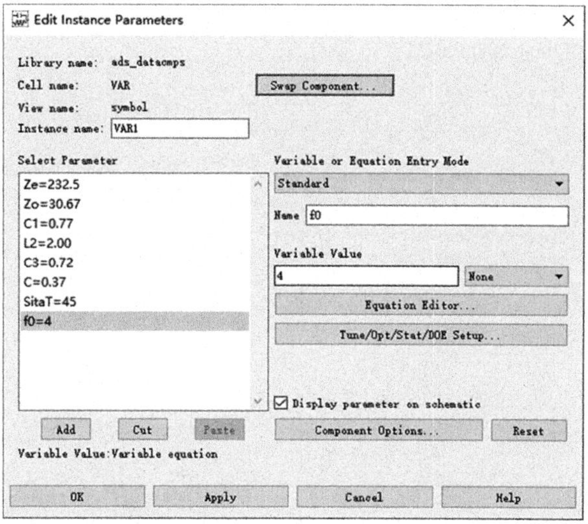

Fig. 5.27 Definitions of all variables in the circuit

on the keyboard to exit. After that, execute menu commands [Insert] → [GROUND] to place three grounds (Or click the [TermG Port Impedance Termination for S-Parameters with Grounded Reference] icon 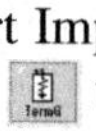 to insert three term ports with the grounded reference). Execute menu commands [Insert] → [Wire] to connect the inserted components. After finishing all the steps, press "Esc" to exit.

(2) Set simulation frequency. Double-click the S-parameter simulator 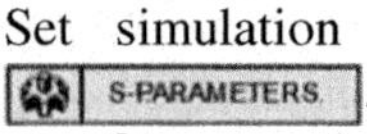. Complete the settings shown in Fig. 5.31, in which the start frequency is 0 GHz, the stop frequency is 10 GHz, and the step-size is

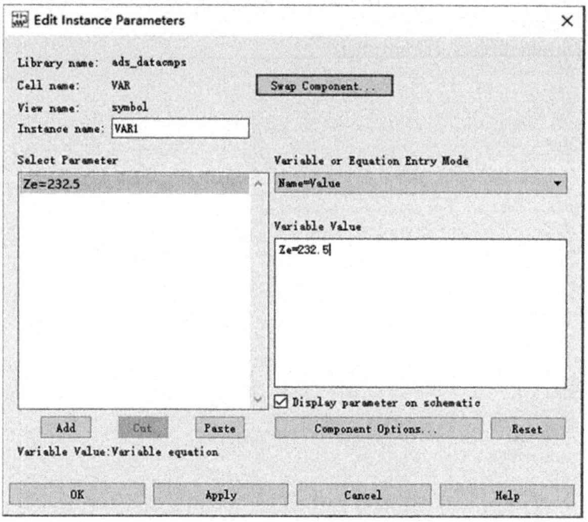

Fig. 5.28 The second method to define the variable Ze

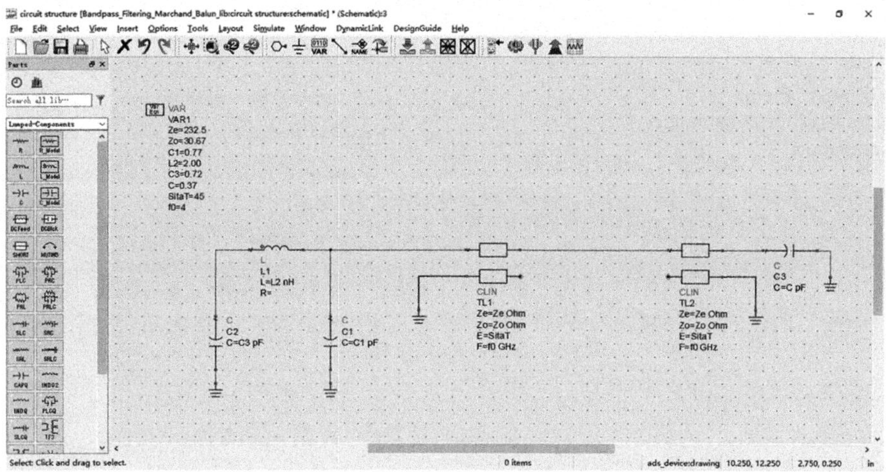

Fig. 5.29 Bandpass filtering Marchand balun circuit model defined by ideal parameters

0.01 GHz, then click the [OK] button. The final ideal parameter simulated circuit model of the bandpass filtering Marchand balun is obtained, as shown in Fig. 5.32.

2. View Simulated Results

(1) Execute menu commands [Simulate] → [Simulate]. After the simulation completes, the data display window will be opened, as shown in Fig. 5.33.
(2) Click the [Rectangular Plot] icon in the [Palette], then click the left mouse button at the blank data display area to open the [Plot Traces &

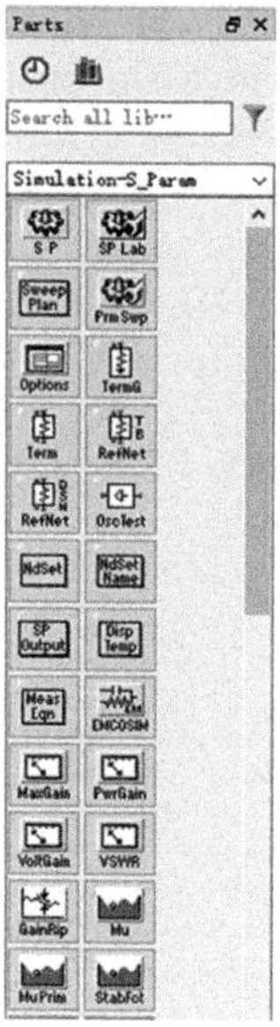

Fig. 5.30 [Simulation-S_Param] list

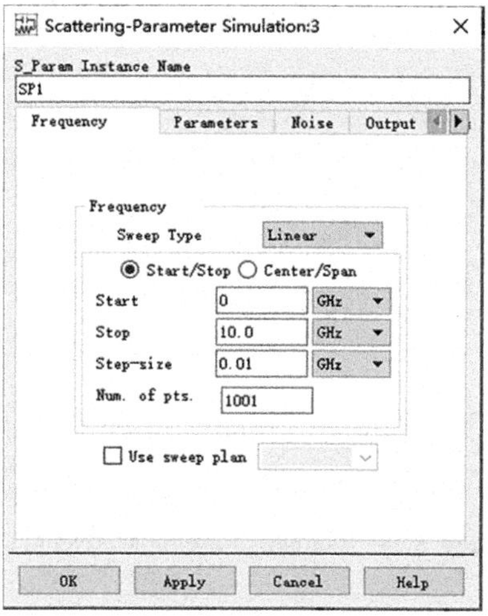

Fig. 5.31 Frequency settings of the S-parameter simulation

Attributes] dialog box shown in Fig. 5.34, where the parameter traces can be set and plotted.

(3) Long press the "Ctrl" key to select S(1,1) and S(2,1), then click the [>>Add>>] button. Choose [dB] in the pop-up dialog box shown in Fig. 5.35, then click the [OK] button. It can be seen that dB(S(1,1)) and dB(S(2,1)) have been added to the [Traces] list box, as shown in Fig. 5.36.

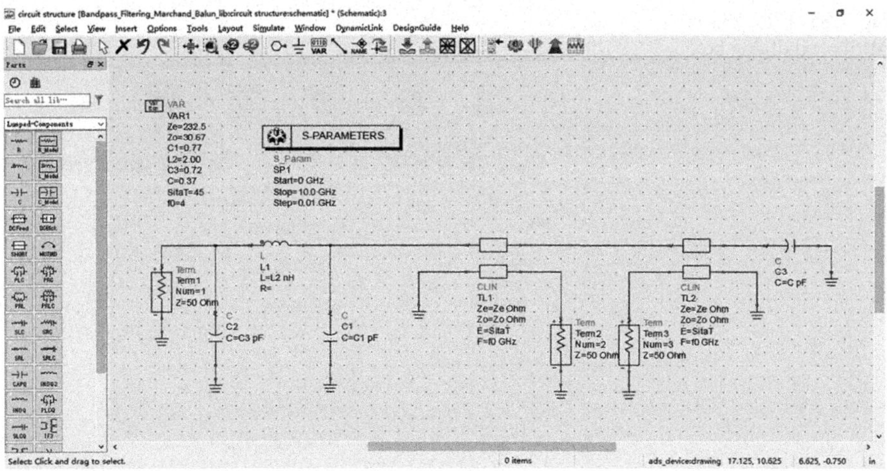

Fig. 5.32 Final ideal parameter simulated circuit model of the bandpass filtering Marchand balun

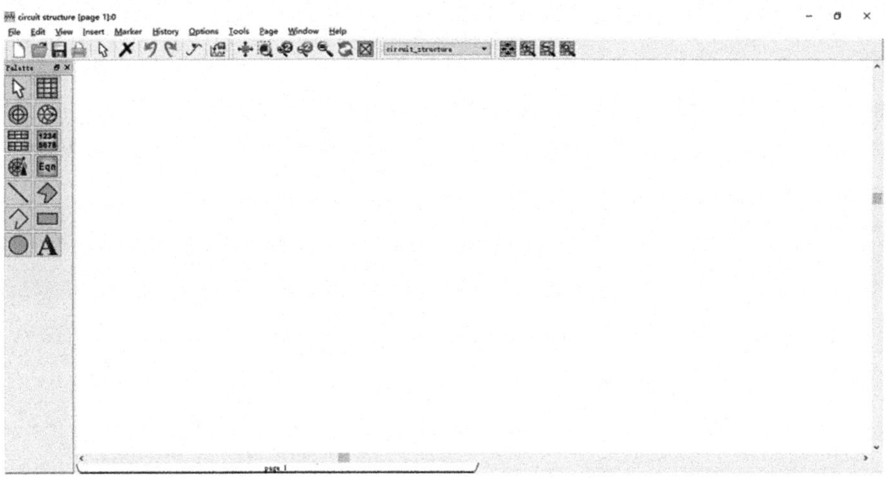

Fig. 5.33 Data display window

(4) Click the [OK] button in Fig. 5.36, the traces of dB(S(1,1)) and dB(S(2,1)) can be displayed in the data display area, the ordinate is the dB value, as shown in Fig. 5.37.

(5) Click the [Equation] icon Eqn in the left [Palette]. Click the left mouse button at the data display area to open the [Eenter Equation] dialog box. Write the equation in Fig. 5.38 under the [Enter equation here] to calculate the magnitude imbalance, then click the [OK] button to close the dialog box. Similarly, the equation in Fig. 5.39 can be entered to calculate the phase imbalance.

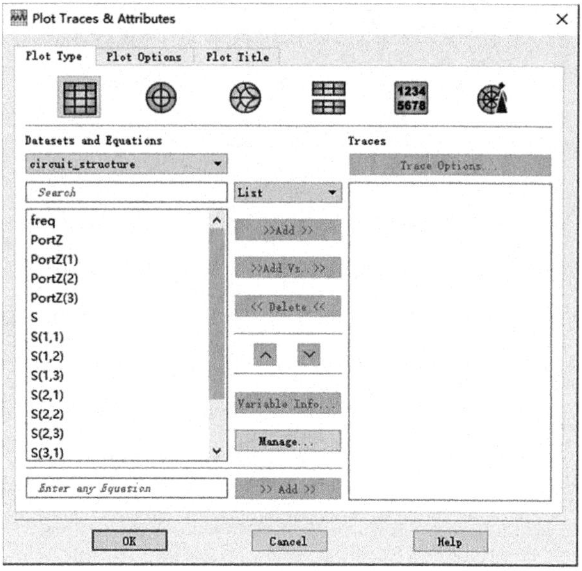

Fig. 5.34 [Plot Traces & Attributes] dialog box

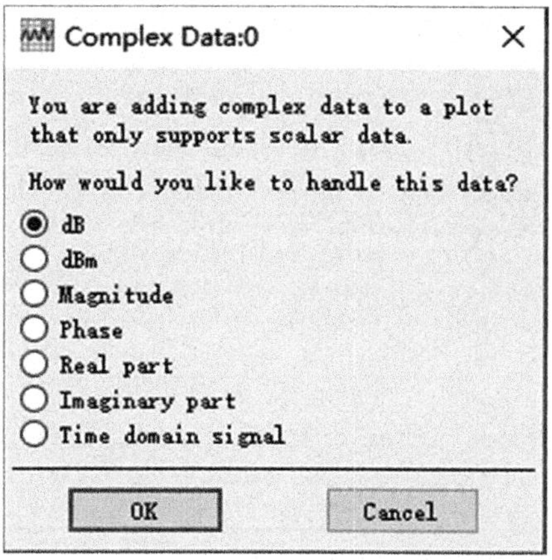

Fig. 5.35 Handling of the data

(6) Click the [Rectangular Plot] icon  in the [Palette] again, then click the left mouse button at the blank data display area to open the [Plot Traces & Attributes] dialog box, where the parameter traces can be set and plotted.
(7) Select "Equations" in the "Datasets and Equations" column, select the defined DiffMagnitude, and click the [>>Add>>] button. Choose [dB] in the pop-up dialog box and click the [OK] button. The DiffMagnitude trace with dB values on the ordinate is displayed in the data display area, as shown in Fig. 5.40.
(8) Similarly, the DiffPhase trace with the unit of on the ordinate can also be displayed, as shown in Fig. 5.41.

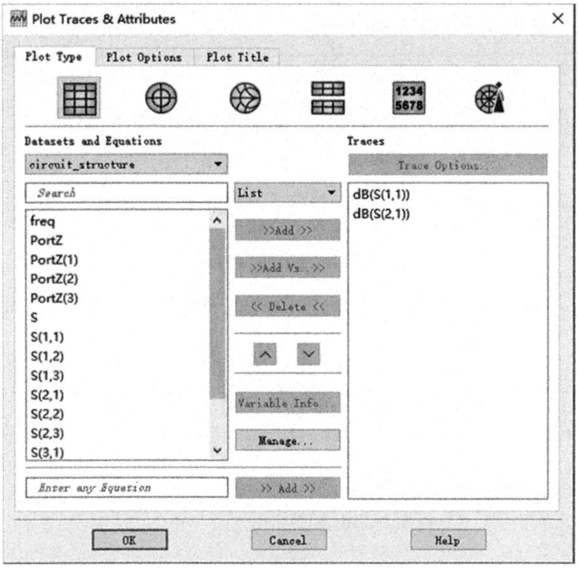

Fig. 5.36 Add of S(1,1) and S(2,1)

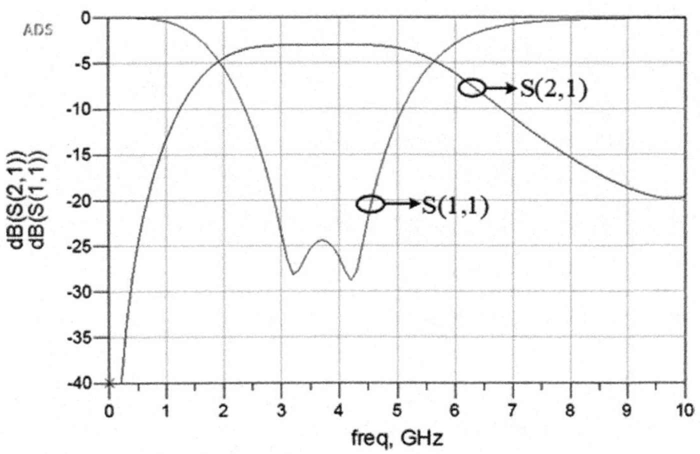

Fig. 5.37 Traces of dB(S(1,1)) and dB(S(2,1))

3. Handle Traces

   Next, traces of dB(S(1,1)) and dB(S(2,1)) are taken as examples to introduce the curves handling. Similar steps can be applied in other traces.

   (1) By changing the frequency position of the marker, the values of any point on the trace can be read. Execute menu commands [Marker] → [New…] to open the dialog box shown in Fig. 5.42. Select a trace and click the left mouse button to insert a marker on it, as shown in Fig. 5.43. Similarly, insert

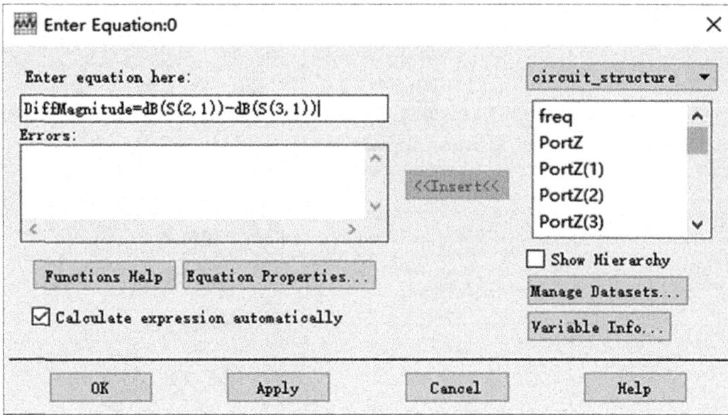

Fig. 5.38 Operation of writing the magnitude imbalance equation

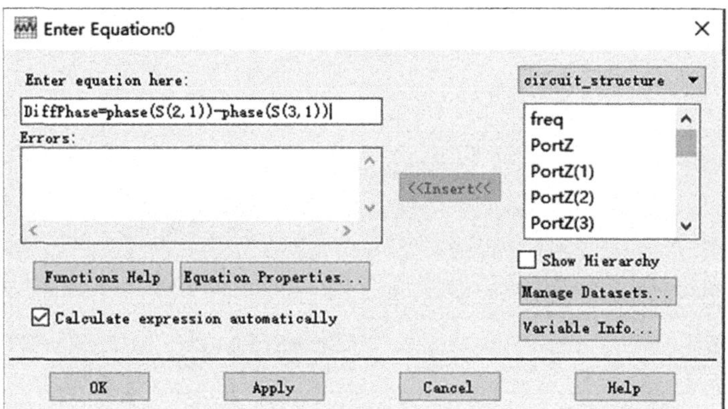

Fig. 5.39 Operation of writing the phase imbalance equation

a marker for another trace dB(S(2,1)). In addition, you can long press the left mouse button to move the position of the data frame.

(2) After selecting the inserted marker, the left and right arrow keys on the keyboard can be used to adjust the position of the abscissa (freq). You can also click the position shown in Fig. 5.44 with the left mouse button to modify the specific frequency you want to view directly. Here, the values of dB(S(1,1)) and dB(S(2,1)) at the center frequency of 4 GHz are displayed, as shown in Fig. 5.45.

(3) The editing function of the data display is introduced below. Take the modification and beautification of the Y axis as examples. Double-click the simulated result figure, then the [Plot Traces & Attributes] dialog box pops up. Click the [Plot Options] tab to deselect the "Auto Scale" (The automatic adjustment scale of the ADS software is not used). Adjust the Y axis scale

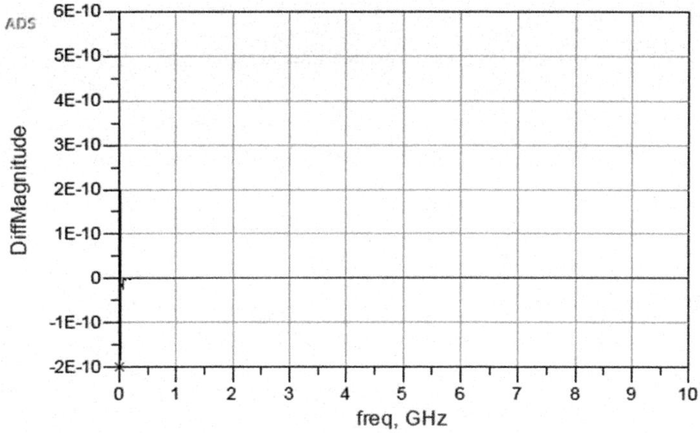

Fig. 5.40 DiffMagnitude trace

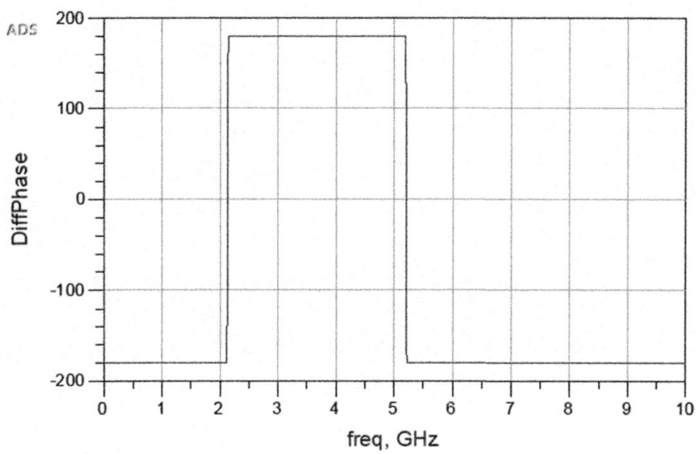

Fig. 5.41 DiffPhase trace

Fig. 5.42 [Insert Marker] dialog box

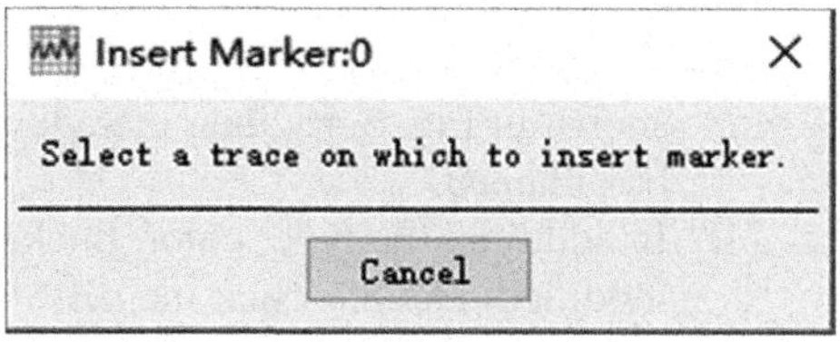

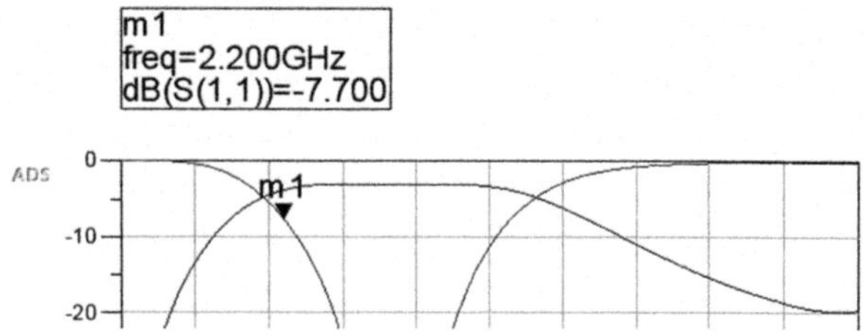

Fig. 5.43 Insertion of marker on the trace

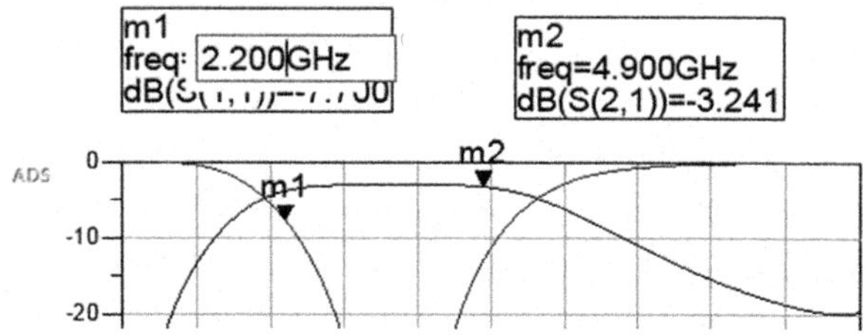

Fig. 5.44 Modification of the abscissa (freq) value

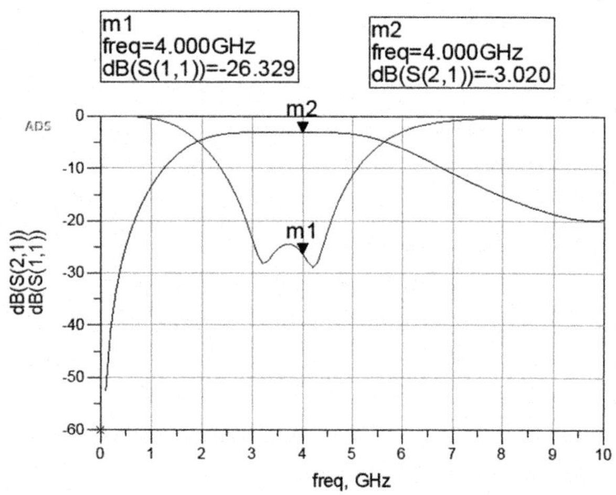

Fig. 5.45 Values of dB(S(1,1)) and dB(S(2,1)) at the center frequency

shown in Fig. 5.46, then the adjusted S-parameter results in Fig. 5.47 can be obtained.

(4) In addition, the type, color, thickness, and symbol of the trace can also be modified. Double-click the dB(S(1,1)) to open the [Trace Options] dialog box and modify the trace, as shown in Fig. 5.48 (Keep the trace color as default). Similarly, modify the dB(S(2,1)) trace shown in Fig. 5.49. The final traces of dB(S(1,1)) and dB(S(2,1)) are shown in Fig. 5.50.

(5) Similarly, modify the traces of DiffMagnitude and DiffPhase. The final traces of DiffMagnitude and DiffPhase are shown in Figs. 5.51 and 5.52.

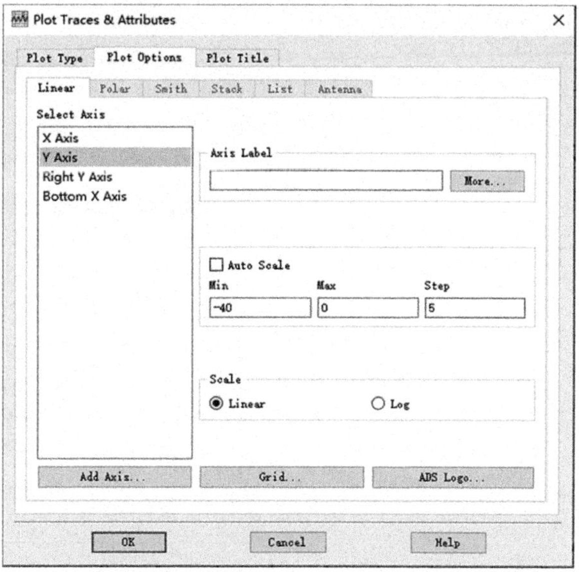

Fig. 5.46 Adjustment of the Y axis scale

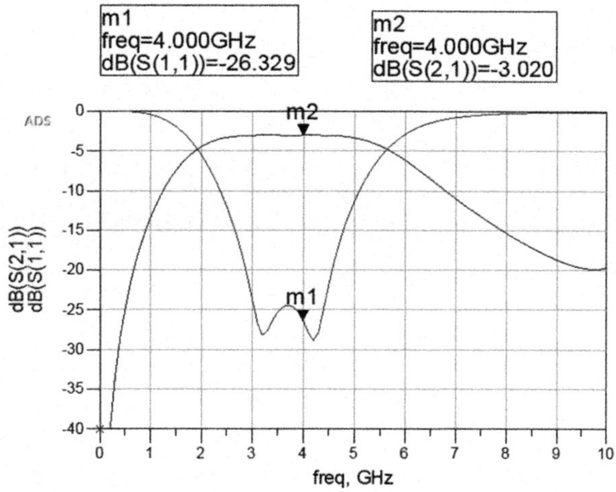

Fig. 5.47 The adjusted S-parameter results

5.4 Layout of Bandpass Filtering Marchand Balun

The performance of the actual circuit often has some differences from the theoretical results, so the influence of the interference, the coupling, and some other factors should be taken into consideration. Therefore, it is necessary to use the ADS software for the layout simulation.

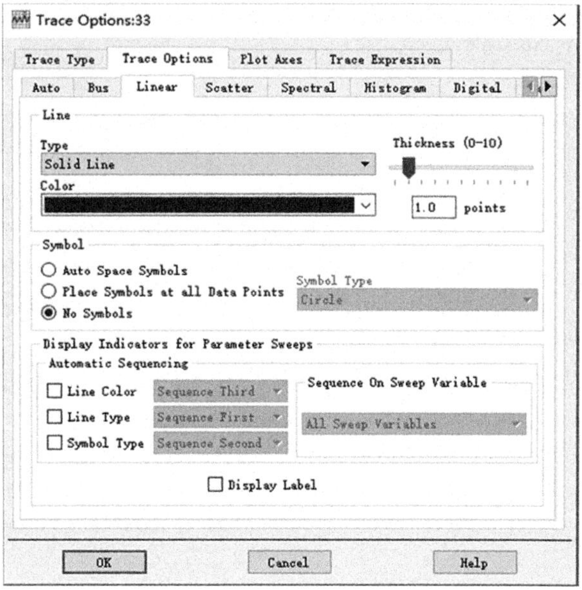

Fig. 5.48 Modification of the dB(S(1,1)) trace

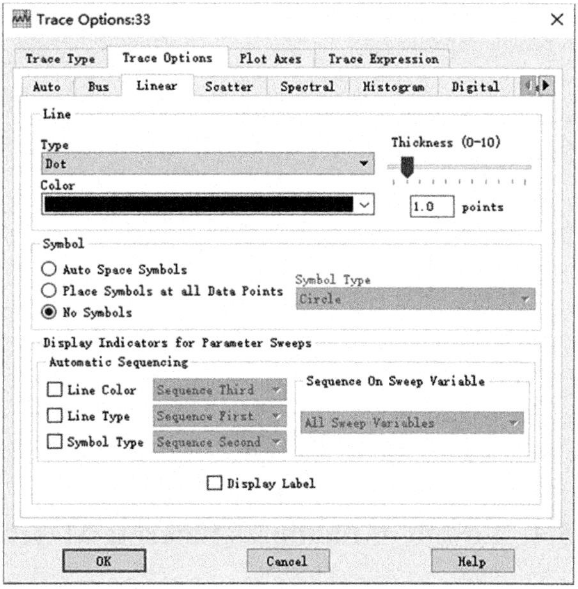

Fig. 5.49 Modification of the dB(S(2,1)) trace

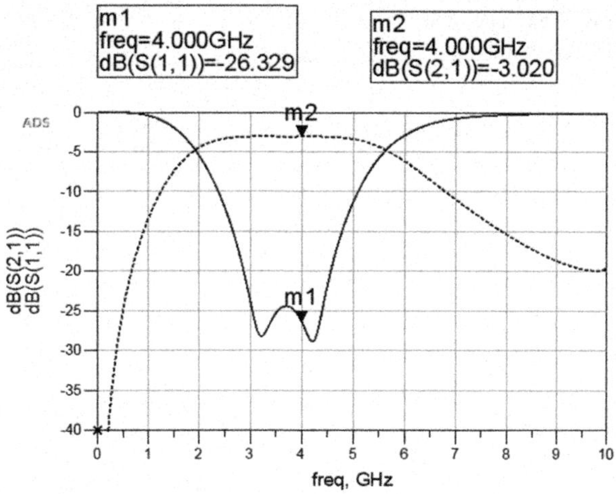

Fig. 5.50 The final dB(S(1,1)) and dB(S(2,1)) traces

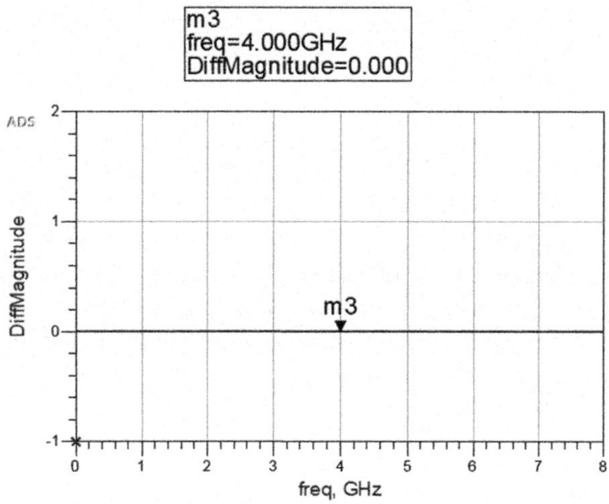

Fig. 5.51 The final DiffMagnitude traces

5.4.1 New Substrate File

All circuit elements are constructed on the gallium arsenide (GaAs) substrate with the thickness of 200 μm, the relative dielectric constant of 12.85, and the loss tangent of 0.006. The 75-nm-thick nickel chromium alloy (NiCr) layer with a block resistance of about 25 Ω/sq is used to realize the thin-film resistance, while the top and bottom 5-μm-thick metals and the 0.2-μm-thick intermediate Si_3N_4 layer construct the MIM capacitor. In addition, an air bridge can be built between the two copper layers to connect the circuits inside the spiral inductor and the peripheral circuits, which can make the layout design more flexible.

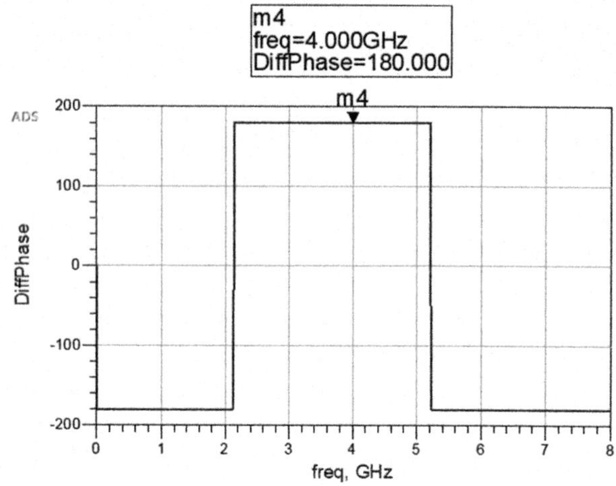

Fig. 5.52 The final DiffPhase traces

Before drawing the layout, the substrate file needs to be set according to the TFIPD technology.

1. New Layout

   Return to the "Bandpass_Filtering_Marchand_Balun_wrk" main interface of the workspace. Execute menu commands [File] → [New] → [Layout…] to open the [New Layout] dialog box in Fig. 5.53. Modify the "Cell" name to "MIM capacitor", then click the [Create Layout] button. After that, the [Choose Layout Technology] dialog box shown in Fig. 5.54 will pop up. Here, select "Standard ADS Layers, 0.001 micron layout resolution", which is 0.001 μm (note that the unit is μm in this chapter), click [Finish], and the new layout window will pop up, as shown in Fig. 5.55.
2. New Substrate

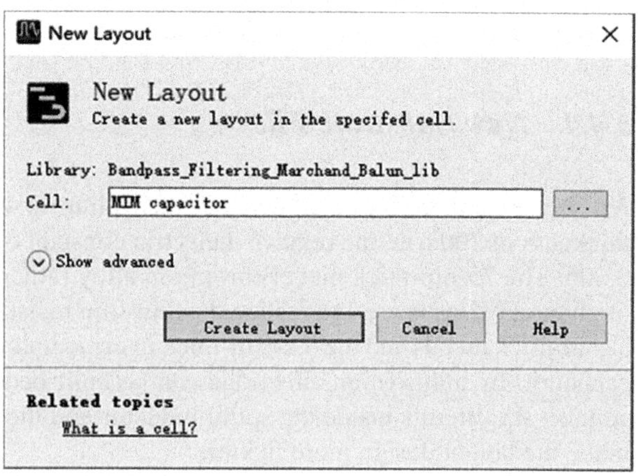

Fig. 5.53 [New Layout] dialog box

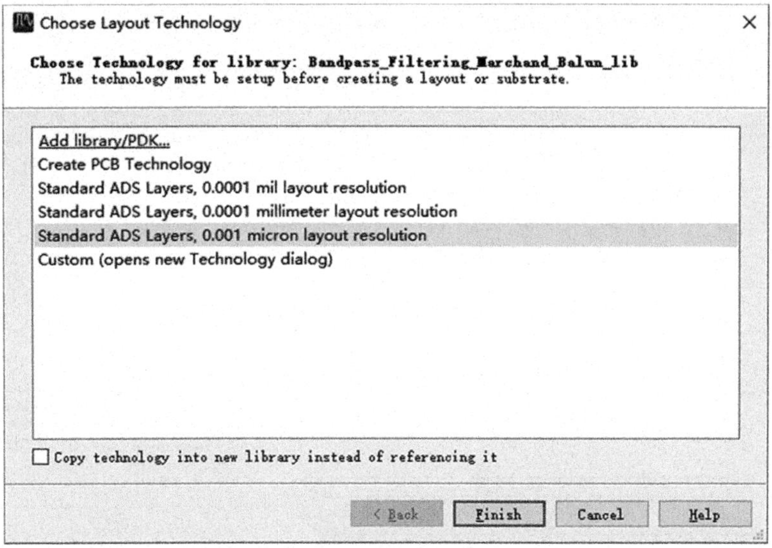

Fig. 5.54 [Choose Layout Technology] dialog box

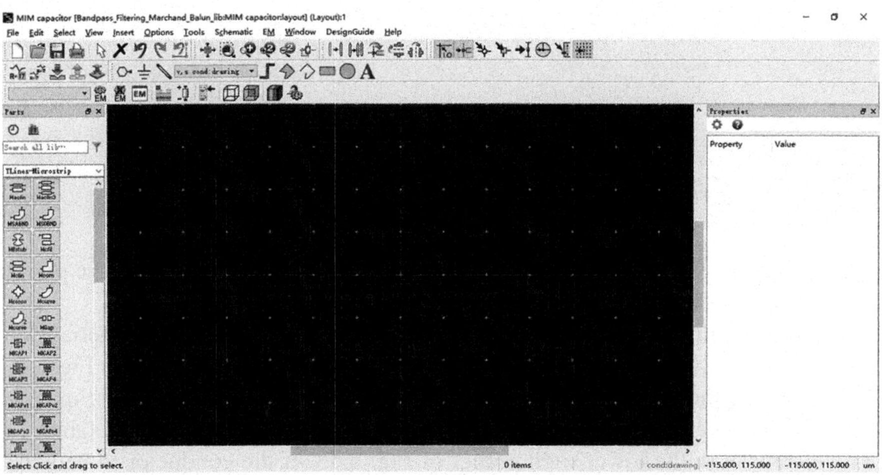

Fig. 5.55 New layout window

(1) Create a new substrate. Execute menu commands [EM] → [Substrate…] in the layout window, then click [OK] in the dialog box which pops up. The [New Substrate] dialog box in Fig. 5.56 is displayed, where "File name" and "Template" can be modified. Keep the default file name here. Because the used TFIPD technology is based on the GaAs substrate, select "100µmGaAs" in the "Template" column and click the [Create Substrate]

Fig. 5.56 [New Substrate] dialog box

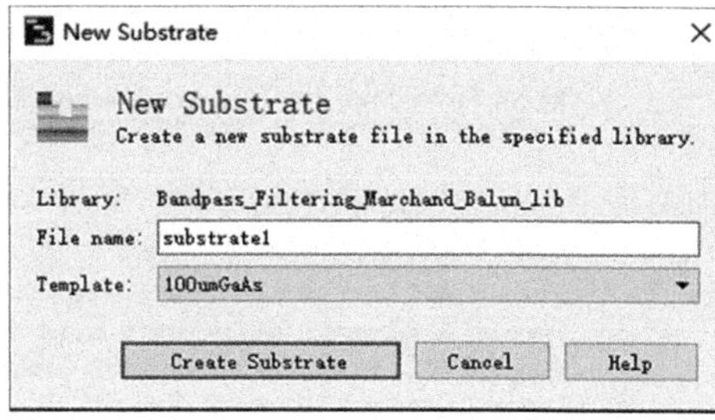

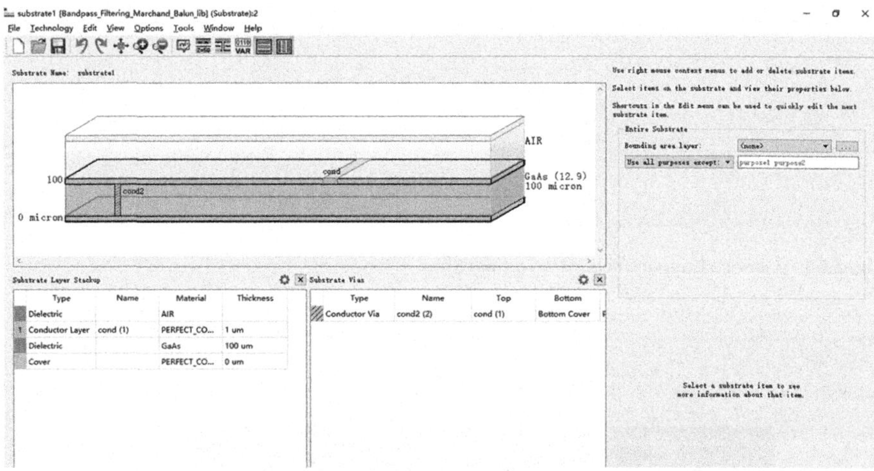

Fig. 5.57 View of all substrate items

button. Then execute menu commands [View] → [View All] in the pop-up window to view all substrate items, as shown in Fig. 5.57.

(2) Add conductors. Execute menu commands [Technology] → [Material Definitions…] to open the [Material Definitions] dialog box shown in Fig. 5.58. Select the [Conductors] tab and define the relevant conductors in Fig. 5.59. The specific steps are listed as follows: Click the [Add From Database…] button in the lower right corner of Fig. 5.58. If the conductor to be added exists in the pop-up [Add Materials From Database] dialog box shown in Fig. 5.60, select the conductor and click [OK] to add it. If the conductor to be added does not exist in the [Add Materials From Database] dialog box, return to the [Material Definitions] dialog box and click the [Add Conductor] button to add a conductor and modify its properties. In addition, you can remove unwanted conductors by clicking the [Remove Conductor] button in the [Material Definitions] dialog box. After finishing all the steps, click the [Apply] button.

(3) Add dielectrics. Select the [Dielectrics] tab and define the relevant dielectrics in Fig. 5.61. The specific steps are listed as follows: Click the [Add From

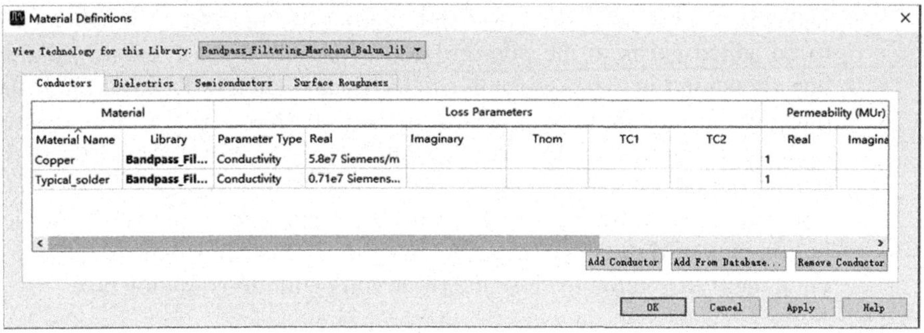

Fig. 5.58 [Material Definitions] dialog box

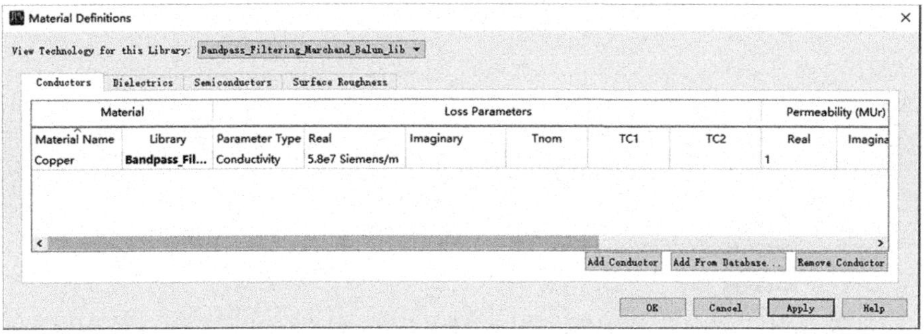

Fig. 5.59 Add of conductors and modification of their properties

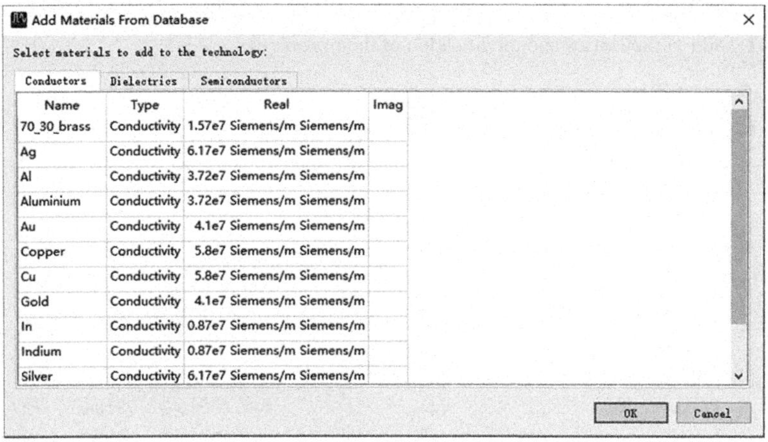

Fig. 5.60 [Conductors] tab of the [Add Materials From Database] dialog box

Database…] button in the lower right corner of Fig. 5.61. If the dielectric to be added exists in the pop-up [Add Materials From Database] dialog box shown in Fig. 5.62, select the dielectric and click [OK] to add it. If the dielectric to be added does not exist in the [Add Materials From Database] dialog box, return to the [Material Definitions] dialog box and click the [Add Dielectric] button to add a dielectric and modify its properties. In addition, you can remove unwanted dielectrics by clicking the [Remove Dielectric] button in the [Material Definitions] dialog box. After finishing all the steps, click the [OK] button to close the [Material Definitions] dialog box.

3. Edit Substrate Items

(1) Select the "cond" layer and click the right mouse button to pop up 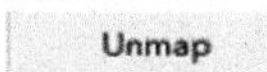

. Click the "Unmap" to delete this layer (Or select the "cond"

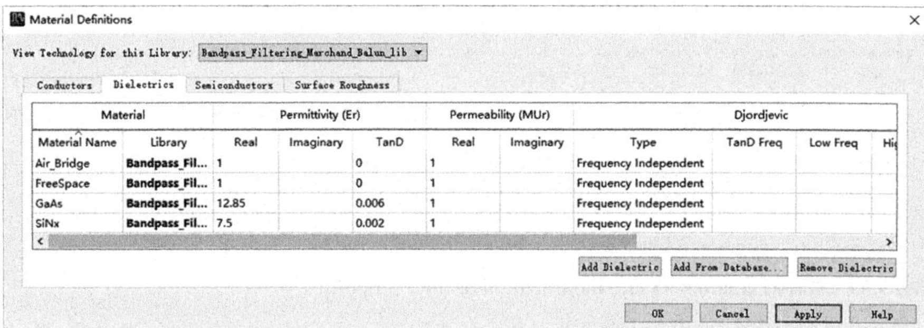

Fig. 5.61 Add of dielectrics and modification of their properties

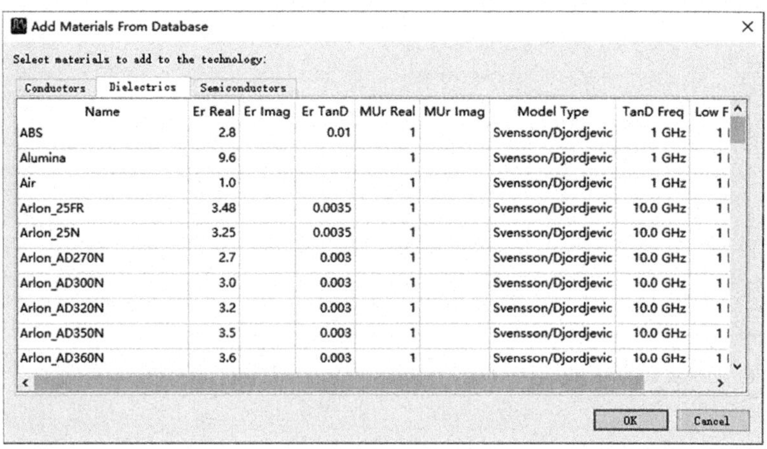

Fig. 5.62 [Dielectrics] tab of the [Add Materials From Database] dialog box

layer, then press "Delete" on the keyboard to delete it). Delete the "cond2" layer in the same way.

(2) Edit substrate layers. Select an existing substrate layer, right-click the mouse, and choose the [Insert Substrate Layer] from the pop-up menu to insert a new substrate layer. Select the substrate layer to be modified to view and modify its properties under the [Substrate Layer] at the right side of the window.

(3) Edit conductor layers. Select the surface of the substrate layer on which you want to insert a conductor layer, right-click the mouse, and choose the [Map Conductor Layer] from the pop-up menu to insert a new conductor layer. Select the conductor layer to be modified to view and modify its properties under the [Conductor Layer] at the right side of the window.

(4) Edit conductor vias. Select the substrate layer into which you want to insert a conductor via, right-click the mouse, and choose the [Map Conductor Via] from the pop-up menu to insert a new conductor via. Select the conductor via to be modified to view and modify its properties under the [Conductor Via] at the right side of the window.

(5) All substrate items are shown in Fig. 5.63. The bottom layer is "Cover". The thickness of the GaAs substrate layer is 200 μm. The thickness of the first SiNx layer is 0.1 μm. For the bond layer, the "Process Role" is selected as "Conductor", the "Material" is selected as "Copper", the "Operation" is selected as "Expand the substrate", the "Position" is selected as "Above interface", and the thickness is 5 μm. The thickness of the second SiNx layer is 0.2 μm. For the text layer, the "Process Role" is selected as "Conductor", the "Material" is selected as "Copper", the "Operation" is selected as "Expand the substrate", the "Position" is selected as "Above interface", and the thickness is 0.5 μm. The thickness of the Air_Bridge layer is 3 μm. For the leads layer, the "Process Role" is selected as "Conductor", the "Material" is selected as "Copper", the "Operation" is selected as "Intrude the substrate", the "Position" is selected as "Above interface", and the thickness is 5 μm. The top layer is the "FreeSpace". For the symbol layer, the "Process Role" is selected as "Conductor Via" and the "Material" is selected as "Copper". For the packages layer, the "Process Role" is selected as "Conductor Via" and the "Material" is selected as "Copper".

(6) Moreover, execute menu commands [Technology] → [Layer Definitions…] to open the [Layer Definitions] dialog box shown in Fig. 5.64, where you can modify the display properties of each layer such as color and pattern, etc. Here, keep all the settings in default.

5.4.2 Layout of MIM Capacitor

To facilitate the layout drawing, select the function key in the box of Fig. 5.65 in the "MIM Capacitor" cell. Execute menu commands [Options] → [Preferences…],

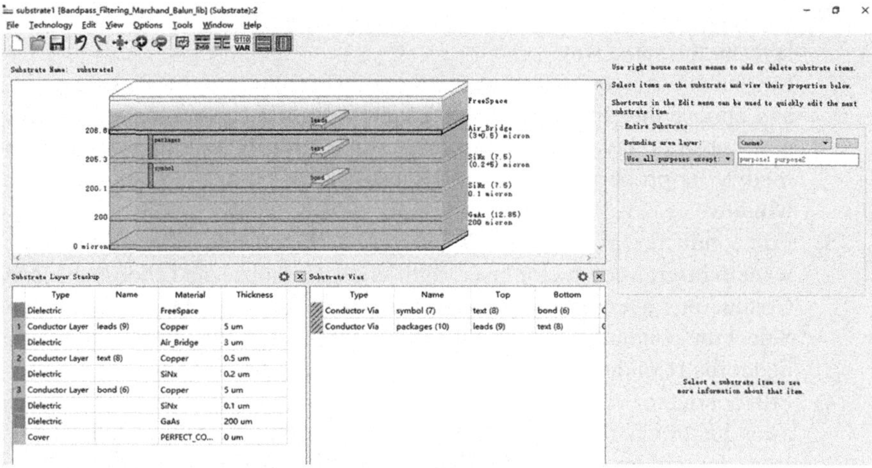

Fig. 5.63 Substrate layer stackup and substrate vias

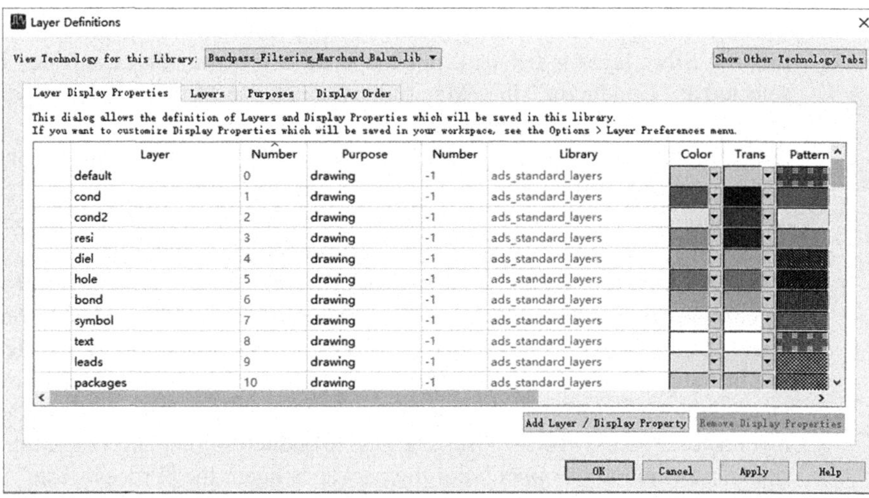

Fig. 5.64 [Layer Display Properties] tab of the [Layer Definitions] dialog box

select the [Grid/Snap] tab in the pop-up [Preferences for Layout] dialog box to modify the display grid in the layout. Set the "Snap Grid Distance (in layout units)", "Snap Grid Per Minor Display Grid", and the "Minor Grid Per Major Display Grid" to appropriate values, as shown in Fig. 5.66. (Or right-click in the layout drawing area and select "<0.1-1-100>" under [Grid Spacing…]; Or use the shortcut key "Ctrl + Shift + 8").

1. Layout of MIM Capacitor

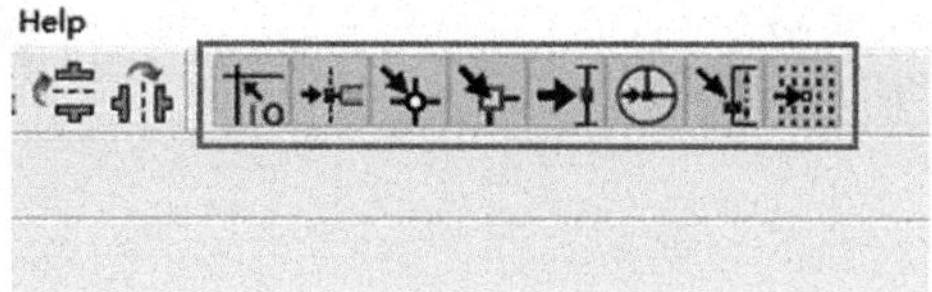

Fig. 5.65 Selection of the function key in the box

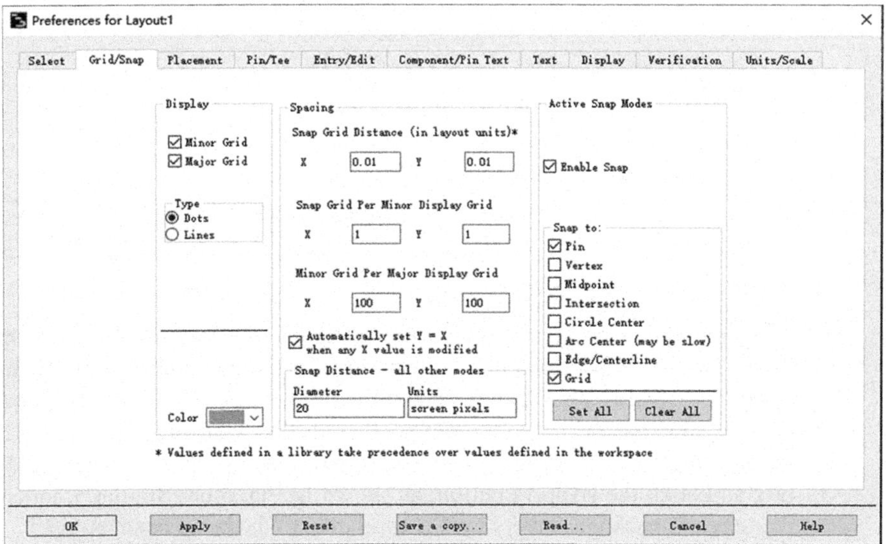

Fig. 5.66 Modification of the display grid in the layout

Two 5-μm-thick copper layers (bond and leads layers) and a 0.2-μm-thick intermediate Si_3N_4 layer are used to construct the MIM capacitor. The area and the thickness of the intermediate dielectric layer determine the value of the MIM capacitor. The MIM capacitor with the capacitance of 0.37 pF is taken as an example to describe its drawing steps in detail below.

(1) MIM capacitor lamination. Execute menu commands [Insert] → [Rectangle] to insert a rectangle into the layout, then press "Esc" to exit. Select the inserted rectangle, select "bond: drawing" in the drop-down menu of the [All Shapes] → [Layer] under [Properties] at the right side of the window, and set the [Rectangles] → [Width] to 35.4 μm, [Height] to 40.4 μm. Execute menu commands [Edit] → [Copy/Paste] → [Copy To Layer…], select [text: drawing] in the pop-up [Copy to Layer] dialog box, as shown in Fig. 5.67. Click the [Apply] button to copy a text layer in the original position. Similarly, copy a leads layer and a packages layer in the original position, respectively. After copying all the layers, click the [Cancel] button to close this dialog box.

Fig. 5.67 [Copy to Layer] dialog box

(2) The indentation of layers. There are different indentations between the layers of the MIM capacitor. Select the leads layer and execute menu commands [Edit] → [Scale/Oversize] → [Oversize…]. Since the leads layer is indented by 1.5 μm compared with the bond layer, enter − 1.5 in the "Oversize(+)/Undersize(−)" column of the pop-up [Oversize] dialog box and click the [Apply] button, as shown in Fig. 5.68. Similarly, indent the text and packages layers by 3 μm relative to the bond layer.

(3) Draw connection parts. Execute menu commands [Insert] → [Rectangle] to insert a rectangle into the layout, then press "Esc" to exit. Select the inserted rectangle, select "bond: drawing" in the drop-down menu of the

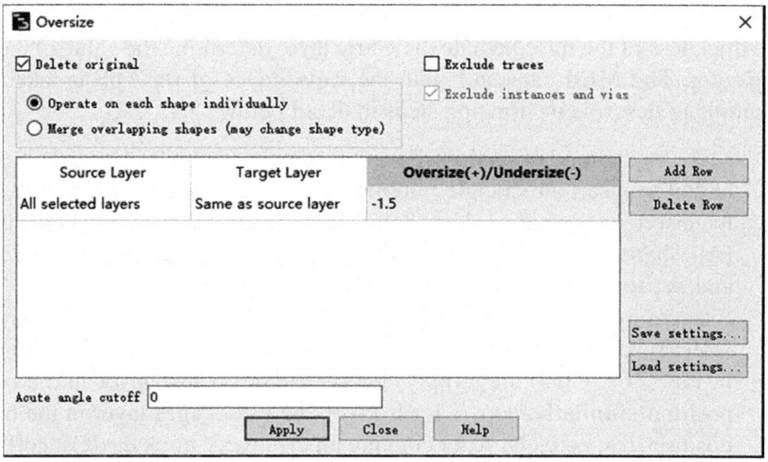

Fig. 5.68 [Oversize] dialog box

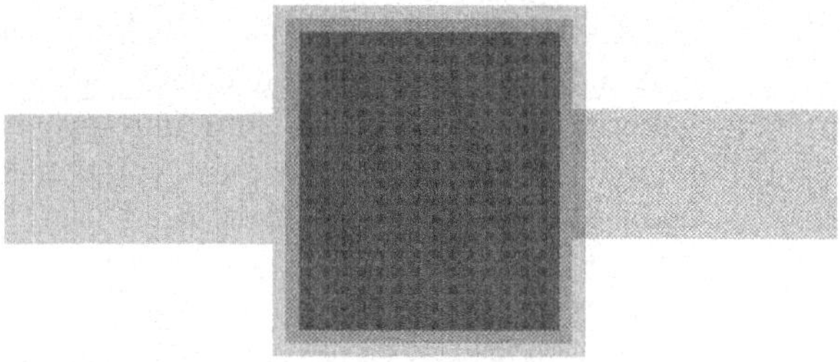

Fig. 5.69 Final layout of a MIM capacitor

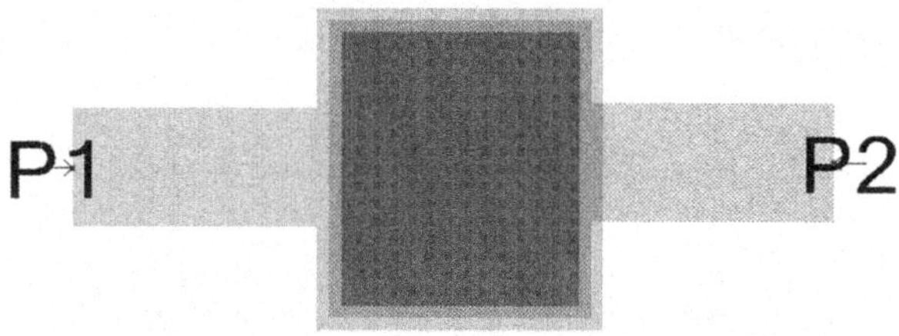

Fig. 5.70 Insertion of pins

[All Shapes] → [Layer] under [Properties] at the right side of the window, and set the [Rectangles] → [Width] to 30 μm, [Height] to 15 μm. Select and long press the modified rectangle by the left mouse button to move it to the middle position of the MIM capacitor and connect it with the original bond layer. Similarly, a leads layer rectangle is inserted in the opposite position and is connected to the original leads layer. The layout of a MIM capacitor has been drawn, as shown in Fig. 5.69.

2. Layout Simulation of MIM Capacitor

(1) Insert ports. Execute menu commands [Insert] → [Pin] and click the left mouse button to insert two pins into the I/O ports of the MIM capacitor, as shown in Fig. 5.70.

(2) Modify simulation settings. Execute menu commands [EM] → [Simulation Settings…], click the [Create EM Setup View] button in the pop-up the [New EM Setup View] dialog box, as shown in Fig. 5.71, and the [EM Setup for simulation] window will pop up, as shown in Fig. 5.72. Select EM simulator first. The second method "Momentum Microwave" is usually used, because this method runs fast and its accuracy can also meet the application requirement (The first method "Momentum RF" is the fastest one but shows the least accuracy. The third method "FEM" has the highest accuracy, but shows the slowest simulation speed, so it is mainly used for some complex three-dimensional structure simulations). Then select the [Frequency Plan] tab to set the simulation frequency. Select "Adaptive" in the "Type" column and set the "Fstart" to 0 GHz, the "Fstop" to 10 GHz, and the "Npts" to 10. Finally, select the [Options] tab. Click "Preprocessor", select the "User specified snap distance" option under the "Heal the layout",

and set the distance to 2.5 μm. Click "Mesh" and select the "Edge Mesh" option. Keep other settings in default. After finishing all the steps, close the window and click the [OK] button to save the settings.

(3) Layout simulation. Execute menu commands [EM] → [Simulate] to simulate the layout of the MIM capacitor. During the simulation, a status window will pop up to display the current item's progress. After the finishing of the simulation, the data display window will be opened automatically. Refer to the steps in Sect. 5.3.2 to view and handle dB(S(1,1)) and dB(S(2,1)) traces, the final results are shown in Fig. 5.73.

3. Co-Simulation of MIM Capacitor

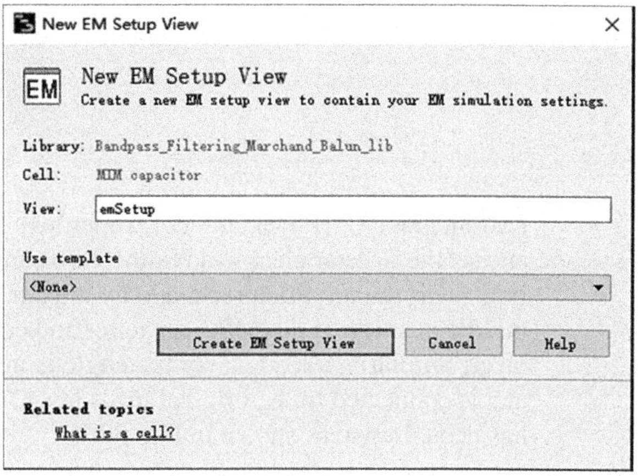

Fig. 5.71 [New EM Setup View] dialog box

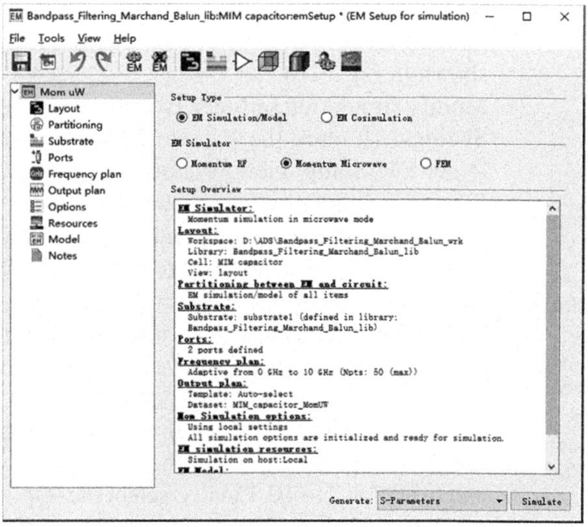

Fig. 5.72 [EM Setup for simulation] window

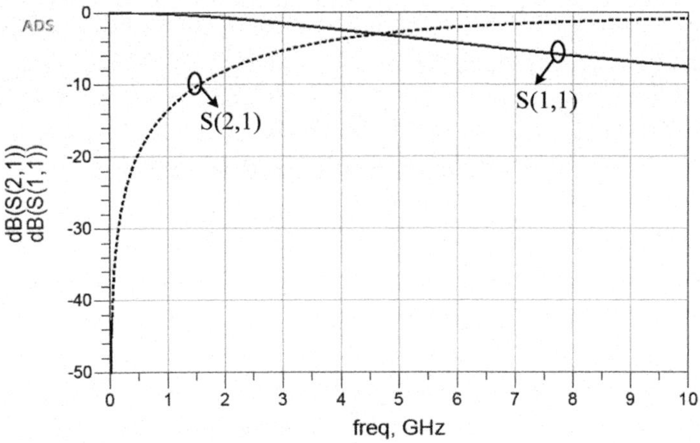

Fig. 5.73 Final simulated results of the drawn MIM capacitor

To verify whether the capacitance of the drawn MIM capacitor is 0.37 pF, the co-simulation of capacitor schematic and layout is required.

(1) Create EM model and symbol of the MIM capacitor. In the layout window, execute menu commands [EM] → [Component] → [Create EM Model And Symbol…], then click the [OK] button in the pop-up dialog boxes. Execute menu commands [Edit] → [Component] → [Update Component Definitions…], then click the [OK] button in the pop-up dialog box to complete the creation of the EM model and the symbol in the MIM capacitor.

(2) Create a new schematic and insert the MIM capacitor component. Return to the "Bandpass_Filtering_Marchand_Balun_wrk" main interface of the workspace. Execute menu commands [File] → [New] → [Schematic…], modify the "Cell" name to "MIM capacitor-cosimulation" in the opened [New Schematic] dialog box, and click the [Create Schematic] button to create a new schematic. Click the [Open the Library Browser] icon at the left side of the schematic window, then select the "MIM capacitor" component under the [Workspace Libraries] in the displayed [Component Library] window, as shown in Fig. 5.74. Click the right mouse button, then click the [Place Component] in the pop-up menu to insert a MIM capacitor component in the schematic. Press "Esc" to exit.

(3) Insert ideal capacitor. Select the [Lumped-Components] from the drop-down menu in the left component palette list and click the [Capacitor] icon to insert a capacitor. Press "Esc" on the keyboard to exit. Double-click the capacitor and modify C = 0.37 pF in the [Edit Instance Parameter] pop-up dialog box (check whether the unit setting is consistent), then click the [OK] button to save the parameter.

(4) Insert the S-parameter simulator, term ports, and grounds. Select the [Simulation-S_Param] from the drop-down menu in the left component

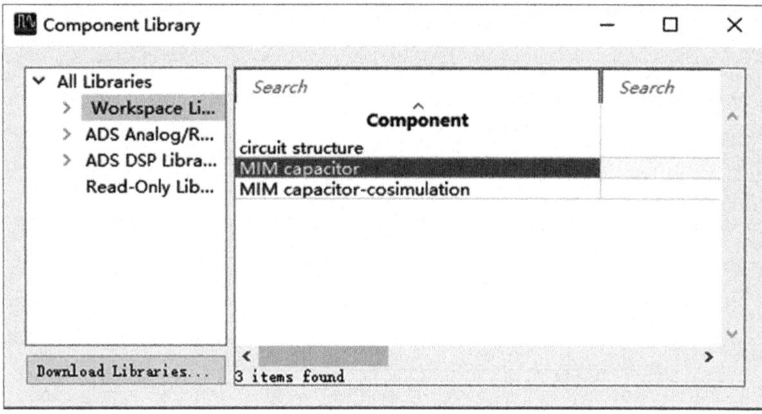

Fig. 5.74 MIM capacitor component in the [Component Library] window

palette list and click the [S-parameter Simulator] icon to insert an S-parameter simulator. Click the [Port Impedance Termination for S-Parameters] icon to insert four term ports, then press "Esc" to exit. Execute menu commands [Insert] → [GROUND] to place four grounds (Or click the [TermG Port Impedance Termination for S-Parameters with Grounded Reference] icon to insert four term ports with the grounded reference). Execute menu commands [Insert] → [Wire] to connect the inserted components. After finishing all the steps, press "Esc" to exit.

(5) Set simulation frequency. Double-click the S-parameter simulator S-PARAMETERS. Set the "Start" to 0 GHz, the "Stop" to 10 GHz, and the "Step-size" to 0.01 GHz, and click the [OK] button. The final capacitor co-simulation circuit is obtained, as shown in Fig. 5.75.

(6) Co-simulation. Execute menu commands [Simulate] → [Simulate]. After the simulation completes, the data display window will be opened automatically. Refer to the steps in Sect. 5.3.2 to view and handle traces of dB(S(1,1)), dB(S(2,1)), dB(S(3,3)), and dB(S(4,3)), the final results are shown in Fig. 5.76. It can be seen that traces dB(S(1,1)) and dB(S(3,3)), traces dB(S(2,1)) and dB(S(4,3)) are almost coincident, indicating that the capacitance of the drawn MIM capacitor is approximately equal to 0.37 pF. If the differences between these traces are large, go back to modify the MIM capacitor layout and repeat the steps above until the error of two traces is in the range of acceptance (within an acceptable range).

5.4.3 Layout of Spiral Inductor

1. Layout Drawing of Spiral Inductor

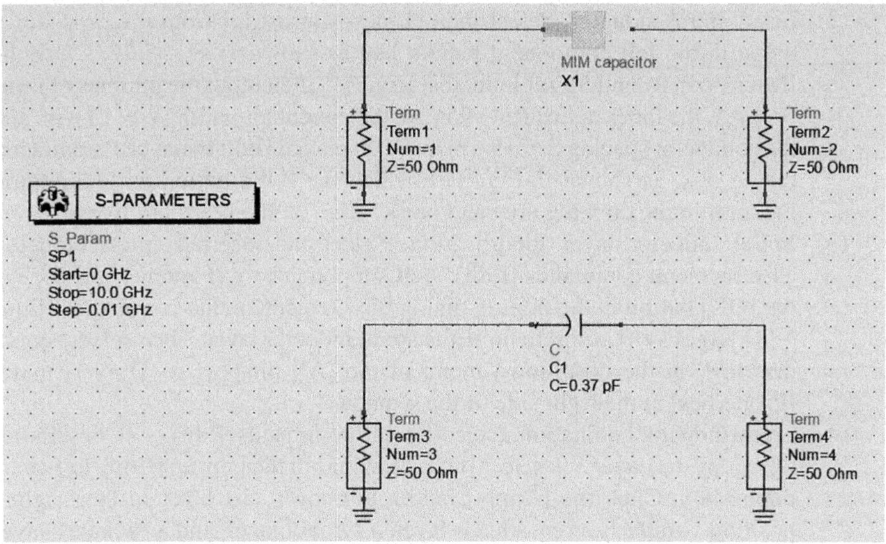

Fig. 5.75 Capacitor co-simulation circuit

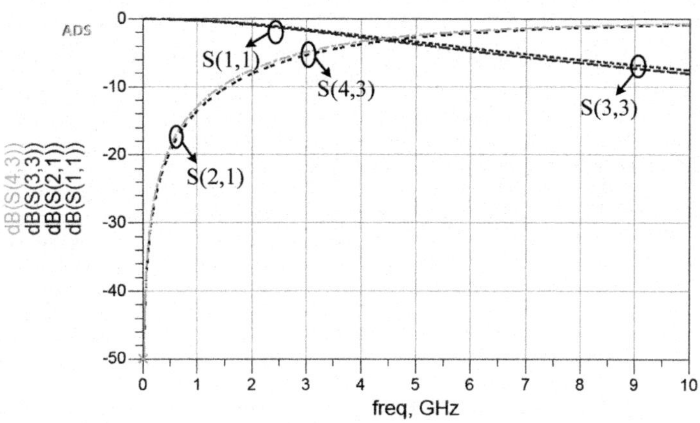

Fig. 5.76 S-parameter results of the capacitor co-simulation

The inductance value of a spiral inductor is related to its inner radius, number of turns, width, and spacing of winding. The spiral inductor with the inductance of 2.00 nH is taken as an example to describe its drawing steps in detail below.

(1) In the "Bandpass_Filtering_Marchand_Balun_wrk" main interface of the workspace, execute menu commands [File] → [New] → [Layout…] and modify the "Cell" name to "spiral inductor" in the opened [New Layout] dialog box. Click the [Create Layout] button, a new layout window will pop up.

(2) Insert spiral inductor. Select the [TLines-Microstrip] from the drop-down menu in the left component palette list, as shown in Fig. 5.77. Click the [Microstrip Round Spiral Inductor] icon , then input the number of turns N = 2.5, the inner radius Ri = 95 μm, the conductor width W = 15 μm, and the conductor spacing S = 15 μm in the displayed [Edit Instance Parameters] dialog box, as shown in Fig. 5.78. Click the [OK] button to add a spiral inductor in the drawing area and press "Esc" on the keyboard to exit.
(3) Spiral inductor layer modification. Select the inserted spiral inductor, execute menu commands [Edit] → [Component] → [Flatten…], and click the [OK] button in the pop-up dialog box. Execute menu commands [Edit] → [Merge] → [Union] to unite the spiral inductor layer. Then select "bond: drawing" in the drop-down menu of the [All Shapes] → [Layer] under [Properties] at the right side of the window.
(4) Spiral inductor lamination. Execute menu commands [Edit] → [Copy/Paste] → [Copy To Layer…], select [text: drawing] in the pop-up [Copy to Layer] dialog box. Click the [Apply] button to copy a text layer in the original position. Similarly, copy a leads layer, a symbol layer, and a packages layer in the original position, respectively. After copying all the layers, click the [Cancel] button to close this dialog box.
(5) Air bridge construction. In order to make circuits inside the spiral inductors connect with the peripheral circuits, air bridges are constructed. Execute menu commands [Insert] → [Rectangle] to insert a rectangle in the layout,

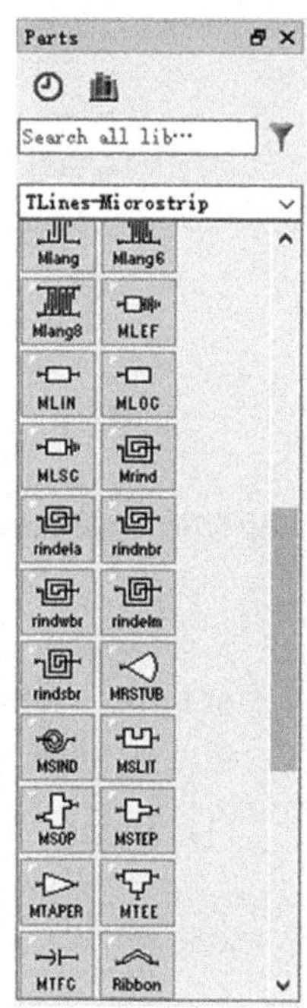

Fig. 5.77
[TLines-Microstrip] list

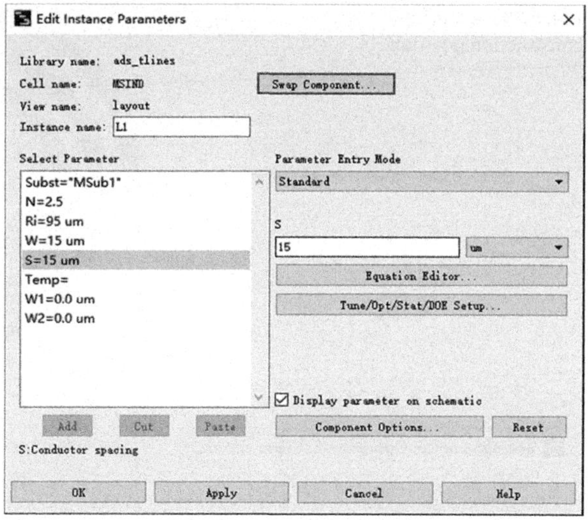

Fig. 5.78 [Edit Instance Parameters] dialog box

then press "Esc" to exit. Select the inserted rectangle, select "cond: drawing" in the drop-down menu of the [All Shapes] → [Layer] under [Properties] at the right side of the window, and set the [Rectangles] → [Width] to an arbitrary value, [Height] to 50 μm. Select and long press the modified rectangle with the left mouse button to move it to the air bridge construction position, as shown in Fig. 5.79. Press the shortcut "Ctrl + A" to select all layers. Execute menu commands [Edit] → [Boolean Logical...] in the pop-up [Boolean Logical Operation Between Layers] dialog box to modify the Boolean logical operation between the bond layer and the cond layer according to Fig. 5.80, then click [Apply] button to complete the operation. Similarly, complete the Boolean operations between the text layer, symbol layer, packages layer, and the cond layer, only retain the top leads metal layer, and click the [Cancel] button to close the dialog box. Then select the cond layer and press "Delete" on the keyboard to delete it. The spiral inductor after the construction of an air bridge is shown in Fig. 5.81.

(6) The indentation of layers. There are different indentations between the layers of spiral inductor. Because the layer has been disconnected after the Boolean logical operation, execute menu commands [View] → [Layer View] → [By Name...]. Then, in the [Layout Layers] dialog box shown in Fig. 5.82, select [text: drawing] to only display the text layer of the spiral inductor. Press the shortcut "Ctrl + A" to select the text layer and execute menu commands [Edit] → [Scale/Oversize] → [Oversize...] to open the dialog box. Since the text layer is indented by 2 μm compared with the bond layer, enter − 2 in the "Oversize(+)/Undersize(−)" column of the pop-up [Oversize] dialog box and click the [Apply] button. Similarly, indent the symbol and packages layers by 2 μm relative to the bond layer.

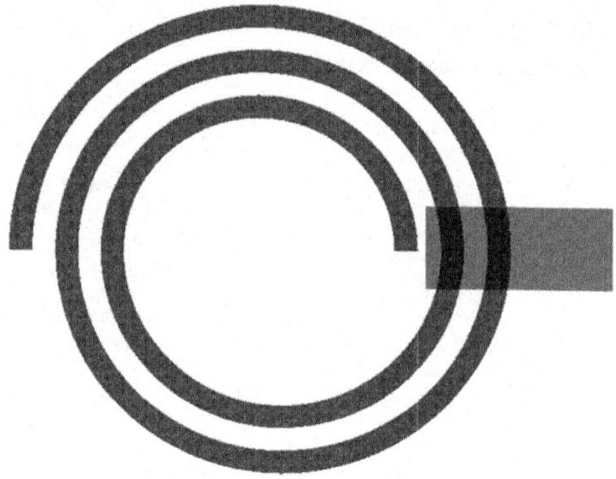

Fig. 5.79 Air bridge construction position

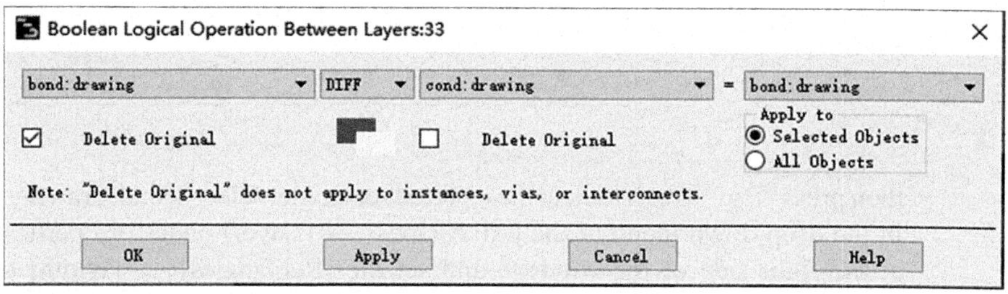

Fig. 5.80 [Boolean Logical Operation Between Layers] dialog box

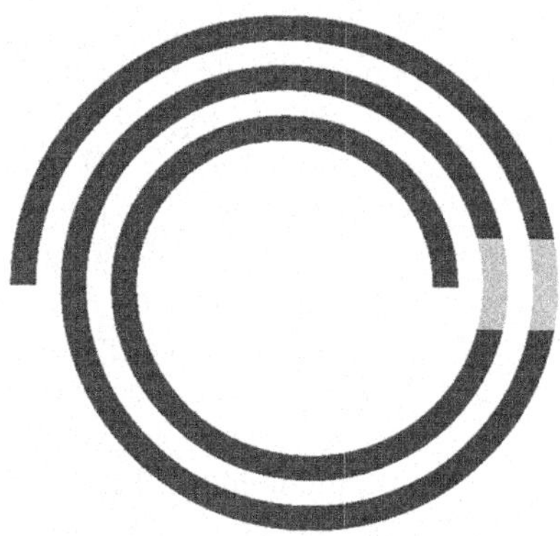

Fig. 5.81 Spiral inductor after the construction of an air bridge

(7) Draw connection parts. Execute menu commands [Insert] → [Rectangle] to insert a rectangle in the layout, then press "Esc" to exit. Select the inserted rectangle, select "bond: drawing" in the drop-down menu of the [All Shapes] → [Layer] under [Properties] at the right side of the window, and set the [Rectangles] → [Height] to 15 μm. [Width] can be set according to the specific size of the air bridge. Select and long press the modified rectangle by the left mouse button to move it to the middle position of the air bridge

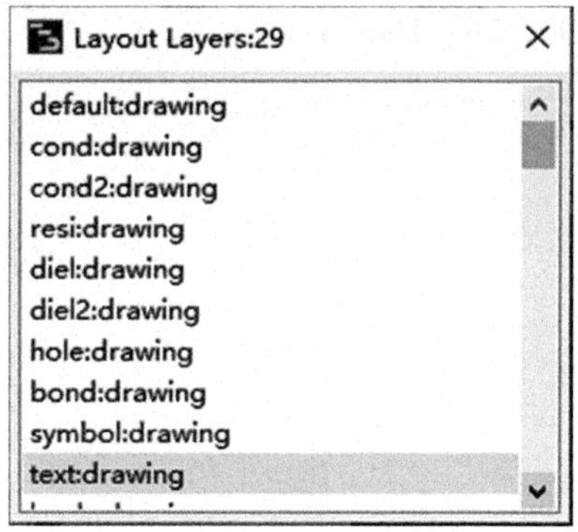

Fig. 5.82 [Layout Layers] dialog box

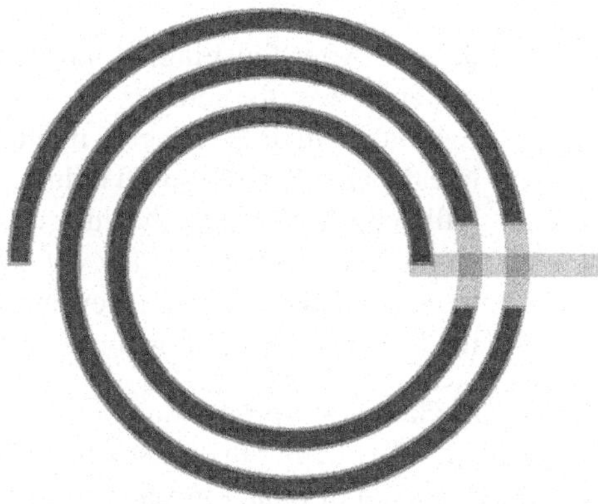

Fig. 5.83 Final layout of a spiral inductor

and connect it with the original bond layer. The layout of a spiral inductor has been drawn, as shown in Fig. 5.83.

2. Layout Simulation of Spiral Inductor

(1) Insert ports. Execute menu commands [Insert] → [Pin] and click the left mouse button to insert two pins into the I/O ports of the spiral inductor, as shown in Fig. 5.84.

(2) Modify simulation settings. Execute menu commands [EM] → [Simulation Settings…], click the [Create EM Setup View] button in the pop-up the [New EM Setup View] dialog box, and the [EM Setup for simulation] window will pop up. Select the second method “Momentum Microwave” of the EM simulator at first. Then select the [Frequency Plan] tab to set the simulation frequency. Select “Adaptive” in the “Type” column and set the “Fstart” to 0 GHz, the “Fstop” to 10 GHz, and the “Npts” to 10. Finally, select the [Options] tab. Click “Preprocessor”, select the “User specified snap distance” option under the “Heal the layout”, and set the distance to 2.5 μm. Click “Mesh” and select the “Edge Mesh” option. Keep other settings in

Fig. 5.84 Insertion of pins

default. After finishing all the steps, close the window and click the [OK] button to save the settings.

(3) Layout simulation. Execute menu commands [EM] → [Simulate] to simulate the layout of the spiral inductor. During the simulation, a status window will pop up to display the current item's progress. After the finishing of the simulation, the data display window will be opened automatically. Refer to the steps in Sect. 5.3.2 to view and handle traces dB(S(1,1)) and dB(S(2,1)), the final results are shown in Fig. 5.85.

3. Co-simulation of Spiral Inductor

To verify whether the inductance of the drawn spiral inductor is 2.00 nH, the co-simulation of inductor schematic and layout is required.

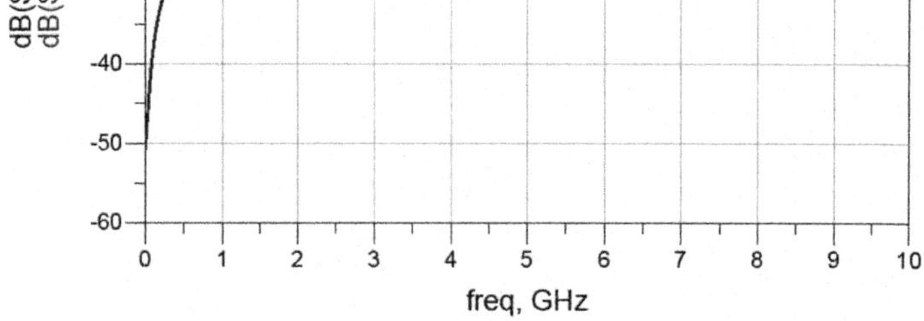

Fig. 5.85 Final simulated results of the drawn spiral inductor

(1) Create EM model and symbol of the spiral inductor. In the layout window, execute menu commands [EM] → [Component] → [Create EM Model And Symbol…], then click the [OK] button in the pop-up dialog boxes. Execute menu commands [Edit] → [Component] → [Update Component Definitions…], then click the [OK] button in the pop-up dialog box to complete the creation of the EM model and the symbol in the spiral inductor.

(2) Create a new schematic and insert the spiral inductor component. Return to the "Bandpass_Filtering_Marchand_Balun_wrk" main interface of the workspace. Execute menu commands [File] → [New] → [Schematic…], modify the "Cell" name to "spiral inductor-cosimulation" in the opened [New Schematic] dialog box, and click the [Create Schematic] button to create a new schematic. Click the [Open the Library Browser] icon at the left side of the schematic window, then select the "spiral inductor" component under the [Workspace Libraries] in the displayed [Component Library] window. Click the right mouse button, then click the [Place Component] in the pop-up menu to insert a spiral inductor component into the schematic. Press "Esc" to exit.

(3) Insert ideal inductor. Select the [Lumped-Components] from the drop-down menu in the left component palette list and click the [Inductor] icon to insert an inductor. Press "Esc" on the keyboard to exit. Double-click the inductor and modify L = 2.00 nH in the [Edit Instance Parameter] pop-up dialog box (check whether the unit setting is consistent), then click the [OK] button to save the parameter.

(4) Insert the S-parameter simulator, term ports, and grounds. Select the [Simulation-S_Param] from the drop-down menu in the left component palette list and click the [S-parameter Simulator] icon to insert an S-parameter simulator. Click the [Port Impedance Termination for S-Parameters] icon to insert four term ports, then press "Esc" to exit. Execute menu commands [Insert] → [GROUND] to place four grounds (Or click the [TermG Port Impedance Termination for S-Parameters with Grounded Reference] icon to insert four term ports with the grounded reference). Execute menu commands [Insert] → [Wire] to connect the inserted components. After finishing all the steps, press "Esc" to exit.

(5) Set simulation frequency. Double-click the S-parameter simulator S-PARAMETERS. Set the "Start" to 0 GHz, the "Stop" to 10 GHz, and the "Step-size" to 0.01 GHz, and click the [OK] button. The final inductor co-simulation circuit is obtained, as shown in Fig. 5.86.

(6) Co-simulation. Execute menu commands [Simulate] → [Simulate]. After the simulation completes, the data display window will be opened automatically. Refer to the steps in Sect. 5.3.2 to view and handle traces of dB(S(1,1)), dB(S(2,1)), dB(S(3,3)), and dB(S(4,3)), the final results are shown in Fig. 5.87. It can be seen that traces dB(S(1,1)) and dB(S(3,3)), traces dB(S(2,1)) and dB(S(4,3)) are almost coincident, indicating that the inductance of the drawn spiral inductor is approximately equal to 2.00 nH. If the differences between these traces are large, go back to modify the spiral

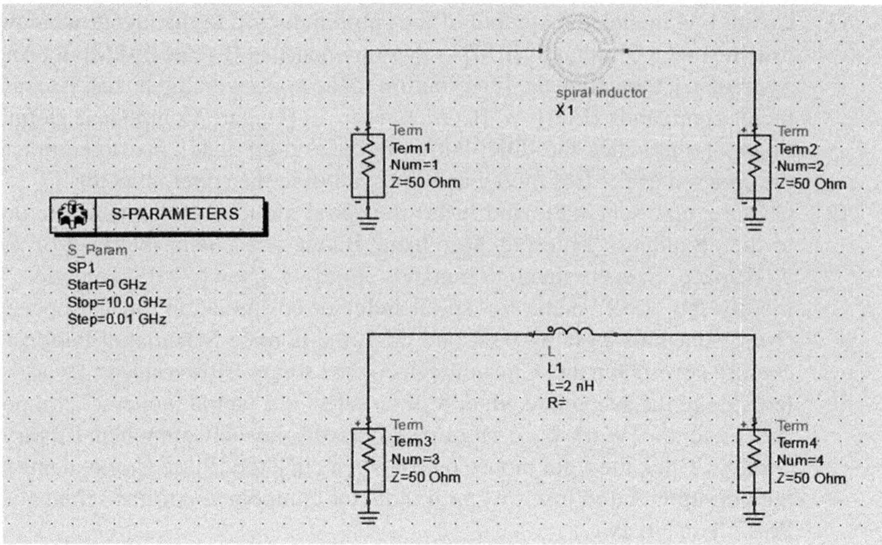

Fig. 5.86 Inductor co-simulation circuit

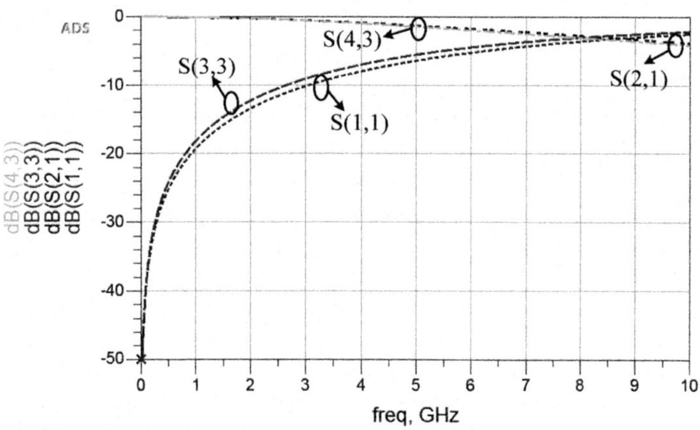

Fig. 5.87 S-parameter results of the inductor co-simulation

inductor layout and repeat the steps above until the error of two traces is in the range of acceptance (within an acceptable range).

5.4.4 *Layout of Spiral Coupled Line*

1. Layout Drawing of Spiral Coupled Line

The spiral coupled line with the odd-mode impedance of 232.5 Ω and the even-mode impedance of 30.67 Ω is taken as an example to describe its drawing steps in detail below.

(1) Create a new layout. In the "Bandpass_Filtering_Marchand_Balun_wrk" main interface of the workspace, execute menu commands [File] → [New] → [Layout...] and modify the "Cell" name to "spiral coupled line" in the opened [New Layout] dialog box. Click the [Create Layout] button, a new layout window will pop up.
(2) Insert spiral microstrip lines. Select the [TLines-Microstrip] from the drop-down menu in the left component palette list. Click the [Microstrip Round Spiral Inductor] icon , then input the number of turns N = 2.75, the inner radius Ri = 125 μm, and the conductor width W = 15 μm in the displayed [Edit Instance Parameters] dialog box. Since the spiral coupled line is formed by two microstrip lines in crossing spiral, the spacing S is modified to 45 μm. Click the [OK] button to add two spiral microstrip lines in the drawing area, then press "Esc" on the keyboard to exit
(3) Layer modification of spiral microstrip line. Select the inserted spiral microstrip lines, execute menu commands [Edit] → [Component] → [Flatten...], and click the [OK] button in the pop-up dialog box. Execute menu commands [Edit] → [Merge] → [Union] to unite the spiral inductor layer. Then select "bond: drawing" in the drop-down menu of the [All Shapes] → [Layer] under [Properties] at the right side of the window.
(4) Spiral microstrip line rotation and movement. Select one of the spiral microstrip lines, click the right mouse button, and select [Rotate] from the pop-up menu to rotate it in 180° clockwise. Long press the rotated spiral microstrip line by the left mouse button to move it to the position shown in Fig. 5.88. Execute the menu command [Edit] → [Move] → [Move Relative...], modify [Delta X] to – 220 (unit: μm) in the pop-up [Move Relative] dialog box, as shown in Fig. 5.89. Click the [OK] button to close this dialog box. Two spiral microstrip lines are wound into the spiral coupled line shown in Fig. 5.90 (the winding spacing is 15 μm).
(5) Spiral coupled line lamination. Press the shortcut "Ctrl + A" to select the inserted spiral coupled line. Execute menu commands [Edit] → [Copy/Paste] → [Copy To Layer...], select [text: drawing] in the pop-up [Copy to Layer] dialog box. Click the [Apply] button to copy a text layer in the original position. Similarly, copy a leads layer, a symbol layer, and a packages layer in the original position, respectively. After copying all the layers, click the [Cancel] button to close this dialog box.
(6) Air bridge construction. In order to make the circuits inside spiral coupled lines connect with the peripheral circuits, air bridges are constructed. Execute menu commands [Insert] → [Rectangle] to insert two rectangles in the layout, then press "Esc" to exit. Select the inserted rectangles successively, select "cond: drawing" in the drop-down menu of the [All Shapes] →

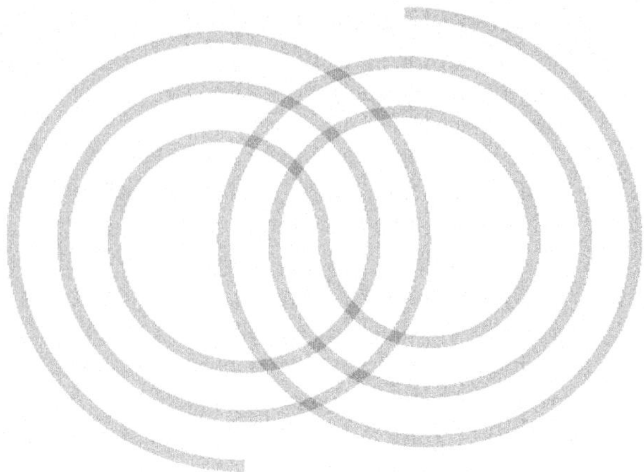

Fig. 5.88 Relative position of the two spiral microstrip lines before the moving

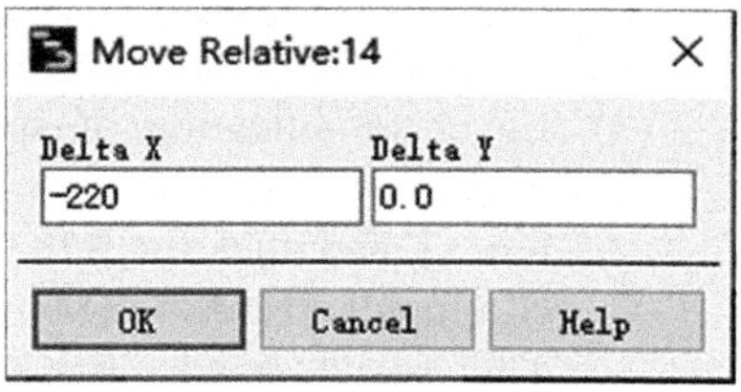

Fig. 5.89 [Move Relative] dialog box

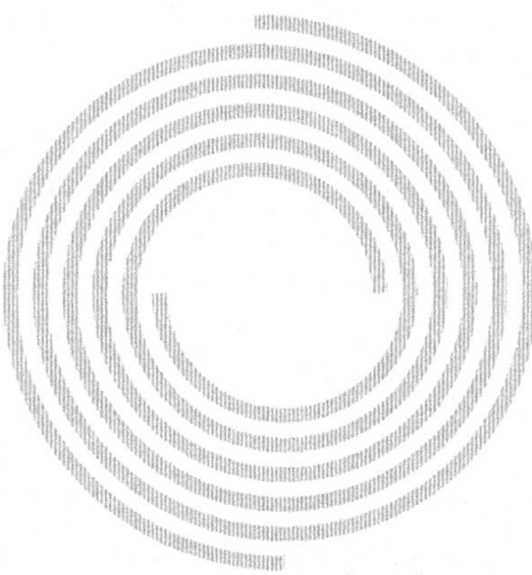

Fig. 5.90 Relative position of the two spiral microstrip lines after the moving

[Layer] under [Properties] at the right side of the window, and set the [Rectangles] → [Width] to an arbitrary value, [Height] to 50 μm. Select and long press the modified rectangle by the left mouse button successively to move them to the air bridge construction position, as shown in Fig. 5.91. Press the shortcut "Ctrl + A" to select all layers. Execute menu commands [Edit] → [Boolean Logical...] in the pop-up [Boolean Logical Operation Between Layers] dialog box to modify the Boolean logical operation between the bond layer and the cond layer, then click the [Apply] button to complete the operation. Similarly, complete the Boolean operations among the text layer, symbol layer, packages layer, and the cond layer, only retain the top leads metal, and click the [Cancel] button to close the dialog box. Then long press the "Ctrl" key to select the two cond layer and press "Delete" on the keyboard to delete them. The spiral coupled line after the construction of two air bridges is shown in Fig. 5.92.

(7) The indentation of layers. There are different indentations between the layers of spiral coupled lines. Since the layer has been disconnected after the Boolean logical operation, execute menu commands [View] → [Layer View] → [By Name...]. Then, in the [Layout Layers] dialog box shown, select [text: drawing] to only display the text layer of the spiral coupled line. Press the shortcut "Ctrl + A" to select the text layer and execute menu commands [Edit] → [Scale/Oversize] → [Oversize...] to open the dialog box. Since the text layer is indented by 2 μm compared with the bond layer,

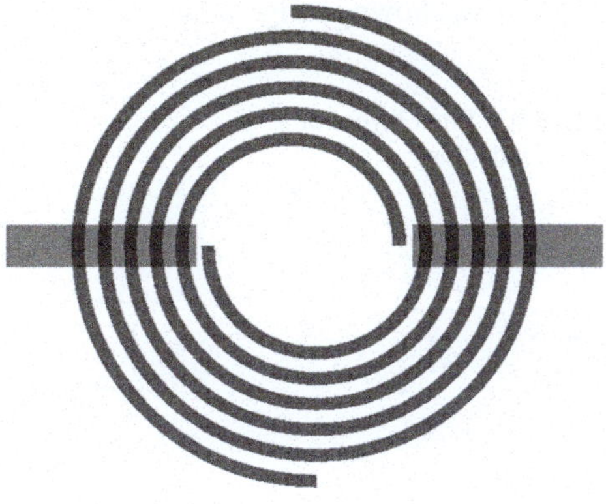

Fig. 5.91 Air bridge construction positions

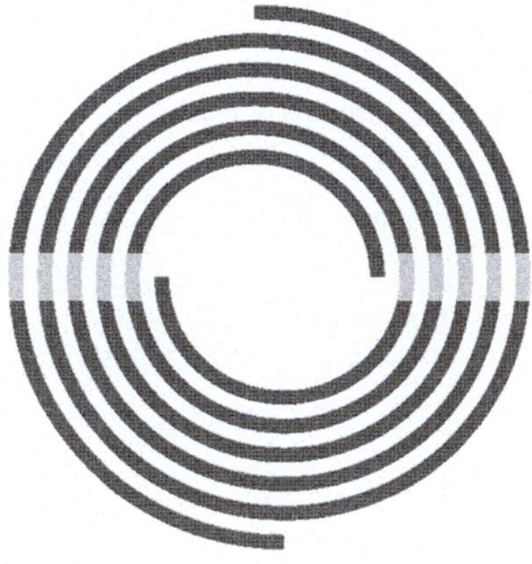

Fig. 5.92 Spiral coupled line after the construction of two air bridges

enter − 2 in the "Oversize(+)/Undersize(−)" column of the pop-up [Oversize] dialog box and click the [Apply] button. Similarly, indent the symbol and packages layers by 2 μm relative to the bond layer.

(8) Draw connection parts. Execute menu commands [Insert] → [Rectangle] to insert two rectangles in the layout, then press "Esc" to exit. Select the inserted rectangle successively, select "bond: drawing" in the drop-down menu of the [All Shapes] → [Layer] under [Properties] at the right side of the window, and set the [Rectangles] → [Height] to 15 μm. [Width] can be set according to the specific size of the air bridges. Select and long press the modified rectangle by the left mouse button to move them to the middle position of the air bridges and connect them with the original bond layers. The layout of spiral coupled line has been drawn, as shown in Fig. 5.93.

2. Layout Simulation of Spiral Coupled Line

(1) Insert ports. Execute menu commands [Insert] → [Pin] and click the left mouse button to insert four pins into the I/O ports of the spiral coupled line, as shown in Fig. 5.94.

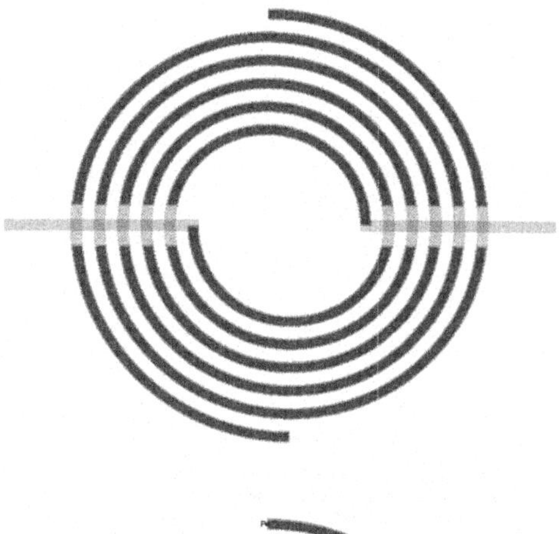

Fig. 5.93 Final layout of a spiral inductor

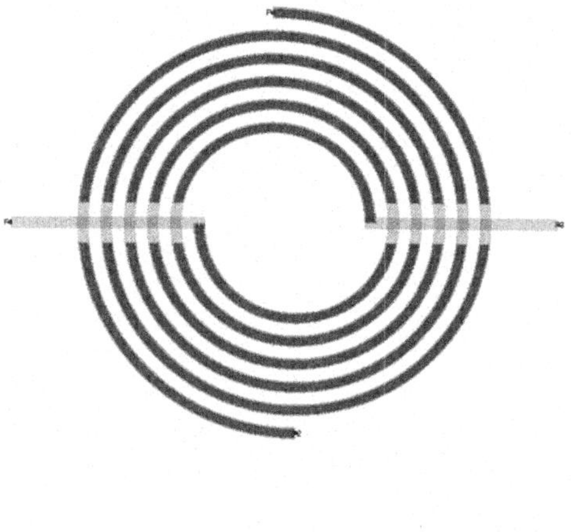

Fig. 5.94 Insertion of pins

(2) Modify simulation settings. Execute menu commands [EM] → [Simulation Settings…], click the [Create EM Setup View] button in the pop-up the [New EM Setup View] dialog box, and the [EM Setup for simulation] window will pop up. Select the second method "Momentum Microwave" of the EM simulator at first. Then select the [Frequency Plan] tab to set the simulation frequency. Select "Adaptive" in the "Type" column and set the "Fstart" to 0 GHz, the "Fstop" to 10 GHz, and the "Npts" to 10. Finally, select the [Options] tab. Click "Preprocessor", select the "User specified snap distance" option under the "Heal the layout", and set the distance to 2.5 μm. Click "Mesh" and select the "Edge Mesh" option. Keep other settings in default. After finishing all the steps, close the window and click the [OK] button to save the settings.

(3) Layout simulation. Execute menu commands [EM] → [Simulate] to simulate the layout of the spiral coupled line. During the simulation, a status window will pop up to display the current item's progress. After the finishing of the simulation, the data display window will be opened automatically. Refer to the steps in Sect. 5.3.2 to view and handle traces dB(S(1,1)) and dB(S(2,1)), the final results are shown in Fig. 5.95.

3. Co-Simulation of Spiral Coupled Line

To verify whether the odd- and even-mode impedances of the drawn spiral coupled line are 232.5 Ω are 30.67 Ω, respectively, and whether the electrical length is 45° at the center frequency of 4 GHz, the co-simulation of the spiral coupled line schematic and layout is required.

(1) Create EM model and symbol of the spiral coupled line. In the layout window, execute menu commands [EM] → [Component] → [Create EM Model And Symbol…], then click the [OK] button in the pop-up

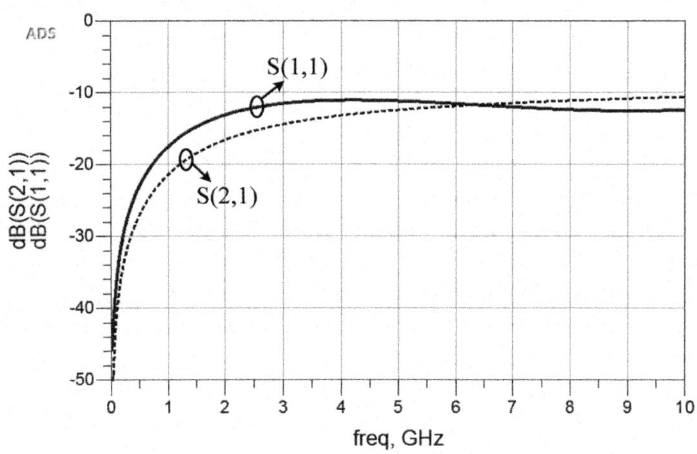

Fig. 5.95 Final simulated results of the drawn spiral coupled line

dialog boxes. Execute menu commands [Edit] → [Component] → [Update Component Definitions…], then click the [OK] button in the pop-up dialog box to complete the creation of the EM model and the symbol in the spiral coupled line.

(2) Create a new schematic and insert the spiral coupled line component. Return to the "Bandpass_Filtering_Marchand_Balun_wrk" main interface of the workspace. Execute menu commands [File] → [New] → [Schematic…], modify the "Cell" name to "spiral coupled line-cosimulation" in the opened [New Schematic] dialog box, and click the [Create Schematic] button to create a new schematic. Click the [Open the Library Browser] icon at the left side of the schematic window, then select the "spiral coupled line" component under the [Workspace Libraries] in the displayed [Component Library] window. Click the right mouse button, then click the [Place Component] in the pop-up menu to insert a spiral coupled line component in the schematic. Press "Esc" to exit.

(3) Insert ideal coupled line. Select the [TLines-Ideal] from the drop-down menu in the left component palette list and click the [Ideal Coupled Transmission Lines] icon to insert a coupled line. Press "Esc" on the keyboard to exit. Double-click the coupled line and modify Ze = 232.5 Ω, Zo = 30.67 Ω, E = 45°, and F = 4 GHz in the [Edit Instance Parameter] pop-up dialog box (check whether the unit setting is consistent), then click the [OK] button to save the parameter.

(4) Insert the S-parameter simulator, term ports, and grounds. Although the coupled line has four ports, there are only two ports connected in the co-simulation circuit. Select the [Simulation-S_Param] from the drop-down menu in the left component palette list and click the [S-parameter Simulator] icon to insert an S-parameter simulator. Click the [Port Impedance Termination for S-Parameters] icon to insert four term ports, then press "Esc" to exit. Execute menu commands [Insert] → [GROUND] to place four grounds (Or click the [TermG Port Impedance Termination for S-Parameters with Grounded Reference] icon to insert four term ports with the grounded reference). Execute menu commands [Insert] → [Wire] to connect the inserted components. After finishing all the steps, press "Esc" to exit.

(5) Set simulation frequency. Double-click the S-parameter simulator S-PARAMETERS. Set the "Start" to 0 GHz, the "Stop" to 10 GHz, and the "Step-size" to 0.01 GHz, and click the [OK] button. The final coupled line co-simulation circuit is obtained, as shown in Fig. 5.96.

(6) Co-simulation. Execute menu commands [Simulate] → [Simulate]. After the simulation completes, the data display window will be opened automatically. Refer to the steps in Sect. 5.3.2 to view and handle traces of dB(S(1,1)), dB(S(2,1)), dB(S(3,3)), and dB(S(4,3)), the final results are shown in Fig. 5.97. It can be seen that traces dB(S(1,1)) and dB(S(3,3)), traces dB(S(2,1)) and dB(S(4,3)) are almost coincident, indicating that the

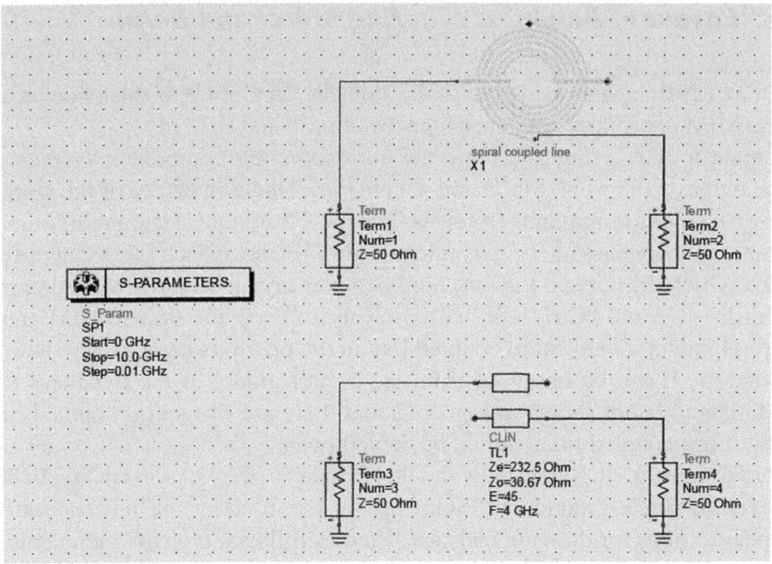

Fig. 5.96 Coupled line co-simulation circuit

odd-mode impedance and even-mode impedance of the drawn spiral coupled line are approximately equal to 232.5 Ω are 30.67 Ω, respectively, and the electrical length is approximately equal to 45° at the center frequency of 4 GHz. If the differences between these traces are large, go back to modify the spiral coupled line layout and repeat the steps above until the error of two traces is in the range of acceptance (within an acceptable range).

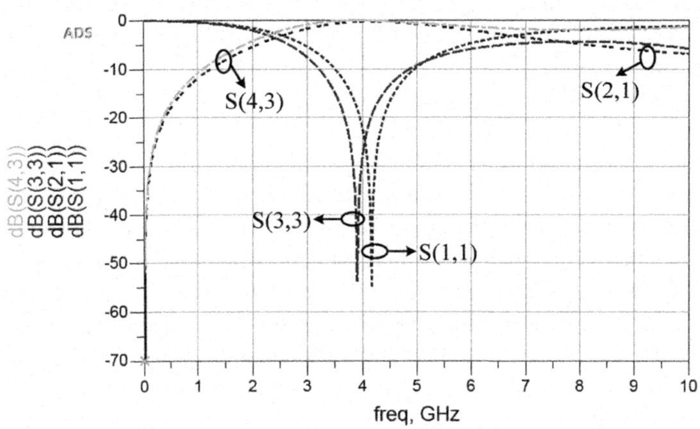

Fig. 5.97 S-parameter results of the coupled line co-simulation

5.4.5 Layout of Bandpass Filtering Marchand Balun

(1) Draw MIM capacitors. Refer to the steps in Sect. 5.4.2 of this chapter to draw the MIM capacitors with capacitances of 0.77 and 0.72 pF.

(2) Create a new layout and copy the corresponding component layouts. In the "Bandpass_Filtering_Marchand_Balun_wrk" main interface of the workspace, execute menu commands [File] → [New] → [Layout…] and modify the "Cell" name to "bandpass filtering marchand balun" in the opened [New Layout] dialog box. Click the [Create Layout] button, a new layout window will pop up. Press "Ctrl + C" and "Ctrl + V" successively to copy the drawn MIM capacitor, spiral inductor, and spiral coupled line in the previous layout to the new layout window. It can be observed from the circuit model of the bandpass filtering Marchand balun shown in Fig. 5.32 that there are two spiral coupled lines, so the spiral coupled lines should be copied twice.

(3) Draw signal pads. Signal pads should be added to the I/O ports in layout because "ground-single-ground" ("GSG") probes are used in the measurement. The specific steps for drawing pads are listed as follows: Execute menu commands [Insert] → [Rectangle] to insert a rectangle in the layout, then press "Esc" to exit. Select the inserted rectangle, select "bond: drawing" in the drop-down menu of the [All Shapes] → [Layer] under [Properties] at the right side of the window. Then set the [Rectangles] → [Width] and [Height] to appropriate values. Execute menu commands [Edit] → [Copy/Paste] → [Copy To Layer…], select [text: drawing] in the pop-up [Copy to Layer] dialog box, and click the [Apply] button to copy a test layer in the original position. Similarly, copy a leads layer, a symbol layer, and a packages layer in the original position, respectively. After copying all the needed layers, click the [Cancel] button to close this dialog box. Then indent the layers. Select the text layer of the pads and microstrip lines, execute menu commands [Edit] → [Scale/Oversize] → [Oversize…]. Since the text layer is indented by 2 μm compared with the bond layer, enter − 2 in the "Oversize(+)/Undersize(−)" column of the pop-up [Oversize] dialog box and click the [Apply] button. Similarly, indent the symbol layer and packages layer by 2 μm relative to the bond layer.

(4) Introduce of a compensation line. It is worth mentioning that spiral coupled lines are not ideal coupled lines in essence, its unequal odd- and even-mode phase velocities can cause the imperfect magnitude and phase imbalances. Therefore, we introduce a compensation line located between Port 3 and the spiral coupled line to optimize the magnitude and phase imbalances. The steps to draw a compensation line are similar to the procedure of drawing the spiral inductor. The specific parameters include the number of turns $N = 1.5$, the inner radius $Ri = 50$ μm, the conductor width $W = 15$ μm, and the conductor spacing $S =$ 15 μm. For more details, please refer to Sect. 5.4.3 of this chapter, which are not repeated here.

(5) Layout adjustment and component connection. The layout is adjusted in the consideration of the circuit size, the layout beautification, and other factors.

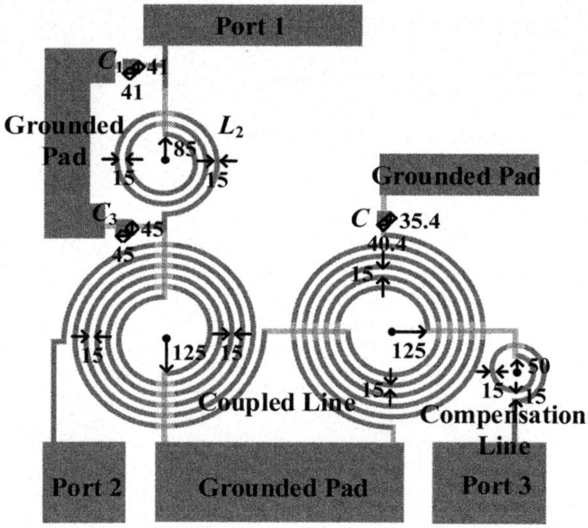

Fig. 5.98 Final layout of the bandpass filtering Marchand balun

According to the simulated circuit model in Fig. 5.32, all components are connected by microstrip lines. The drawing of microstrip line is similar to that of pad. For more details, please refer to step (3), which is not repeated here. The final layout of the bandpass filtering Marchand balun is obtained after the optimizations in several times, as shown in Fig. 5.98 (unit: μm)

5.4.6 Layout Simulation

1. Layout Simulation of Bandpass Filtering Marchand Balun

   (1) Insert pins. Execute menu commands [Insert] → [Pin], then click the left mouse button to insert pins into the I/O and grounded ports of the bandpass filtering Marchand balun, as shown in Fig. 5.99. Long press the "Ctrl" key and select all pins, then select "leads: drawing" in the drop-down menu of the [All Shapes] → [Layer] under [Properties] at the right side of the window.
   (2) Modify simulation settings. Execute menu commands [EM] → [Simulation Settings…], click the [Create EM Setup View] button in the pop-up the [New EM Setup View] dialog box, and the [EM Setup for simulation] window will be displayed. Select the second method "Momentum Microwave" of the EM simulator at first. Then select the [Frequency Plan] tab to set the simulation frequency. Select "Adaptive" in the "Type" column and set the "Fstart" to 0 GHz, the "Fstop" to 10 GHz, and the "Npts" to 10. Finally, select the [Options] tab. Click "Preprocessor", select the "User specified snap distance" option under the "Heal thc layout", and set the distance

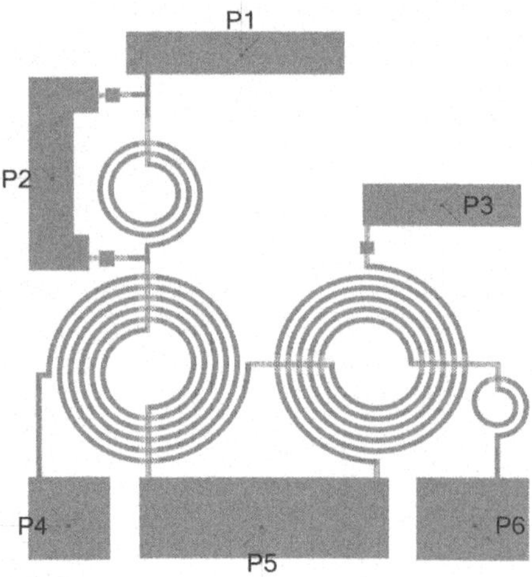

Fig. 5.99 Insertion of pins

to 2.5 μm. Click "Mesh" and select the "Edge Mesh" option. Keep other settings in default. After finishing all the steps, close the window and click the [OK] button to save the settings.

(3) Layout simulation. Execute menu commands [EM] → [Simulate] to simulate. During the simulation, a status window will pop up to display the current item's progress. After the finishing of the simulation, the data display window will be opened automatically. The whole simulation is usually long. Since the layout is not grounded, the simulated results are not used to evaluate the performance of the bandpass filtering Marchand balun. After the simulation is finished, close the data display window.

2. Co-Simulation of Bandpass Filtering Marchand Balun

In order to evaluate the performance of the bandpass filtering Marchand balun, the co-simulation of the bandpass filtering Marchand balun is carried out.

(1) Create EM model and symbol of the bandpass filtering Marchand balun. In the layout window, execute menu commands [EM] → [Component] → [Create EM Model And Symbol…], then click the [OK] button in the pop-up dialog boxes. Execute menu commands [Edit] → [Component] → [Update Component Definitions…], then click the [OK] button in the pop-up dialog box to complete the creation of the EM model and symbol in the bandpass filtering Marchand balun.

(2) Create a new schematic and insert the bandpass filtering Marchand balun component. Return to the "Bandpass_Filtering_Marchand_Balun_wrk" main interface of the workspace. Execute menu commands [File] → [New] → [Schematic…], modify the "Cell" name to "bandpass filtering Marchand balun-cosimulation" in the opened [New Schematic] dialog box, and click the [Create Schematic] button to create a new schematic. Click the [Open

the Library Browser] icon at the left side of the schematic window, then select the "bandpass filtering Marchand balun" component under the [Workspace Libraries] in the displayed [Component Library] window. Click the right mouse button, then click the [Place Component] in the pop-up menu to insert a bandpass filtering Marchand balun component to the schematic. Press "Esc" to exit.

(3) Insert the S-parameter simulator, term ports, and grounds. Select the [Simulation-S_Param] from the drop-down menu in the left component palette list and click the [S-parameter Simulator] icon to insert an S-parameter simulator. Click the [Port Impedance Termination for S-Parameters] icon to insert three term ports, then press "Esc" to exit. Execute menu commands [Insert] → [GROUND] to place six grounds (Or click the [TermG Port Impedance Termination for S-Parameters with Grounded Reference] icon to insert three term ports with the grounded reference, then execute menu commands [Insert] → [GROUND] to place three grounds). Execute menu commands [Insert] → [Wire] to connect all the inserted components. After finishing all the steps, press "Esc" to exit.

(4) Set simulation frequency. Double-click the S-parameter simulator S-PARAMETERS. Set the "Start" to 0 GHz, the "Stop" to 8 GHz, and the "Step-size" to 0.01 GHz, and click the [OK] button. The final bandpass filtering Marchand balun co-simulation circuit is obtained, as shown in Fig. 5.100.

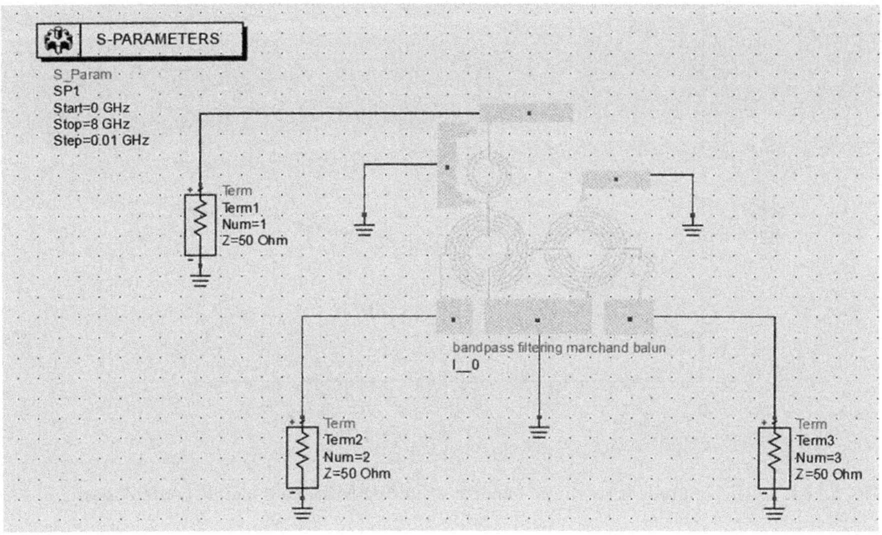

Fig. 5.100 Bandpass filtering Marchand balun co-simulation circuit

(5) Co-simulation. Execute menu commands [Simulate] → [Simulate]. After the simulation completes, the data display window will be opened automatically. Refer to the steps in Sect. 5.3.2 to view and handle traces of dB(S(1,1)), dB(S(2,1)), DiffMagnitude, and DiffPhase. The final results are shown in Figs. 5.101, 5.102 and 5.103.

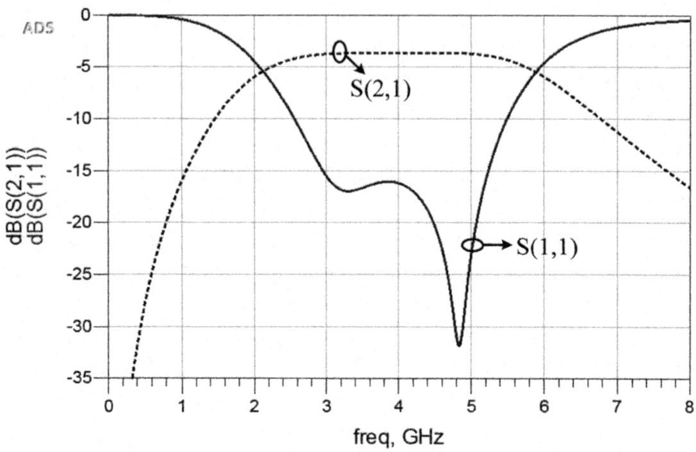

Fig. 5.101 Co-simulation of dB(S(1,1)) and dB(S(2,1)) traces

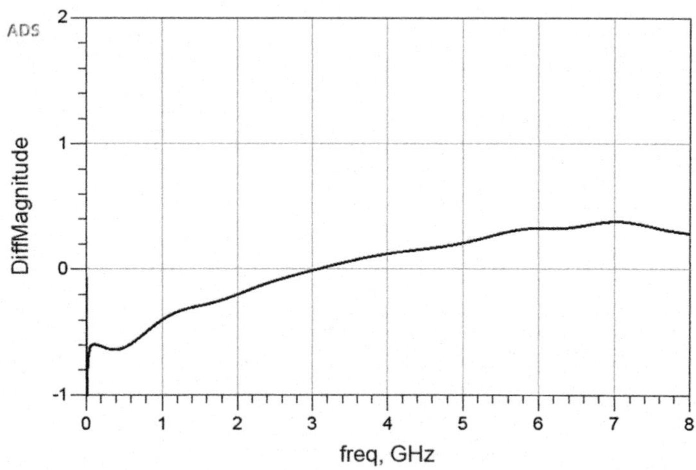

Fig. 5.102 DiffMagnitude trace of the bandpass filtering Marchand balun co-simulation

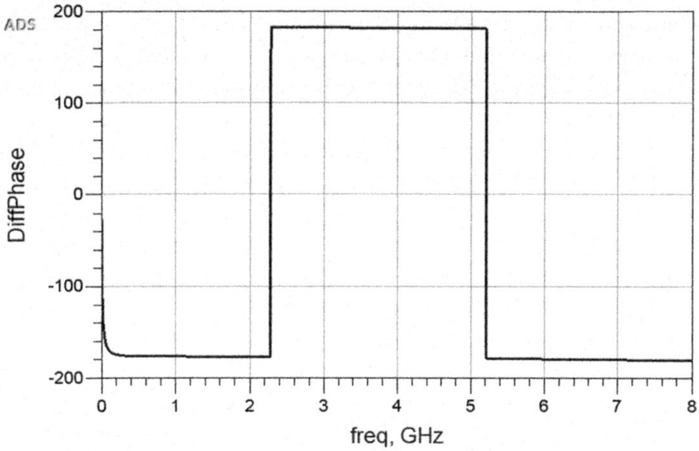

Fig. 5.103 DiffPhase trace of the bandpass filtering Marchand balun co-simulation

5.5 Package and Measurement

5.5.1 Chip Package

After the manufacture, the final bandpass filter Marchand Balun chips are obtained. In this chapter, the bandpass filter Marchand balun chip is packaged using the wire bonding technology. When currents pass through the Au wires, magnetic fields in the surrounding space can be excited, so Au wires can be viewed as inductors. An Au wire, which has length of 400 μm long and width of 1.5 mil, is equivalent to a 0.6 nH inductor. If a design is sensitive to parasitic inductors, the designer can use multiple parallel Au wires to reduce the values of inductances. The characteristic impedances of the used CPWGs are 50 Ω.

1. Evaluation of Package Influence

   Since the packaging influences are not considered in the design of the bandpass filter Marchand balun, the influences of ideal inductor with the inductance of 0.1 nH and ideal CPWG with the characteristic impedance of the 50 Ω on the performance of bandpass Marchand balun are evaluated in this section, which is generated in the process of the packaging.

   (1) Create a new schematic and insert the bandpass filtering Marchand balun component. Return to the "Bandpass_Filtering_Marchand_Balun_wrk" main interface of the workspace. Execute menu commands [File] → [New] → [Schematic…], modify the "Cell" name to "bandpass filtering Marchand balun-packaged" in the opened [New Schematic] dialog box, and click the [Create Schematic] button to create a new schematic. Click the [Open the Library Browser] icon at the left side of the schematic window and select the "bandpass filtering Marchand balun" component under the [Workspace

Libraries] in the displayed [Component Library] window. Click the right mouse button and click the [Place Component] in the pop-up menu to insert a bandpass filtering Marchand balun component to the schematic. Press "Esc" to exit.

(2) Insert equivalent inductors of wire bonds. Select the [Lumped-Components] from the drop-down menu in the left component palette list and click the [Inductor] icon to insert an inductor. Press "Esc" on the keyboard to exit. Double-click the capacitor and modify L = 0.1 nH in the [Edit Instance Parameter] pop-up dialog box, then click the [OK] button to save the parameter settings. Use the same steps to insert five inductors of 0.1 nH (Or select the inserted inductor, then press "Ctrl + C" and "Ctrl + V" to copy and paste).

(3) Insert CPWGs and substrate parameters. Select the [TLines-Waveguide] from the drop-down menu in the left component palette list shown in Fig. 5.104, then click the [Coplanar Waveguide Ground Plane] icon to insert a CPWG. Press "Esc" on the keyboard to exit. Double-click the CPWG and input W = 1000 μm, G = 500 μm, and L = 9000 μm in the [Edit Instance Parameter] pop-up dialog box, then click the [OK] button to save the parameters (check whether the unit settings are consistent), as shown in Fig. 5.105. Use the same steps to insert two CPWGs (Or select the inserted CPWG and press "Ctrl + C" and "Ctrl + V" to copy and paste). Click the [Coplanar Waveguide Substrate] icon to insert a CPWG substrate model. Double-click the icon to edit the substrate parameters of the CPWG according to Fig. 5.106, then click the [OK] button to save the parameter settings.

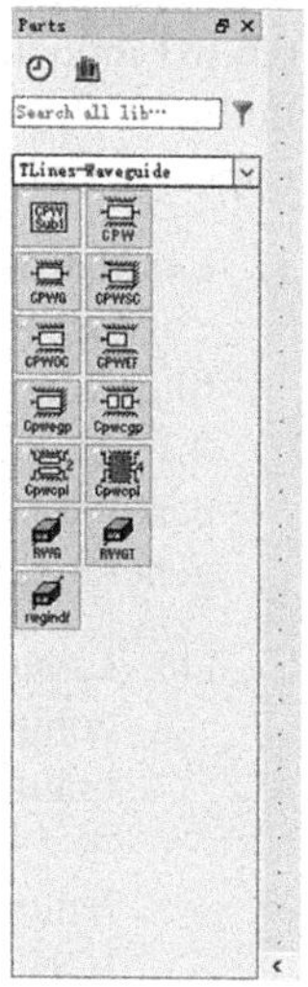

Fig. 5.104 [TLines-Waveguide] list

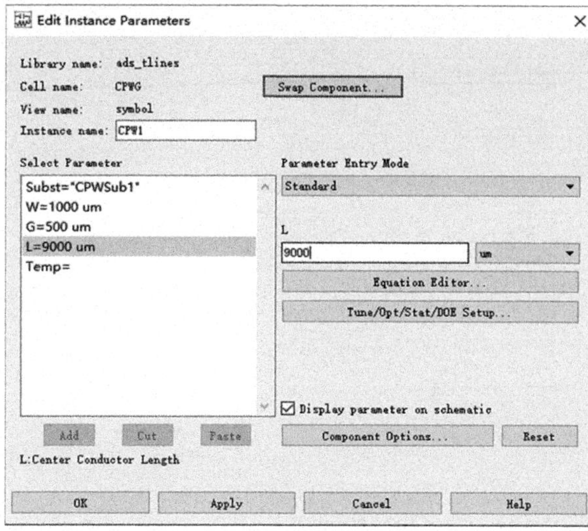

Fig. 5.105 [Edit Instance Parameters] dialog box of CPWG

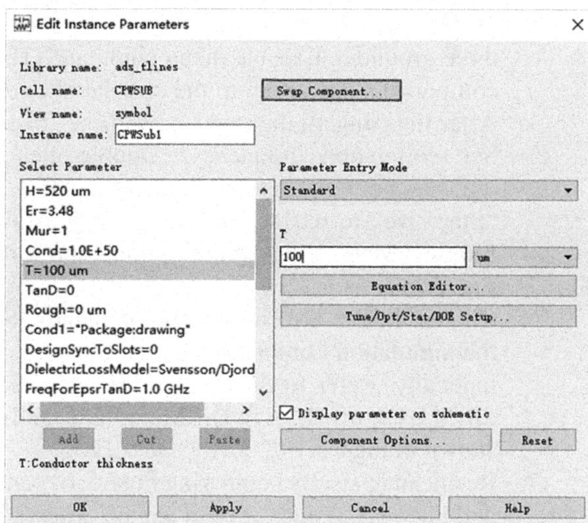

Fig. 5.106 [Edit Instance Parameters] dialog box of CPWG substrate

(4) Insert the S-parameter simulator, term ports, and grounds. Select the [Simulation-S_Param] from the drop-down menu in the left component palette list and click the [S-parameter Simulator] icon to insert an S-parameter simulator. Click the [Port Impedance Termination for S-Parameters] icon to insert three term ports, then press "Esc" to exit. Execute menu commands [Insert] → [GROUND] to place six grounds (Or click the [TermG Port Impedance Termination for S-Parameters with Grounded Reference] icon to insert three term ports with the grounded reference, then execute menu commands [Insert] → [GROUND] to place

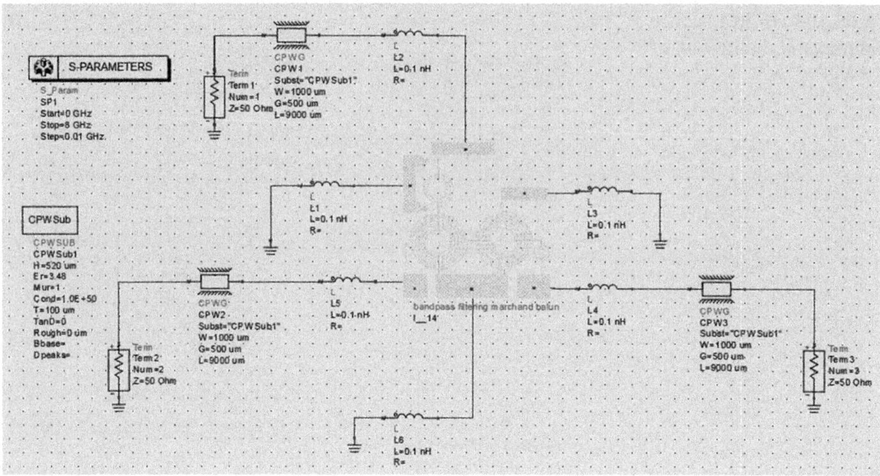

Fig. 5.107 Bandpass filtering Marchand balun co-simulation circuit after considering package influences

three grounds). Execute menu commands [Insert] → [Wire] to connect all components according to the simulated circuit model shown in Fig. 5.32. After finishing all the steps, press "Esc" to exit.

(5) Set simulation frequency. Double-click the S-parameter simulator 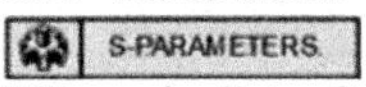

. Set the "Start" to 0 GHz, the "Stop" to 8 GHz, and the "Step-size" to 0.01 GHz, and click the [OK] button. The final bandpass filtering Marchand balun co-simulation circuit is obtained, as shown in Fig. 5.107.

(6) Co-simulation. Execute menu commands [Simulate] → [Simulate]. After the simulation completes, the data display window will be opened automatically. Refer to the steps in Sect. 5.3.2 to view and handle traces of dB(S(1,1)), dB(S(2,1)), DiffMagnitude, and DiffPhase, the final results are shown in Figs. 5.108, 5.109 and 5.110.

(7) Result analysis. By comparing Figs. 5.101 and 5.108, Figs. 5.102 and 5.109, and Figs. 5.103 and 5.110, it can be observed that Au wires and CPWGs in the chip package do not have unacceptable influences on the performance of bandpass filter Marchand balun.

2. Chip Package

As shown in Fig. 5.111, a bandpass filtering Marchand balun chip is fixed on the PCB by Au wires. To be specific, ports are linked to CPWGs while grounded pads are attached to the ground. The function of the CPWGs is to extend the ports of the bandpass filtering Marchand balun chip outward, so that the SMA connectors can be fixed easier at the edges of the PCB.

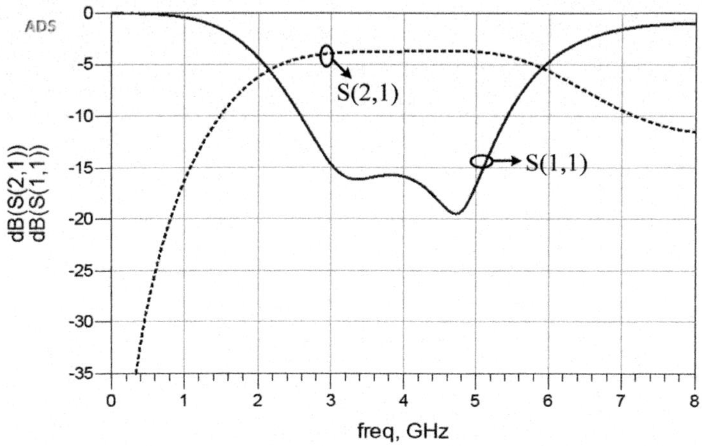

Fig. 5.108 dB(S(1,1)) and dB(S(2,1)) traces of the bandpass filtering Marchand balun co-simulation after considering package influences

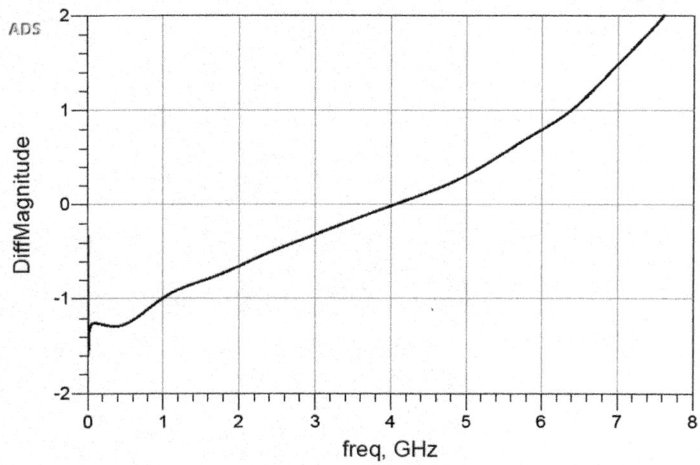

Fig. 5.109 DiffMagnitude trace of the bandpass filtering Marchand balun co-simulation after considering package influences

5.5.2 Chip Measurement

1. Measured Results

   The measurement of the bandpass filtering Marchand balun chip fixed on the PCB by the wire bonding is completed by using the vector network analyzer of ROHDE & SCHWARZ ZVA8. The simulated and measured results of the S-parameters, magnitude imbalances, and phase differences are shown in Figs. 5.112, 5.113 and 5.114.

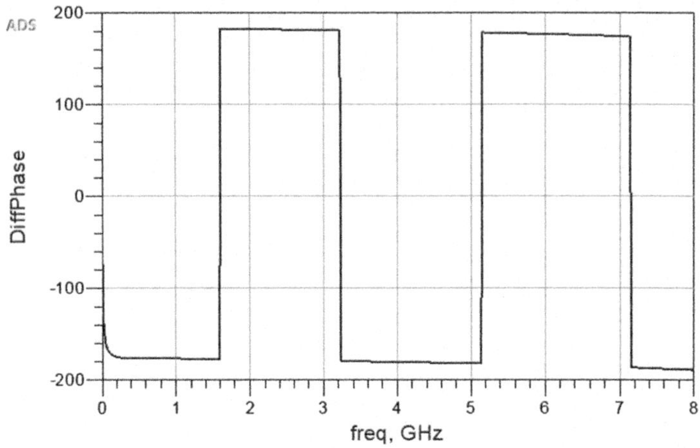

Fig. 5.110 DiffPhase trace of the bandpass filtering Marchand balun co-simulation after adding package influences

Fig. 5.111 Package of the bandpass filter Marchand balun chip

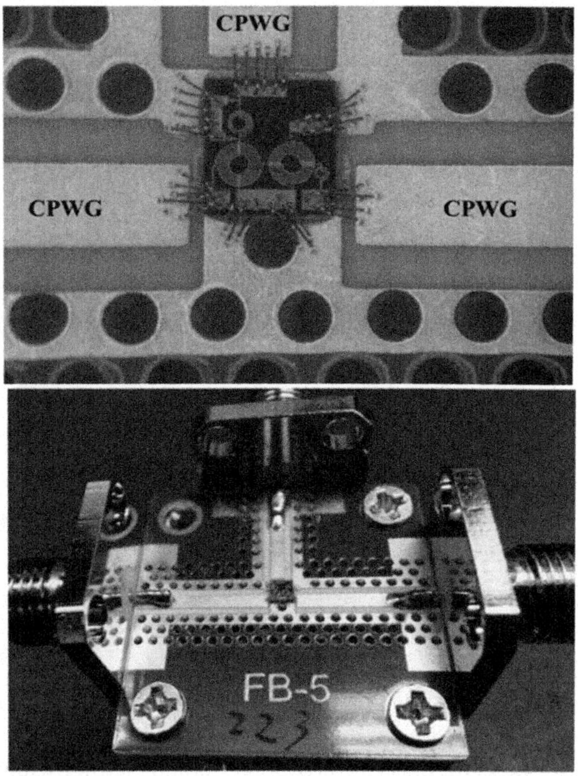

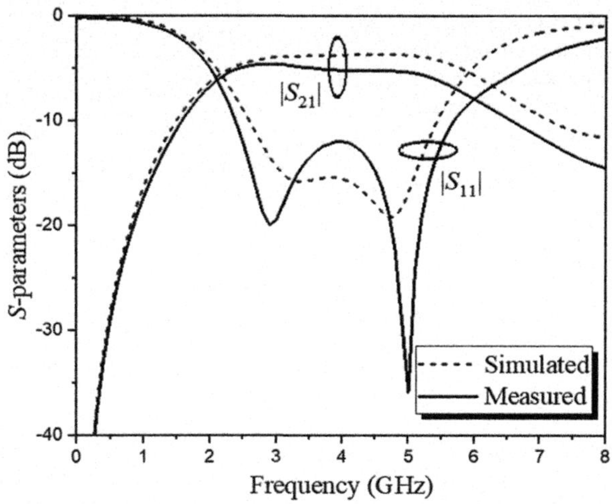

Fig. 5.112 EM simulated and measured results of the bandpass filtering Marchand balun chip of $|S_{11}|$ and $|S_{21}|$

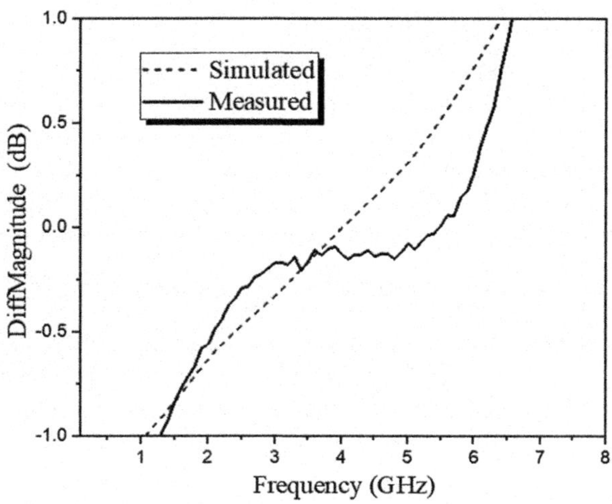

Fig. 5.113 EM simulated and measured results of the bandpass filtering Marchand balun chip of magnitude imbalances

2. Result Analysis

   It can be seen from Figs. 5.112, 5.113 and 5.114 that the measured results are mostly consistent with the simulated results. The deviations can be caused by the mechanical error and inaccurate dielectric constants of industrial materials. Within the passband, the measured in-band return loss at the unbalanced port is greater than 10.00 dB from 2.39 to 5.68 GHz. The in-band maximum magnitude and phase imbalances are 0.35 dB and 5.26°, respectively. In addition, the out-band performance is also great. It is only disappointing that the effect of parasitic parameters, especially the parasitic capacitors, is more obvious in the high band, which leads to the serious signal leakage.

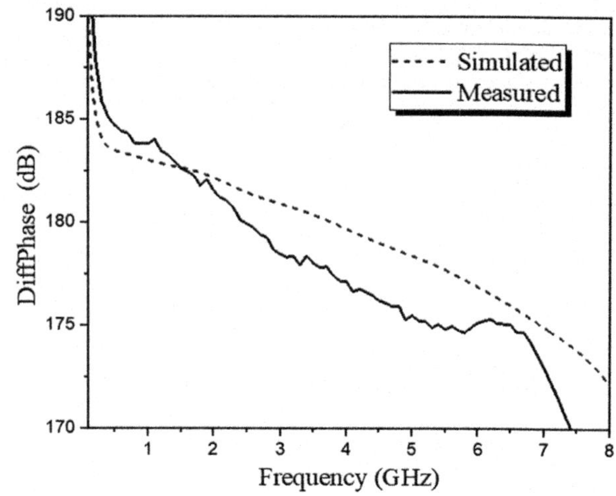

Fig. 5.114 EM simulated and measured results of the bandpass filtering Marchand balun chip of phase differences

5.5.3 Data Processing

All the drawn data figures in this book are completed by the Origin software [5, 6]. In this section, the S-parameter, magnitude imbalances, and phase differences of the bandpass filtering Marchand Balun chip in this chapter are taken as examples to introduce the detailed data processing using the Origin software. The figures in other chapters are drawn in the same way, which are not repeated in this section.

1. New Project and Generate the primary figure

   (1) Double-click the Origin shortcut icon to start the ADS software. Enter the main interface [OriginPro 2019b], as shown in Fig. 5.115.

   (2) Select the data to be processed and paste it into the [Book1] table, which is created automatically.

> • Note
>
> Pasting data is not allowed in the first four lines, including "Long Name", "Units", "Comments", and "F(x)=".

Enter names in the [Long Name] and units in the [Units] of each column. Since the two frequency columns of co-simulation and test data are different, two X columns need to be set. The specific steps are listed as follows: Select the test frequency column (column F) and execute menu commands [column] → [Set As] → [X] (Or select the test frequency column, right-click and select [Set As] → [X]). The final data to be processed is shown in Fig. 5.116.

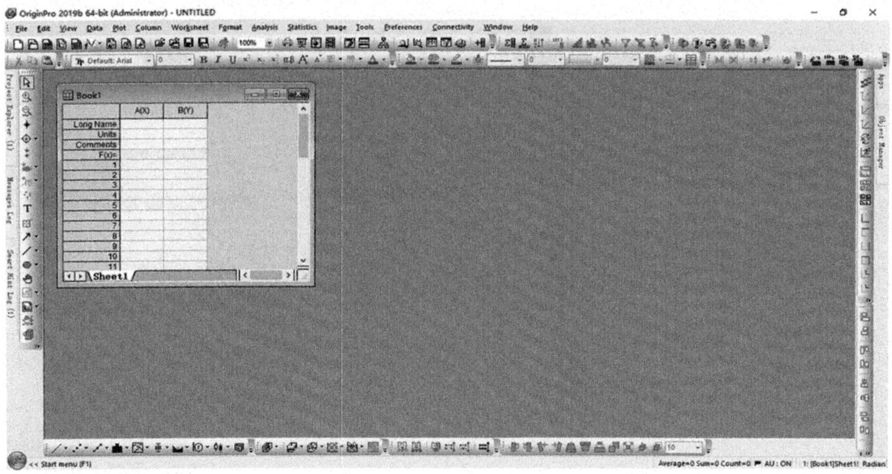

Fig. 5.115 Main interface of Origin

	A(X1)	B(Y1)	C(Y1)	D(Y1)	E(Y1)	F(X2)	G(Y2)	H(Y2)	I(Y2)	J(Y2)
Long Name	frequency	S11-Simulated	S21-Simulate	DiffMagnitude-Simulated	DiffPhase-Simulated	frequency	S11-Measured	S21-Measured	DiffMagnitude-Measured	DiffPhase-Measured
Units	GHz	dB	dB	dB	Degree	GHz	dB	dB	dB	Degree
Comments										
F(x)=										
1	0	-7.20974E-10	-285.47282	-0.10961	0	0	-0.15367	-76.34157	20.35827	59.18027
2	0.01	-0.00574	-93.51673	-0.64862	257.82986	0.1	-0.11754	-56.73033	-1.98079	193.29342
3	0.02	-0.00757	-83.21992	-1.08026	224.17901	0.2	-0.1524	-44.7481	-1.77097	187.97298
4	0.03	-0.00809	-76.59269	-1.20088	210.43493	0.3	-0.18077	-37.78113	-1.74979	185.89139
5	0.04	-0.00836	-71.75041	-1.24509	203.20656	0.4	-0.21723	-32.83663	-1.62929	185.09799
6	0.05	-0.00856	-67.94751	-1.26359	198.79335	0.5	-0.25438	-29.02536	-1.57646	184.71734
7	0.06	-0.00875	-64.82066	-1.27132	195.83166	0.6	-0.28711	-25.90786	-1.52759	184.44145
8	0.07	-0.00895	-62.16754	-1.27392	193.71054	0.7	-0.32307	-23.24465	-1.44332	184.33075
9	0.08	-0.00917	-59.86447	-1.27399	192.11758	0.8	-0.4036	-21.03716	-1.36698	183.82985
10	0.09	-0.00941	-57.83048	-1.27286	190.87724	0.9	-0.52748	-19.04687	-1.27637	183.85657
11	0.1	-0.00967	-56.00976	-1.27127	189.88376	1	-0.68838	-17.31297	-1.1732	183.83376
12	0.11	-0.00997	-54.3622	-1.26966	189.06985	1.1	-0.86695	-15.77226	-1.12412	184.04758
13	0.12	-0.01029	-52.85801	-1.26827	188.39085	1.2	-1.03954	-14.33399	-1.05579	183.46855
14	0.13	-0.01064	-51.47448	-1.26724	187.81598	1.3	-1.29104	-13.02157	-0.99084	183.30125
15	0.14	-0.01102	-50.19388	-1.26661	187.3234	1.4	-1.58349	-11.83282	-0.92539	182.98948
16	0.15	-0.01144	-49.00211	-1.26641	186.89718	1.5	-1.93701	-10.7183	-0.83702	182.63154
17	0.16	-0.01189	-47.88774	-1.26661	186.52543	1.6	-2.35092	-9.70019	-0.76921	182.48628
18	0.17	-0.01238	-46.84143	-1.26715	186.19908	1.7	-2.83572	-8.80608	-0.71482	182.28688
19	0.18	-0.01291	-45.8554	-1.268	185.91106	1.8	-3.44669	-7.99532	-0.66449	181.77965
20	0.19	-0.01347	-44.92312	-1.26908	185.65576	1.9	-4.15416	-7.26155	-0.58063	182.09689
21	0.2	-0.01407	-44.03906	-1.27034	185.42869	2	-5.03006	-6.65682	-0.55921	181.55758
22	0.21	-0.01471	-43.19849	-1.27171	185.22615	2.1	-6.07001	-6.12162	-0.486	181.21558
23	0.22	-0.0154	-42.39734	-1.27315	185.04508	2.2	-7.31044	-5.68215	-0.44132	181.05486
24	0.23	-0.01613	-41.63206	-1.27459	184.88291	2.3	-8.65961	-5.32423	-0.37317	180.72943
25	0.24	-0.0169	-40.89957	-1.276	184.73745	2.4	-10.30244	-5.05289	-0.33905	180.12526
26	0.25	-0.01772	-40.19716	-1.27733	184.60684	2.5	-12.11874	-4.8603	-0.28941	179.94744

Fig. 5.116 Data to be processed

(3) To facilitate the switch between the data table window and the following graph window, execute menu commands [View] → [Project Explorer] (or click the [Project Explorer] on the left, then click the [Auto Hide] icon) to display the project explorer.

(4) Long press the "Ctrl" key, then select A, B, C, F, G, and H six columns to draw the S-parameter curves. Execute menu commands [Plot] (or in the bottom menu bar shown in Fig. 5.117), you can see the various built-in two-dimensional drawing templates. Execute menu commands [Plot] → [Basic 2D] → [Line + Symbol] (or click the [Line + Symbol] icon in the menu bar), then the S-parameter line shown in Fig. 5.118 whose name is Graph1 in default can pop up.

Fig. 5.117 Built-in two-dimensional drawing templates

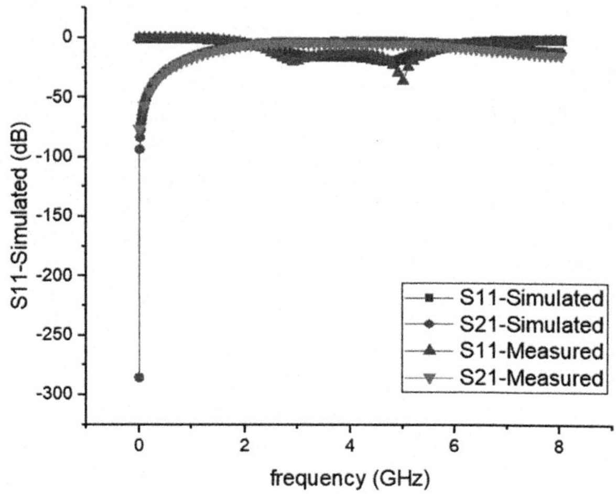

Fig. 5.118 S-parameter curves

(5) Double-click Book1 in the project explorer and return to the data table window. Long press the "Ctrl" key to select A, D, F, and I four columns, then refer to step (4) to draw the DiffMagnitude curves whose name is Graph2 in default, as shown in Fig. 5.119. Similarly, return to the table data window again. Long press the "Ctrl" key to select A, E, F, and J four columns, then refer to step (4) to draw the DiffPhase curves whose name is Graph3 in default, as shown in Fig. 5.120.

2. Beautification of curves

(1) Double-click Graph1 in the project explorer to open the generated S-parameter figure. Double-click one line in the figure, then the [Plot Details] dialog box shown in Fig. 5.121 can pop up. Select the first line under "Layerl" on the left side of the window. Select the [Group] tab and select the "Independent" under the [Edit Mode]. Select the "Line" tab. Select the "Short Dash" in the [Style] column, set the line Width to 2.5, and select "Red" in the [Color] column, as shown in Fig. 5.122. Select the [Symbol] tab and select "None" in the [Symbol Color] column, as shown in Fig. 5.123. And then, select the other three lines under "Layer1" at the left side of the window successively to do similar settings. After setting all the lines, click the [OK] button to save. The beautified S-parameter figure shown in Fig. 5.124 can be obtained.

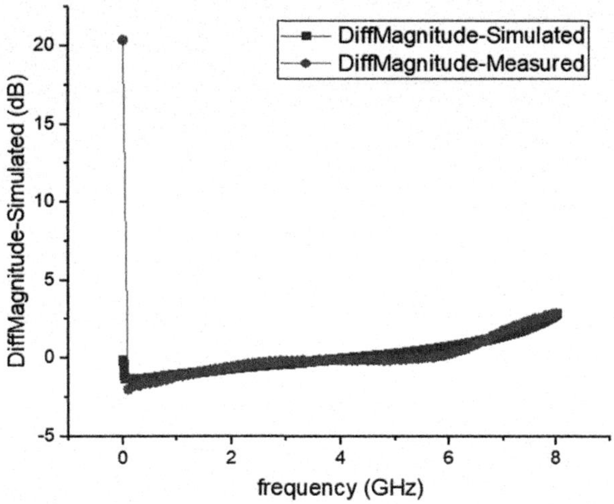

Fig. 5.119 DiffMagnitude curves

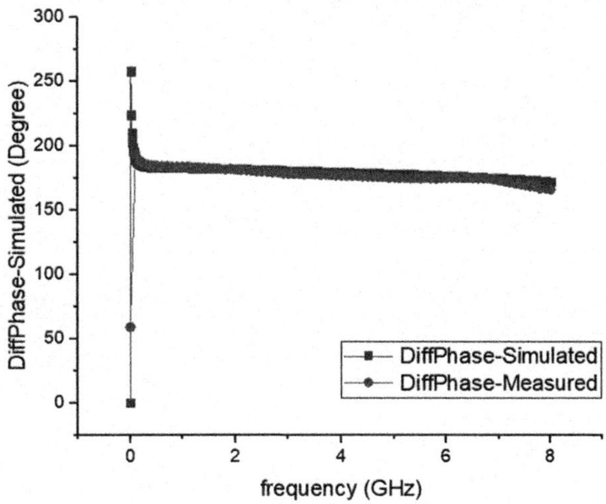

Fig. 5.120 DiffPhase curves

(2) Double-click Graph2 in the project explorer to open the generated magnitude imbalances figure. Refer to Step (1) to beautify lines, then the beautified DiffMagnitude curves shown in Fig. 5.125 can be obtained. Similarly, double-click Graph3 in the project explorer to open the generated phase differences figure. Refer to Step (1) to beautify lines, then the beautified DiffPhase curves shown in Fig. 5.126 can be obtained.

3. Beautification of Display Window

(1) Double-click Graph1 in the project explorer to reopen the S-parameter curves. Double-click the X axis, then the [X Axis] dialog box can pop up, as shown in Fig. 5.127. Select the [Scale] tab and modify the range of the

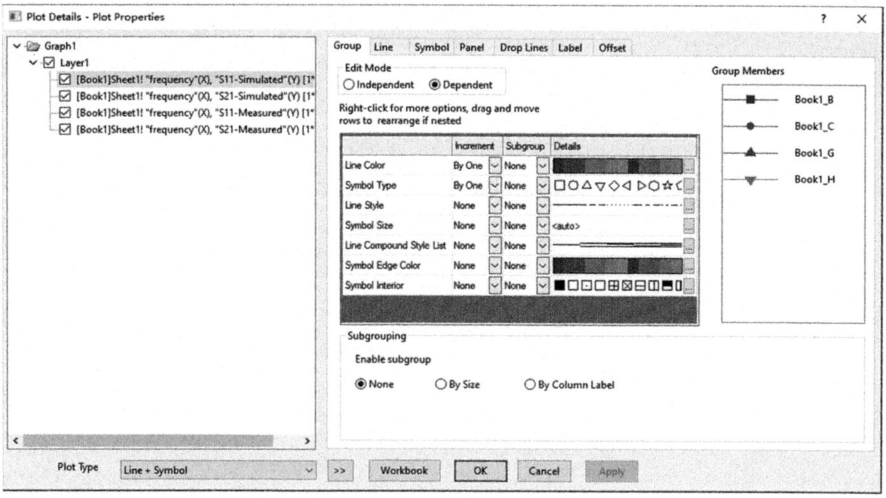

Fig. 5.121 [Plot Details] dialog box ([Group] tab)

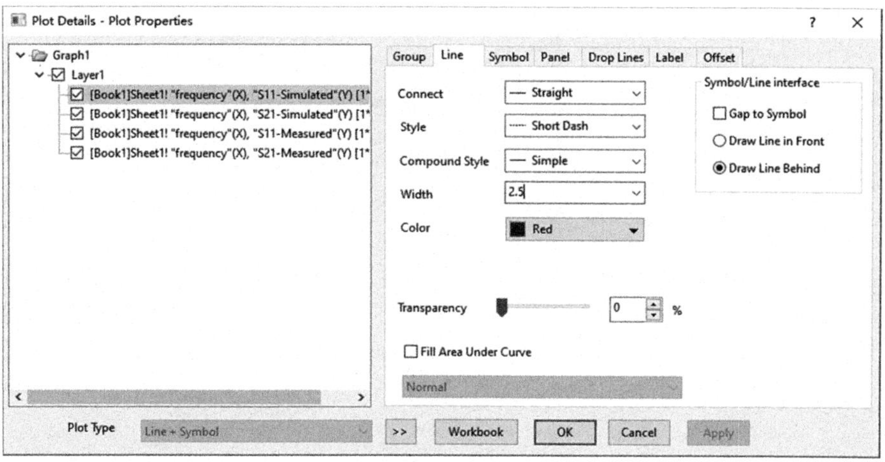

Fig. 5.122 [Plot Details] dialog box ([Line] tab)

horizontal and vertical axes. Firstly, select [Horizontal] to modify the X axis. Set its starting value ("From") to 0, ending value ("To") to 8, [Major Ticks] → [Type] to "By Increment", delta value ("Value") to 1, [Minor Ticks] → [Type] to "By Counts", and quantitative value ("Count") to 1. All settings are shown in Fig. 5.128. Similarly, select [Vertical] and modify the Y axis as shown in Fig. 5.129.

(2) Click the [Line and Ticks] tab to modify lines and ticks. Select [Top], then select [Show Line and Ticks] to add the top line. Select "None" in the

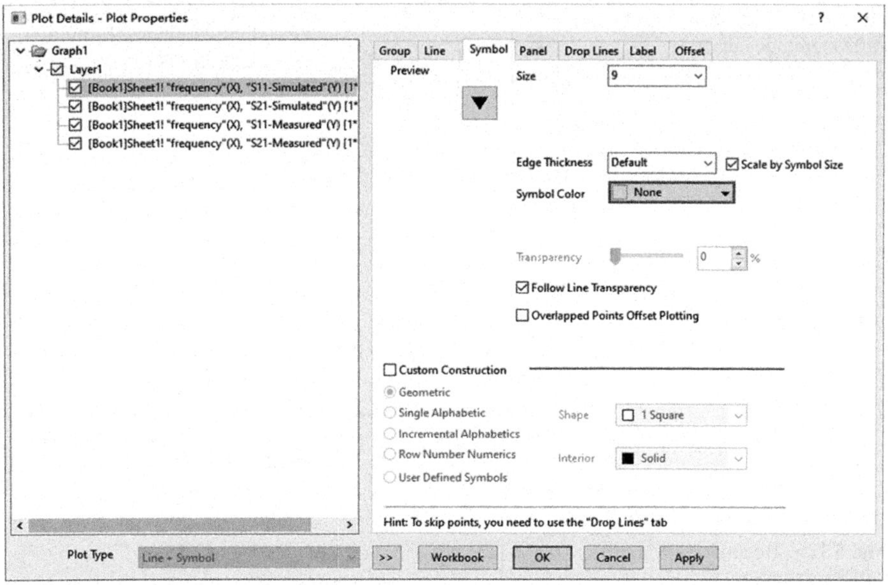

Fig. 5.123 [Plot Details] dialog box ([Symbol] tab)

Fig. 5.124 Beautified S-parameter figure

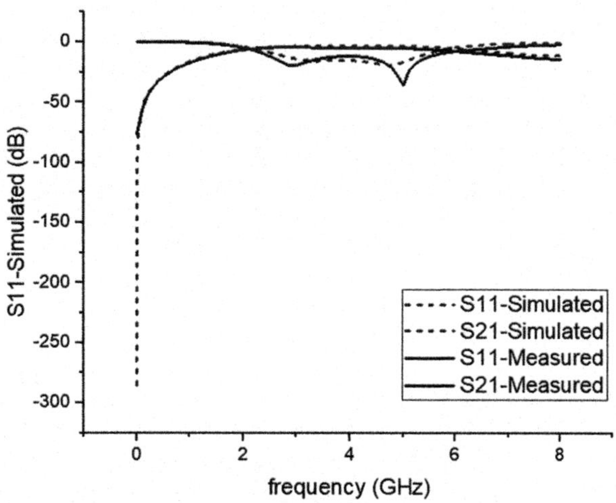

[Major Ticks] → [Style] column, which indicates that the major ticks are not displayed. Select "None" in the [Minor Ticks] → [Style] column, which indicates that the minor ticks are not displayed. Keep other settings in default, as shown in Fig. 5.130. Similarly, select [Right] to add the right line and modify the settings in the same way. After completing all settings, click the [OK] button to save. The obtained S-parameter figure is shown in Fig. 5.131.

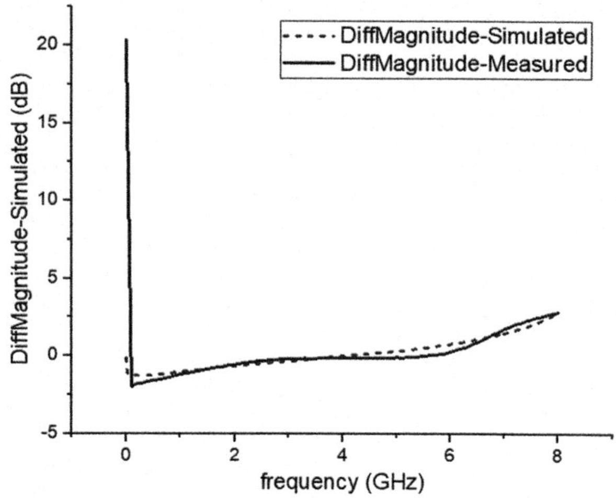

Fig. 5.125 Beautified DiffMagnitude curves

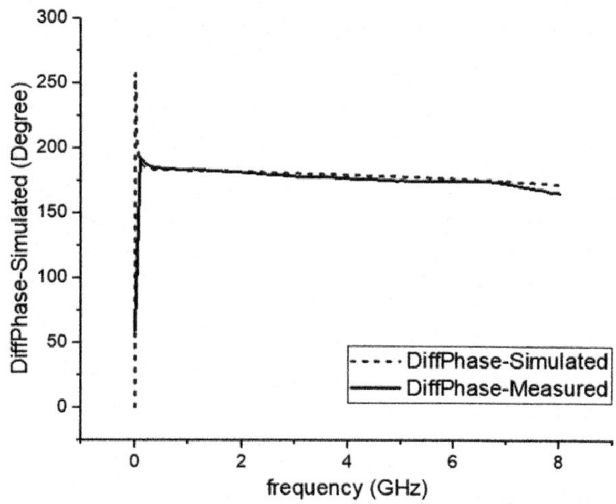

Fig. 5.126 Beautified DiffPhase curves

(3) Select the text object of lines and click the right mouse button to select [Properties…] from the pop-up menu. The [Text Object] dialog box shown in Fig. 5.132 can pop up. Select the [Text] tab to set the text annotation, font type, and size of each line. Or use N B I U x^2 x_2 αβ Ā Ã Å tools to set the text object (for example, x^2, x_2 can be used to add upper and lower scripts, etc.); Set the text object of the lines, as shown in Fig. 5.133. Select the [Frame] tab to modify the frame of the text object. Select "Shadow" under the [Frame] drop-down menu and keep other settings by default, as shown in Fig. 5.134. After completing all the settings, click the [OK] button to save. In addition, long press the text object with the left mouse button to

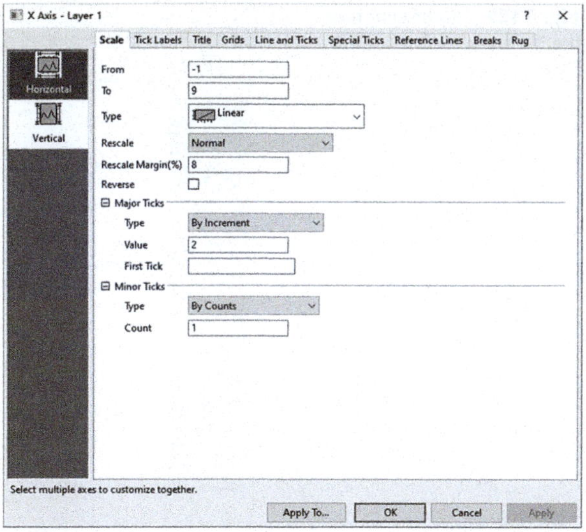

Fig. 5.127 [X Axis] dialog box

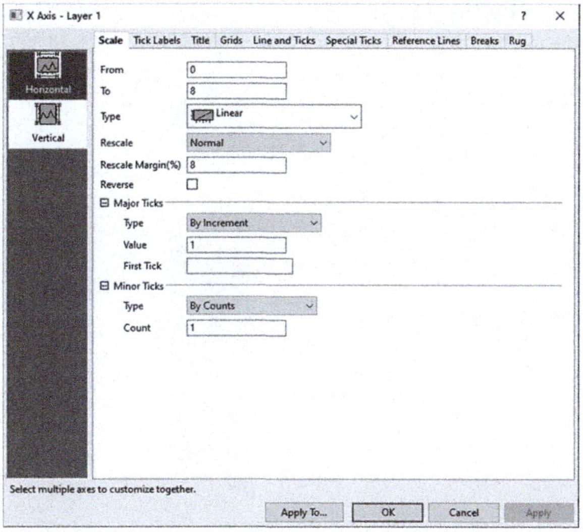

Fig. 5.128 Modification of the X axis

move its position. The S-parameter figure after the modification of the text object is shown in Fig. 5.135.

(4) Select the text object of X axis, click the right mouse button to select [Properties…] from the pop-up menu, and the [Text Object] dialog box can pop up. In the [Text] tab, set the text annotation to "Frequency (GHz)", the font type to "Times New Roman", and the size to 26. Click the [OK] button to save all the settings. Similarly, select the text object of Y axis, click the right mouse button to select [Properties…] from the pop-up menu, and the [Text

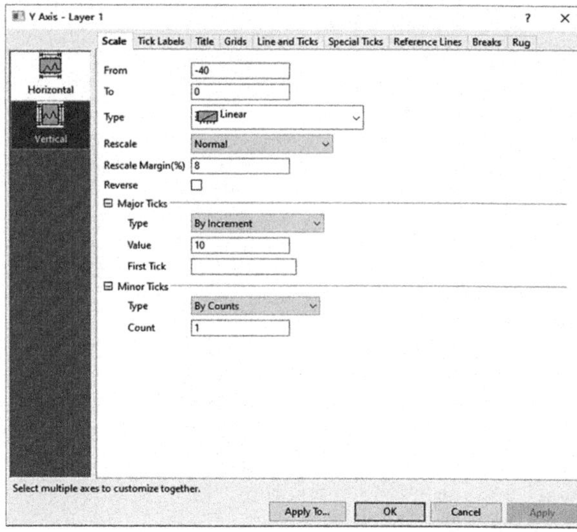

Fig. 5.129 Modification of the Y axis

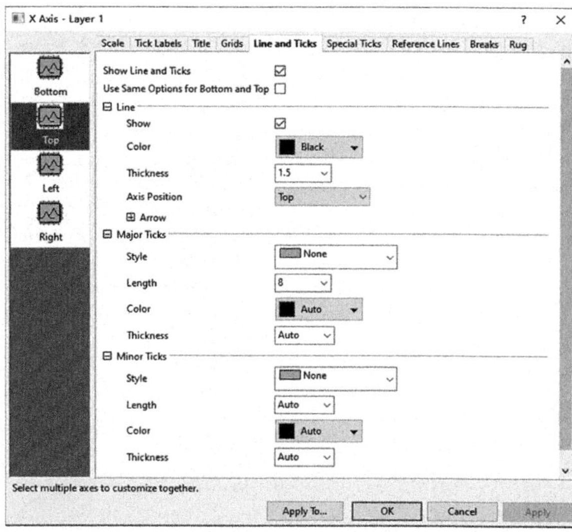

Fig. 5.130 Modification of the top line

Object] dialog box will pop up. In the [Text] tab, set the text annotation to "S-parameters (dB)", the font type to "Times New Roman", the size to 26, and "S" to italics. Click the [OK] button to save all the settings. The S-parameter figure with abscissa information is shown in Fig. 5.136.

(5) Insert identification of lines. Click the [Circle Tool] icon  in the left toolbar as shown in Fig. 5.137 to insert a circle in the drawing area. Select the inserted circle, click the right mouse button to select [Properties...] from the pop-up menu, and the [Object Properties] dialog box will pop up,

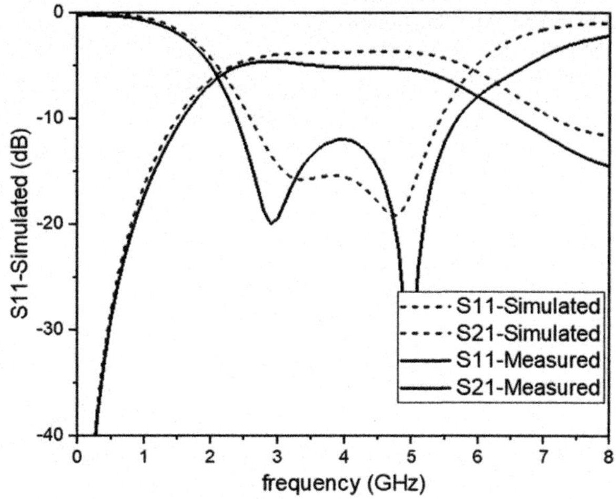

Fig. 5.131 S-parameter curves after modifications of axes, lines, and ticks

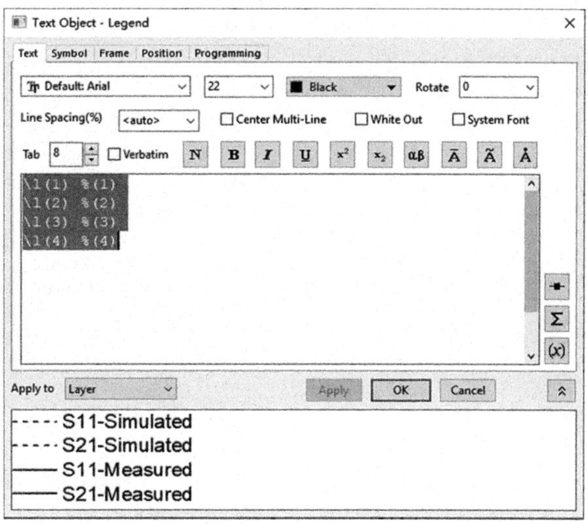

Fig. 5.132 [Text Object] dialog box

as shown in Fig. 5.138. Select the [Border] tab and set the "Width" column to 2.5. Select the [Fill] tab, select the fill color "None" in the [Fill Color] column, and keep other settings in default, as shown in Fig. 5.139. After completing all the settings, click the [OK] button to save. Long press the inserted circle by the left mouse button to move it to the position of S_{11}.

(6) Click the [Text Tool] icon 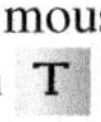 in the left toolbar, click the left mouse button in the drawing area to insert a text box, and enter "|S11|". Select the text box, click the right mouse button to select [Properties…] from the pop-up menu, the [Text Object] dialog box will pop up. Select the [Text] tab, then set the font type to "Times New Roman", size to 26, "S" to the italics, and "11"

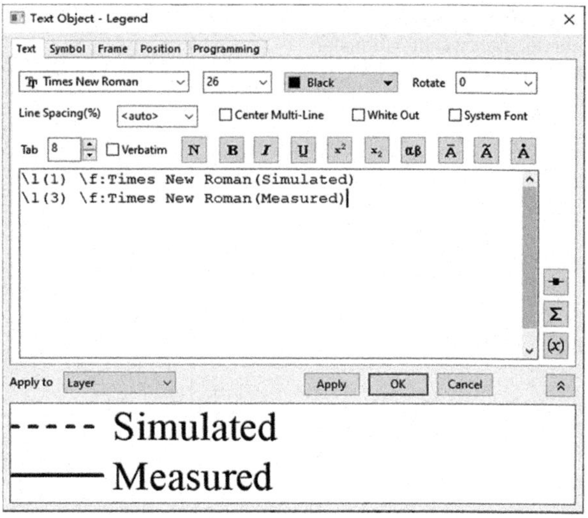

Fig. 5.133 Modification of the text object information

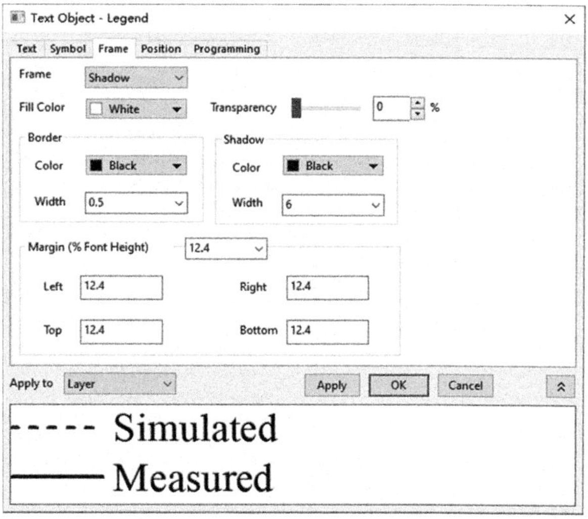

Fig. 5.134 Modified frame of the text object

to the subscript. After completing all the settings, click the [OK] button to save. Then long press the text box by the left mouse button to move it to the vicinity of the circle in step (5) to identify the S_{11}.

(7) Similarly, insert circle and text box for the identification of S_{21}. The final S-parameter curves is shown in Fig. 5.140.

(8) Similarly, double-click Graph2 and Graph3 project explorer successively, refer to the steps mentioned above to complete the beautification of the display windows of the DiffMagnitude and DiffPhase figures. The final DiffMagnitude and DiffPhase figures are shown in Figs. 5.141 and 5.142.

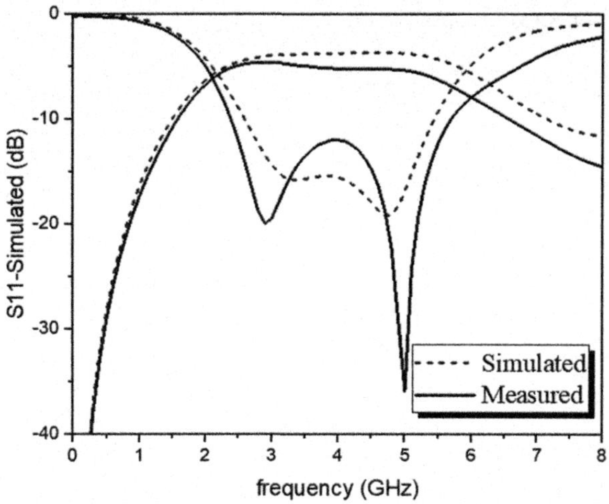

Fig. 5.135 S-parameter figure after the modification of the text object

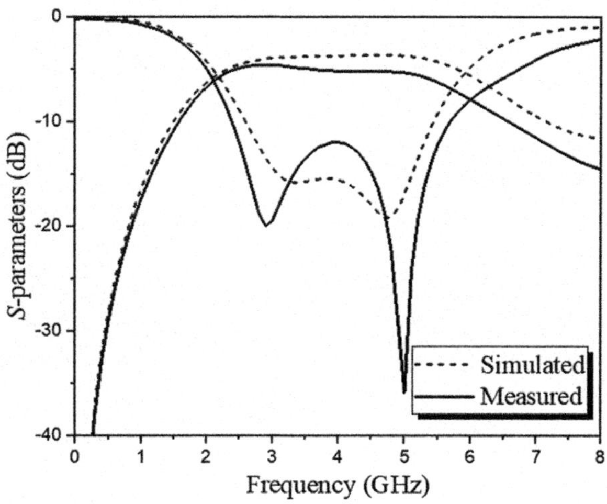

Fig. 5.136 S-parameter figure after the modification of horizontal and vertical coordinate text object

4. Export Graphs

   After the completion of the figure drawing, the figure can be exported in a variety of types. Execute menu commands [File] → [Export Graphs…], the [Export Graphs] dialog box shown in Fig. 5.143 will pop up. Set the "Image Type", the "File Name(s)", and the "Path", etc., then click the [OK] button and the chosen figure can be exported successfully.

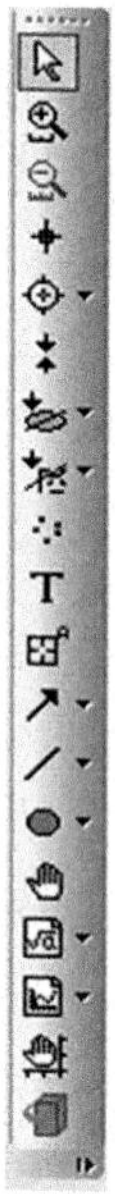

Fig. 5.137 Toolbar options

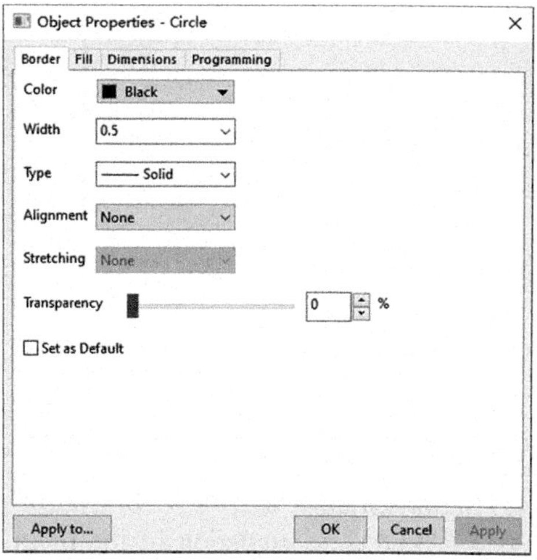

Fig. 5.138 [Object Properties] dialog box

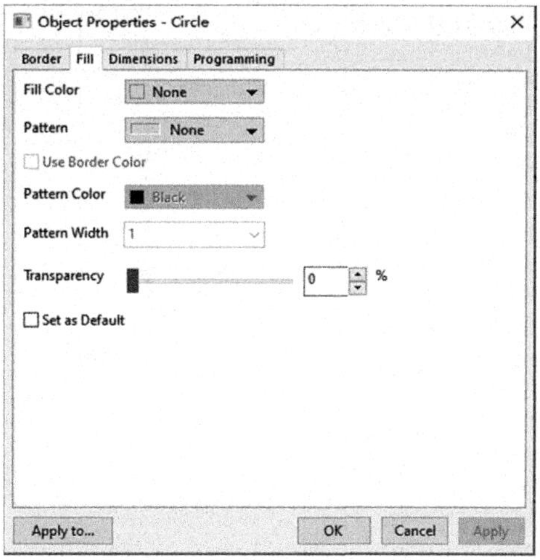

Fig. 5.139 The fill color modification of the inserted circle

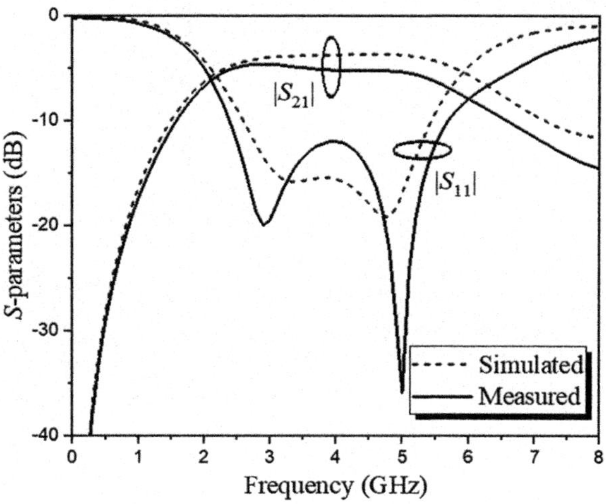

Fig. 5.140 Final S-parameter curves

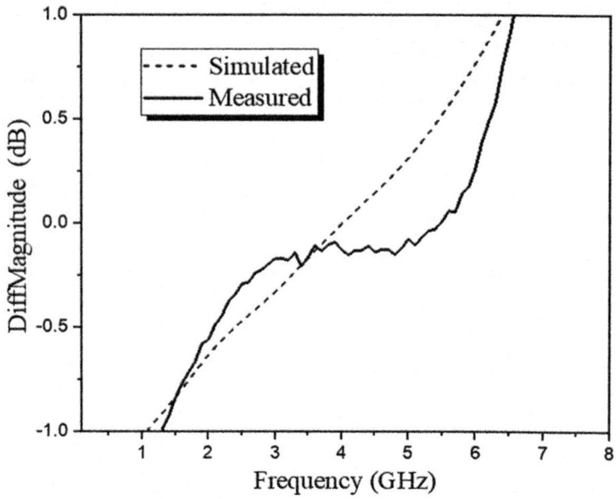

Fig. 5.141 Final DiffMagnitude figure

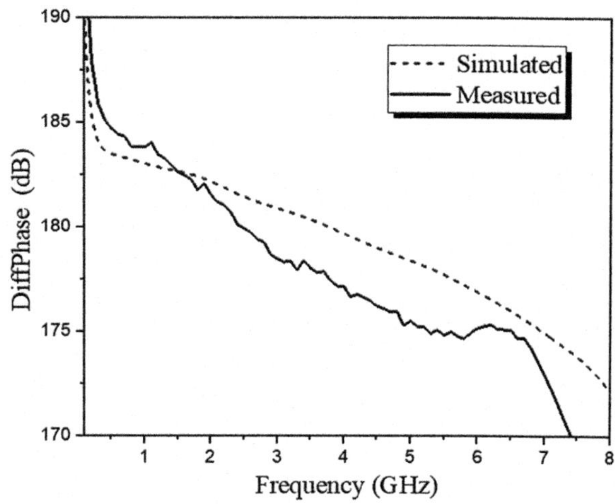

Fig. 5.142 Final DiffPhase figure

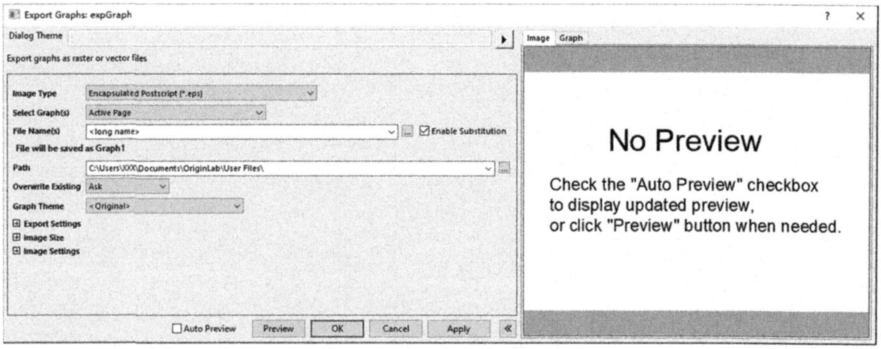

Fig. 5.143 [Export Graphs] dialog box

References

1. X. Xie, Y. Xu, L. Xia, *Microwave Integrated Circuits* (Publishing House of Electronics Industry, Beijing, 2018)
2. Q. Gu, J. Xiang, X. Yuan, *Microwave Integrated Circuit Design* (People's Posts and Telecommunications Press, Beijing, 1978)
3. R. Li, *Key Issues in RF/RFIC Circuit Design* (Higher Education Press, Beijing, 2007)
4. Y. Yang, Y. Wu, Z. Zhuang, W. Wang, C. Wang, An ultraminiaturized bandpass filtering Marchand balun chip with spiral coupled lines based on GaAs integrated passive device technology. IEEE Trans. Plasma Sci. **48**(9), 3067–3075 (2020)
5. A. Fang, W. Ye, *Origin 8.0 Practical Guide* (China Machine Press, Beijing, 2009)
6. J. Zhang, *Origin 9.0 Technology Mapping and Data Analysis Super Learning Manual* (People's Posts and Telecommunications Press, Beijing, 2014)

The manufacturer's authorised representative in the EU is Springer Nature Customer Service Centre GmbH, Europaplatz 3, 69115 Heidelberg, Germany. If you have any concerns regarding our products, please contact ProductSafety@springernature.com

Printed and bound by CPI Group (UK) Ltd, Croydon, CR0 4YY
01/10/2026
02225868-0007